M000268525

Bayesian Networks

STATISTICS IN PRACTICE
Series Advisory Editors

Marian Scott
University of Glasgow, UK

Stephen Senn
University of Glasgow

Founding Editor

Vic Barnett
Nottingham Trent University, UK

Statistics in Practice is an important international series of texts which provide detailed coverage of statistical concepts, methods and worked case studies in specific fields of investigation and study.

With sound motivation and many worked practical examples, the books show in down-to-earth terms how to select and use an appropriate range of statistical techniques in a particular practical field within each title's special topic area.

The books provide statistical support for professionals and research workers across a range of employment fields and research environments. Subject areas covered include medicine and pharmaceutics; industry, finance and commerce; public services; the earth and environmental sciences, and so on.

The books also provide support to students studying statistical courses applied to the above areas. The demand for graduates to be equipped for the work environment has led to such courses becoming increasingly prevalent at universities and colleges.

It is our aim to present judiciously chosen and well-written workbooks to meet everyday practical needs. Feedback of views from readers will be most valuable to monitor the success of this aim.

A complete list of titles in this series appears at the end of the volume.

Bayesian Networks
A Practical Guide to
Applications

Edited by

Dr Olivier Pourret

Electricité de France, France

Mr. Patrick Naim

ELSEWARE, France

Dr Bruce Marcot

USDA Forest Service, USA

John Wiley & Sons, Ltd

Copyright © 2008   John Wiley & Sons Ltd, The Atrium, Southern Gate, Chichester,
West Sussex PO19 8SQ, England

Telephone (+44) 1243 779777

Email (for orders and customer service enquiries): cs-books@wiley.co.uk
Visit our Home Page on www.wileyeurope.com or www.wiley.com

All Rights Reserved. No part of this publication may be reproduced, stored in a retrieval system or
transmitted in any form or by any means, electronic, mechanical, photocopying, recording, scanning or
otherwise, except under the terms of the Copyright, Designs and Patents Act 1988 or under the terms of a
licence issued by the Copyright Licensing Agency Ltd, 90 Tottenham Court Road, London W1T 4LP, UK,
without the permission in writing of the Publisher. Requests to the Publisher should be addressed to the
Permissions Department, John Wiley & Sons Ltd, The Atrium, Southern Gate, Chichester, West Sussex
PO19 8SQ, England, or emailed to permreq@wiley.co.uk, or faxed to (+44) 1243 770620.

This publication is designed to provide accurate and authoritative information in regard to the subject matter
covered. It is sold on the understanding that the Publisher is not engaged in rendering professional services.
If professional advice or other expert assistance is required, the services of a competent professional should
be sought.

Other Wiley Editorial Offices

John Wiley & Sons Inc., 111 River Street, Hoboken, NJ 07030, USA

Jossey-Bass, 989 Market Street, San Francisco, CA 94103-1741, USA

Wiley-VCH Verlag GmbH, Boschstr. 12, D-69469 Weinheim, Germany

John Wiley & Sons Australia Ltd, 42 McDougall Street, Milton, Queensland 4064, Australia

John Wiley & Sons (Asia) Pte Ltd, 2 Clementi Loop #02-01, Jin Xing Distripark, Singapore 129809

John Wiley & Sons Canada Ltd, 6045 Freemont Blvd, Mississauga, Ontario, L5R 4J3, Canada

Wiley also publishes its books in a variety of electronic formats. Some content that appears
in print may not be available in electronic books.

Library of Congress Cataloging-in-Publication Data

Pourret, Olivier.
  Bayesian networks : a practical guide to applications / Olivier Pourret and Patrick Naim.
    p. cm.
  Includes bibliographical references and index.
  ISBN 978-0-470-06030-8 (cloth)
  1. Bayesian statistical decision theory. 2. Mathematical models. I. Naïm. Patrick. II. Title.
  QA279.5.P68 2008
  519.5′42 – dc22
                                     2007045556

British Library Cataloguing in Publication Data

A catalogue record for this book is available from the British Library

ISBN: 978-0-470-06030-8

Typeset in 10/12pt Times by Laserwords Private Limited, Chennai, India
Printed and bound in Great Britain by TJ International, Padstow, Cornwall
This book is printed on acid-free paper responsibly manufactured from sustainable forestry
in which at least two trees are planted for each one used for paper production.

Contents

Foreword

When we, in the late 1980s, worked in a European ESPRIT project on what later became the MUNIN network, expert systems and neural networks were the predominant new artificial intelligence techniques. However, we felt that the most important ingredient of medical diagnosis, *causality with inherent uncertainty*, could not be captured by these techniques.

Rather than trying to model the experts we felt that we should go back to the ancient scientific tradition of modeling the domain, and the new challenge was to incorporate causal uncertainty. We called our models *causal probabilistic networks* (CPNs). They are now called Bayesian networks. The task, we thought, is quite simple: determine a CPN through dialogues with the experts. The rest is just mathematics and computer power.

We were wrong in two ways. It is not 'just' mathematics and computer power. But even worse, to determine a CPN through dialogues with the experts is much more intriguing than we anticipated. Over the two decades since the revival of Bayesian networks, several books have addressed the first problem. Although the need is widely recognized, no book has so far focused on the second problem.

This book meets the demand for an aid in developing Bayesian network models in practice. The authors have done a great job in collecting a large sample of Bayesian network applications from a wide range of domains.

Each chapter tells a story about a particular application. However, they do more than that. By studying the various chapters, the reader can learn very much about how to collaborate with domain experts and how to combine domain knowledge with learning from databases. Furthermore, the reader will be presented to a long list of advantages, problems and shortcomings of Bayesian network modeling and inference.

The sample also reflects the two sides of Bayesian network. On the one hand, a Bayesian network is a causal probabilistic network. On the other hand, a Bayesian network is a way of decomposing a large joint probability distribution. In some of the applications, causality is an important part of the model construction, and in other applications, causality is not an issue.

I hope that this book will be studied by everyone who is about to model a domain containing causality with inherent uncertainty: this book will teach him/her if and how to use Bayesian networks.

Finn V. Jensen
Aalborg University

Preface

The spectacular improvements of the technologies to produce, transmit, store and retrieve information are leading to a paradox: in many circumstances, making the best use of the available information is much more difficult today than a few decades ago. Information is certainly abundant and easily accessible, but at the same time (and to a large extent, consequently) often inconsistent, contradictory, and of uncertain traceability and reliability. The process of interpreting information remains an essential one, because uninterpreted information is nothing else than noise, but becomes more and more delicate. To mention only one domain covered in this book, striking examples of this phenomenon are the famous criminal cases which remain unsolved, despite the accumulation over years of evidences, proofs and expert opinions.

Given this challenge of optimally using information, it is not surprising that a gain of interest for statistical approaches has appeared in many fields in recent years: the purpose of statistics is precisely to convert information into a usable form.

Bayesian networks, named after the works of Thomas Bayes (ca. 1702–1761) on the theory of probability, have emerged as the result of mathematical research carried out in the 1980s, notably by Judea Pearl at UCLA, and from that time on, have proved successful in a large variety of applications.

This book is intended for users, and also *potential* users of Bayesian networks: engineers, analysts, researchers, computer scientists, students and users of other modeling or statistical techniques. It has been written with a dual purpose in mind:

- highlight the versatility and modeling power of Bayesian networks, and also discuss their limitations and failures, in order to help potential users to assess the adequacy of Bayesian networks to their needs;

- provide practical guidance on constructing and using of Bayesian networks.

We felt that these goals would be better achieved by presenting real-world applications, i.e., models actually in use or that have been at least calibrated, tested, validated, and possibly updated from real-world data – rather than demonstration models, prototypes, or hypothetical models. Anyone who has constructed and used models to solve practical problems has learned that the process is never as straightforward as in textbook cases, due to some ever-present difficulties: unability of the model to capture some features of the problem, missing input data, untractability (model size/computing time), and non-validable results. Our aim in the book is, also, to identify and document the techniques invented by practitioners to overcome or reduce the impact of these difficulties.

Besides a brief theoretical introduction to Bayesian networks, based on some basic, easily reproducible, examples (Chapter 1), the substance of this book is 20 application chapters (Chapters 2–21), written by invited authors.

In selecting the applications, we strove to achieve the following objectives:

1. cover the major types of applications of Bayesian networks: diagnosis, explanation, forecasting, classification, data mining, sensor validation, and risk analysis;

2. cover as many domains of applications as possible: industry (energy, defense, robotics), computing, natural and social sciences (medicine, biology, ecology, geology, geography), services (banking, business, law), while ensuring that each application is accessible and attractive for nonspecialists of the domain;

3. invite 'famous names' of the field of Bayesian networks, but also authors who are primarily known as experts of their field, rather than as Bayesian networks practitioners; find a balance between corporate and academic applications;

4. describe the main features of the most common Bayesian network software packages.

Chapter 22 is a synthesis of the application chapters, highlighting the most promising fields and types of applications, suggesting ways that useful lessons or applications in one field might be used in another field, and analysing, in the perspective of artificial intelligence, where the field of Bayesian networks as a whole is heading.

A companion website for this book can be found at: www.wiley.com/go/pourret

Contributors

OLUFIKAYO ADERINLEWO Department of Civil and Environmental Engineering University of Delaware, Newark DE 19716, USA

NII O. ATTOH-OKINE Department of Civil and Environmental Engineering University of Delaware, Newark DE 19716, USA

PHILIPPE BAUMARD Professor, University Paul Cézanne, France, IMPGT 21, rue Gaston Saporta, 13100 Aix-en-Provence, France

NICOLÁS H. BELTRÁN Electrical Engineering Department, University of Chile Av. Tupper 2007, Casilla 412-3, Santiago, Chile

ALEX BIEDERMANN The University of Lausanne – Faculté de Droit et des Sciences Criminelles – École des Sciences Criminelles, Institut de Police Scientifique, 1015 Lausanne-Dorigny, Switzerland

ANDREA BOBBIO Dipartimento di Informatica, Università del Piemonte Orientale, Via Bellini 25 g, 15100 Alessandria, Italy

ROONGRASAMEE BOONDAO Faculty of Management Science, Ubon Rajathanee University Warinchumrab, Ubon Ratchathani 34190, Thailand

LUIS M. DE CAMPOS Departamento de Ciencias de la Computación e Inteligencia Artificial, E.T.S.I. Informática y de Telecomunicaciones, Universidad de Granada, 18071, Granada, Spain

E.J.M. CARRANZA International Institute for Geo-information Science and Earth Observation (ITC), Enschede, The Netherlands

DANIELE CODETTA-RAITERI Dipartimento di Informatica, Università del Piemonte Orientale, Via Bellini 25 g, 15100 Alessandria, Italy

ERIK DAHLQUIST Department of Public Technology, Mälardalen University, S-721 78 Västerås, Sweden

DAVID C. DANIELS 8260 Greensboro Drive, Suite 200, McLean, VA 22102, USA

MANUEL A. DUARTE-MERMOUD Electrical Engineering Department, University of Chile Av. Tupper 2007, Casilla 412-3, Santiago, Chile

ESBEN EJSING Nykredit, Kalvebod Brygge 1-3, 1780 København V, Denmark

JUAN M. FERNÁNDEZ-LUNA Departamento de Ciencias de la Computación e Inteligencia Artificial, E.T.S.I. Informática y de Telecomunicaciones, Universidad de Granada, 18071, Granada, Spain

GIOVANNI FUSCO Associate Professor of Geography and Planning, University of Nice-Sophia Antipolis, UMR 6012 ESPACE, France

LUCAS R. HOPE Clayton School of Information Technology, Monash University, Australia

LINWOOD D. HUDSON 8260 Greensboro Drive, Suite 200, McLean, VA 22102, USA

JUAN F. HUETE Departamento de Ciencias de la Computación e Inteligencia Artificial, E.T.S.I. Informática y de Telecomunicaciones, Universidad de Granada, 18071, Granada, Spain

PABLO H. IBARGÜENGOYTIA Instituto de Investigacions Eléctricas, Av. Reforma 113, Cuernavaca, Morelos, 62490, México

JÖRG KALWA Atlas Elektronik GmbH, Germany

UFFE B. KJÆRULFF Aalborg University, Denmark

KEVIN B. KORB Clayton School of Information Technology, Monash University, Australia

KATHRYN B. LASKEY Dept. of Systems Engineering and Operations Research, George Mason University, Fairfax, VA 22030-4444, USA

SUZANNE M. MAHONEY Innovative Decisions, Inc., 1945 Old Gallows Road, Suite 215, Vienna, VA 22182, USA

ANDERS L. MADSEN HUGIN Expert A/S, Gasværksvej 5, 9000 Aalborg, Denmark

CARLOS MARTÍN Departamento de Ciencias de la Computación e Inteligencia Artificial, E.T.S.I. Informática y de Telecomunicaciones, Universidad de Granada, 18071, Granada, Spain

SHIGERU MASE Department of Mathematical and Computing Sciences, Tokyo Institute of Technology, O-Okayama 2-12-1, W8-28, Meguro-Ku, Tokyo, 152-8550, Japan

STEFANIA MONTANI Dipartimento di Informatica, Università del Piemonte Orientale, Via Bellini 25 g, 15100 Alessandria, Italy

PATRICK NAÏM Elseware, 26–28 rue Danielle Casanova, 75002 Paris, France

ANN E. NICHOLSON Clayton School of Information Technology, Monash University, Australia

JULIETA NOGUEZ Tecnológico de Monterrey, Campus Ciudad de México, México D.F., 14380, México

AGNIESZKA ONIŚKO Faculty of Computer Science, Białystok Technical University, Białystok, 15–351, Poland

THOMAS T. PERLS Geriatric Section, Boston University School of Medicine, Boston, MA 02118, USA

LUIGI PORTINALE Dipartimento di Informatica, Università del Piemonte Orientale, Via Bellini 25 g, 15100 Alessandria, Italy

ALOK PORWAL Center for Exploration Targeting, University of Western Australia, Crawley, WA and Department of Mines and Geology, Rajasthan, Udaipur, India

OLIVIER POURRET Electricité de France, 20 Place de la Défense, 92050, Paris la Défense, France

ALFONSO E. ROMERO Departamento de Ciencias de la Computación e Inteligencia Artificial, E.T.S.I. Informática y de Telecomunicaciones, Universidad de Granada, 18071, Granada, Spain

PAOLA SEBASTIANI Department of Biostatistics, Boston University School of Public Health, Boston MA 02118, USA

J. DOUG STEVENTON Ministry of Forests and Range, British Columbia, Canada

L. ENRIQUE SUCAR Instituto Nacional de Astrofísica, Óptica y Electrónica, Tonantzintla, Puebla, 72840, México

FRANCO TARONI The University of Lausanne – Faculté de Droit et des Sciences Criminelles – École des Sciences Criminelles, Institut de Police Scientifique, 1015 Lausanne-Dorigny, Switzerland

CHARLES R. TWARDY Clayton School of Information Technology, Monash University, Australia

SUNIL VADERA University of Salford, Salford M5 4WT, United Kingdom

PERNILLE VASTRUP Nykredit, Kalvebod Brygge 1-3, 1780 København V, Denmark

SERGIO H. VERGARA Consultant for ALIMAR S.A. Mar del Plata 2111, Santiago, Chile

BRYAN S. WARE 8260 Greensboro Drive, Suite 200, McLean, VA 22102, USA

GALIA WEIDL IADM, University of Stuttgart, Pfaffenwaldring 57, 70550 Stuttgart, Germany

EDWARD J. WRIGHT Information Extraction and Transport, Inc., 1911 North Fort Myer Drive, Suite 600, Arlington, VA 22209, USA

1

Introduction to Bayesian networks

Olivier Pourret

Electricité de France, 20 Place de la Défense, 92050, Paris la Défense, France

1.1 Models

1.1.1 Definition

The primary objective of this book is to discuss the power and limits of Bayesian networks for the purpose of constructing real-world models. The idea of the authors is not to extensively and formally expound on the formalism of mathematical models, and then explain that these models have been – or may be – applied in various fields; the point of view is, conversely, to explain why and how some recent, real-world problems have been modeled using Bayesian networks, and to analyse what worked and what did not.

Real-world problems – thus the starting point of this chapter – are often described as *complex*. This term is however seldom defined. It probably makes more sense to say that human cognitive abilities, memory, and reason are limited and that reality is therefore difficult to understand and manage. Furthermore, in addition to the biological limitations of human capabilities, a variety of factors, either cultural (education, ideology), psychological (emotions, instincts), and even physical (fatigue, stress) tend to distort our judgement of a situation.

One way of trying to better handle reality – in spite of these limitations and biases – is to use representations of reality called *models*. Let us a introduce a basic example.

Bayesian Networks: A Practical Guide to Applications   Edited by O. Pourret, P. Naïm, B. Marcot
© 2008 John Wiley & Sons, Ltd

Example 1. Consider an everyday life object, such as a DVD recorder. The life cycle of the device includes its phases of design, manufacture, marketing, sale, use, possibly break down/repair, and disposal. The owner of a DVD recorder is involved in a temporally delimited part of its life cycle (i.e., when the object is in his/her living-room) and has a specific need: being able to use the device. The user's instruction manual of a DVD recorder is a description of the device, written in natural language, which exclusively aims at explaining how the device is to be operated, and is expressly dedicated to the owner: the manual does not include any internal description of the device.

In this example, the user's instruction manual is a *model* of the DVD recorder.

The 20 application chapters of this book provide numerous examples of models: models of organizations (Japanese electrical companies), of facts (criminal cases), of individuals (students in a robotics course, patients suffering from liver disorders), of devices (a sprinkler system), of places (potentially 'mineralized' geographic areas in India), of documents (texts of the parliament of Andalusia), of commodities (Chilean wines), or of phenomenons (crime in the city of Bangkok, terrorism threats against US military assets). These parts of reality are material or immaterial: we will use the word 'objects' to refer to them.

These objects, which are delimited in time and space, have only one common point: at some stage of their life cycle (either before they actually occurred in reality, or in 'real-time' when they occurred, or after their occurrence) they have been modeled, and Bayesian networks have been employed to construct the model.

Example 1 suggests that the purpose of a model is to satisfy a need of some person or organization having a particular interest in one or several aspects of the object, but not in a comprehensive understanding of its properties. Using the terminology of corporate finance, we will refer to these individuals or groups of individuals with the word *stakeholders*. Examples of stakeholders include users, owners, operators, investors, authorities, managers, clients, suppliers, competitors. Depending on the role of the stakeholder, the need can be to:

- document, evaluate, operate, maintain the object;

- explain, simulate, predict, or diagnose its behavior;

- or – more generally – make decisions or prepare action regarding the object.

The very first benefit of the model is often to help the stakeholder to explicitly state his need: once a problem is explicitly and clearly expressed, it is sometimes not far from being solved.

The construction of a model involves the intervention of at least one human *expert* (i.e., someone with a practical or theoretical experience of the object), and is sometimes also based upon direct, uninterpreted observations of the object. Figure 1.1 illustrates this process: in Example 1, the object is the DVD recorder; the stakeholder is the owner of the device, who needs to be able to perform the installation, connections, setup and operation of the DVD recorder; the model is the user's instruction manual, which is based on the knowledge of some expert

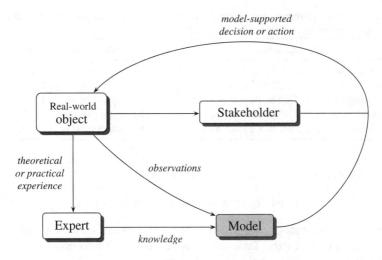

Figure 1.1   Construction and use of a model.

(i.e., the designer), and describes the object from the stakeholder's point of view, in an understandable language; the model-supported action is the use of the device.

Based on the experience of the models discussed in this book, we may agree on the following definition – also fully consistent with Example 1.

Definition 2 (Model)   A model *is a representation of an object, expressed in a specific language and in a usable form, and intended to satisfy one or several need(s) of some stakeholder(s) of the object.*

1.1.2   Use of a model: the inference process

Definition 2 specifies that models are written in a *usable* form. Let us analyse how models are used, i.e., explicitly state the *model-supported decision or action* arrow shown in Figure 1.1.

When the model includes an evaluation of the stakeholder's situation, or a recommendation of decision or action, then the stakeholder makes his decision on the basis of the evaluation or follows the recommendation.

However, most models require – prior to their use – to be adapted to the specific situation of the stakeholder, by the integration of input data.

In Example 1, the input data are the type of the device, the information displayed by the device, and the actions already carried out by the user. The output information is the next action to be done.

Models are thus used to produce information (evaluations, appropriate decisions or actions) on the basis of some input information, considered as valid. This process is called *inference*.

Table 1.1   The inference process: given some input data, what can be inferred from the knowledge of the melting point of gold?

Input data	Inferred information
The ring is of solid gold. Temperature is 1000°C	The ring won't melt.
Temperature is 1000°C. The ring melts.	The ring is not of gold.
Temperature is 1100°C. The ring does not melt.	The ring is not of gold.
Temperature is 1100°C. The ring melts.	The ring is possibly of gold.
The ring is of solid gold. It does not melt.	The temperature is lower than T_m.

For example, if we assume that the statement

$$\text{The melting point of gold is } T_m = 1064.18°C \qquad (1.1)$$

is true, then it constitutes a model which can be used in numerous ways, depending on the available input data: Table 1.1 shows examples of information that can be inferred using the model, on the basis of some input data.

The use of real-world models is not always as straightforward as in the example of Table 1.1. For example, the model user may have some prior belief regarding whether the ring is of gold or not. Also, whether the rings melts or does not melt might be difficult to tell; finally, the temperature might not be known with a high level of precision. In such cases, the use of the model will not produce definitive 'true' statement, but just modify one's assessment of the situation. For instance, if the ring is *believed* not to be of gold, the temperature *estimated* at 1100°C, and the ring *seems* not to melt, then the model output is that the ring is most unlikely of gold. If the uncertainties in the input data can be quantified with probabilities, then the use of the model increases the probability that the ring is not of gold. This is an example of *probabilistic* inference.

1.1.3   Construction

Definition 2 is extremely general: the way a model is constructed obviously depends on several factors, such as the nature of the object, the stakeholder's need(s), the available knowledge and information, the time and resources devoted to the model elaboration, etc. Nevertheless, we may identify two invariants in the process of constructing a model.

1.1.3.1   Splitting the object into elements

One of the precepts of Descartes in his famous *Discourse on the Method* is 'to divide each of the difficulties under examination into as many parts as possible and as might be necessary for its adequate solution' [126].

Indeed, modeling an object implies splitting it into elements and identifying a number of aspects or attributes that characterise the elements.

Deriving a collection of attributes from one single object could at first glance appear as a poor strategy, but this really helps to simplify the problem of satisfying the stakeholder's need: on one hand, each of the attributes is easier to analyze than the whole object; on the other hand, only the attributes which are relevant for the need of the stakeholder are taken into consideration.

1.1.3.2   Saying how it works: the modeling language

To allow inference, the model must include a description of how the elements interact and influence each other. As said in Definition 2, this involves the use of a specific language, which is either textual (natural language, formal rules), mathematical (equations, theorems), or graphical (plans, maps, diagrams).

The well-known consideration 'A good drawing is better than a long speech' also applies to models. Figures are more easily and quickly interpreted, understood and memorized than words. Models which are represented or at least illustrated in a graphical form tend to be more popular and commonly used. It is possible to admit that, throughout history, most successful or unsuccessful attempts of mankind to overcome the complexity of reality have involved, at some stage, a form a graphical representation. Human paleolithic cave paintings – although their interpretation in terms of hunting magic is not beyond dispute – may be considered as the first historical models, in the sense of Definition 2.

1.2   Probabilistic vs. deterministic models

1.2.1   Variables

During the modeling process, the exact circumstances in which the model is going to be used (especially, what input data the model will process) are, to a large extent, unknown. Also, some of the attributes remain unknown when the model is used: the attributes which are at some stage unknown are more conveniently described by *variables*.

In the rest of the chapter, we therefore consider an object which is characterized by a collection of numerical or symbolic variables, denoted $X_1, X_2, \ldots, X_n$. To simplify the formalism, we suppose that the domain of each of the X_j variables, denoted $\mathcal{E}_j$, is discrete.

One may basically distinguish two kinds of variables. The first category is the variables whose values are specified (typically by the stakeholder) at some stage of the model use. Such variables typically describe:

- some aspects of the context: the values of such variables are defined prior to the use of the model and do not change afterwards (in Example 1: which version of the DVD recorder is being installed?);

- some aspects of the object which are directly 'controlled' by the stakeholder:

  - attributes of the object the stakeholder can observe (in Example 1: what is displayed on the control screen of the device?);

- decisions or actions he or she could carry out (in Example 1: what button should be pressed?).

The second category of variables are those which are not directly or not completely controlled – although possibly influenced – by the stakeholder's will (in Example 1: Is the device well setup/out of order?).

At any stage of the practical use of a model, the variables under control have their value fixed, and do not explicitly behave as variables anymore. We may therefore suppose without any loss of generality that the model only comprises variables which are not under control.

1.2.2   Definitions

A deterministic model is a collection of statements, or rules regarding the X_i variables. A sentence (in natural language) such as

$$\text{Elephants are grey} \tag{1.2}$$

is a deterministic model which can be used to identify the race of an African mammal, on the basis of its colour. This model can be considered as a more elegant and intuitive expression of an equation such as

```
colour(elephant)= grey,
```

or of the following formal rule:

```
if animal=elephant then colour=grey.
```

Also, if X_1 and X_2 are variables that correspond to the race and colour of a set of mammals, then the model can be converted in the formalism of this chapter:

$$\text{if } X_1 = \text{'elephant'}, \text{ then } X_2 = \text{'grey'}. \tag{1.3}$$

If the number of variables and the number of possible values of each of them are large, then the object can theoretically reside in a considerable number of states. Let us suppose however that all of these configurations can be enumerated and analyzed. Then the probabilistic modeling of the object consists in associating to any object state (or set of states), a *probability*, i.e., a number between 0 and 1, quantifying how plausible the object state (or set of states) is. We thus define the *joint probability distribution* of the set of variables $X_1, X_2, \ldots, X_n$, denoted

$$\mathbb{P}(X_1, X_2, \ldots, X_n).$$

The domain of this function is the Cartesian product $\mathcal{E}_1 \times \cdots \times \mathcal{E}_n$ and its range is the interval $[0;1]$.

1.2.3 Benefits of probabilistic modeling

1.2.3.1 Modeling power

As far as modeling capability is concerned, probabilistic models are undeniably more powerful than deterministic ones. Indeed, a deterministic model may always be considered as a particular or simplified case of probabilistic model. For example, the model of sentence (1.2) above is a particular case of a slightly more complicated, probabilistic one:

$$x\% \text{ of elephants are grey} \tag{1.4}$$

where $x = 100$. This model can also be written using a conditional probability:

$$\mathbb{P}(X_2 = \text{'grey'} \mid X_1 = \text{'elephant'}) = x. \tag{1.5}$$

Incidentally, the probabilistic model is a more appropriate representation of reality in this example, since, for instance, a rare kind of elephant is white.

1.2.3.2 The power of doubt – exhaustiveness

Doubt is a typically human faculty which can be considered as the basis of any scientific process. This was also pointed out by Descartes, who recommended 'never to accept anything for true which is not clearly known to be such; that is to say, carefully to avoid precipitancy and prejudice, and to comprise nothing more in one's judgement than what was presented to one's mind so clearly and distinctly as to exclude all ground of doubt.' The construction of a probabilistic model requires the systematic examination of all possible values of each variable (each subset $\mathcal{E}_j$), and of each configuration of the object (i.e., each element of $\mathcal{E}_1 \times \cdots \times \mathcal{E}_n$). This reduces the impact of cultural and psychological biases and the risk to forget any important aspect of the object. Furthermore, it is hard to imagine a more precise representation of an object: each of the theoretically possible configurations of the object is considered, and to each of them is associated one element of the infinite set $[0;1]$.

1.2.3.3 Usability in a context of partial information

In many circumstances, probabilistic models are actually much easier to use than deterministic ones. Let us illustrate this with an example.

Example 3. A hiker has gone for a walk in a forest, and brings back home some flashy coloured mushrooms. He wants to decide whether he will have them for dinner, or not. Instead of consulting an encyclopedia of mushrooms, he phones a friend, with some knowledge of the domain. His friend tells him that:

75% of mushrooms with flashy colours are poisonous.

In this example, a deterministic model, such as an encyclopedia of mushrooms, would certainly help identify the exact nature of the mushrooms, but this requires an extensive examination, and takes some time. The probabilistic model provided by the hiker's friend is more suitable to satisfy his need, i.e., make a quick decision for his dinner, than the deterministic one. In fact, if the hiker wants to use the only available information 'the mushroom is flashy-coloured', then a form of probabilistic reasoning – possibly a very basic one – is absolutely necessary.

1.2.4   Drawbacks of probabilistic modeling

In spite of its benefits listed in the previous paragraph, the joint probability distribution $\mathbb{P}(X_1, X_2, \ldots, X_n)$ is rarely employed *per se*. The reason is that this mathematical concept is rather unintuitive and difficult to handle.

Firstly, it can be graphically represented only if $n = 1$ or 2. Even in the in the simplest nontrivial case $n = p = 2$ (illustrated in Figure 1.2), the graphical model is rather difficult to interpret. When $n \geq 3$, no graphical representation is possible, which, as mentioned above, restrains the model usability.

Secondly, the joint probability distribution gives rise to a phenomenon of combinatorial explosion. For instance, if each variable takes on p different values ($p \geq 1$), then the joint probability distribution has to be described by the probabilities of p^n potential configurations of the object, i.e., ten billion values if $n = p = 10$.

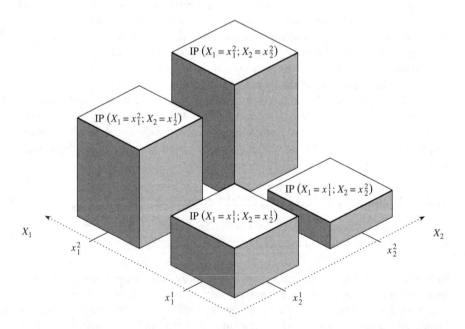

Figure 1.2 Representation of the joint probability distribution of a pair of random variables (X_1, X_2).

1.3   Unconditional and conditional independence

1.3.1   Unconditional independence

Following Descartes's precept of dividing the difficulties, one may try to split the set of n variables into several subsets of smaller sizes which can relevantly be analyzed separately.

Suppose for example that the set of n variables may be divided into two subsets of sizes j and $n - j$ such as:

$$\mathbb{P}(X_1, X_2, \ldots, X_n) = \mathbb{P}\left(X_1, \ldots, X_j\right) \mathbb{P}\left(X_{j+1}, \ldots, X_n\right). \quad (1.6)$$

Then the modeling problem can be transformed into two simpler ones. One can derive the joint probability of subset $X_1, \ldots, X_j$, then that of subset $X_{j+1}, \ldots, X_n$, and use Equation (1.6) to obtain the complete model.

The equality of two functions expressed by Equation (1.6) means that the subsets of variables $(X_1, \ldots, X_j)$ and $(X_{j+1}, \ldots, X_n)$ are independent, or – to avoid confusion with a concept which is defined below – *unconditionally* independent. This means that any information regarding the $(X_1, \ldots, X_j)$ subset (for instance, '$X_1 = 7$' or '$X_1 + X_2 > 3$') does not change the probability distribution of the second subset $(X_{j+1}, \ldots, X_n)$.

However, unconditional independence between two subsets of variables is very unlikely to happen in real-world models. If it does happen, the initial definition of the object is not relevant: in such a case, it makes more sense to construct two separate models.

1.3.2   Conditional independence

A more common – or at least much more reasonably acceptable in real-world models – phenomenon is the so-called 'conditional independence'. Let us introduce this concept by two examples.

1.3.2.1   The lorry driver example

Example 4. A lorry driver is due to make a 600-mile trip. To analyze the risk of his falling asleep while driving, let us consider whether (1) he sleeps 'well' (more than seven hours) on the night before and (2) he feels tired at the beginning of the trip.

In this example, there are obvious causal relationships between the driver's sleep, his perceived fatigue, and the risk of falling asleep: the three variables are dependent. Let us suppose however that we know that the lorry driver feels tired at the beginning of the trip. Then knowing whether this is due to a bad sleep the previous night, or to any other reason is of no use to evaluate the risk. Similarly, if the lorry driver does *not* feel tired at the beginning of the trip, one may then consider that the quality of his sleep on the night before has no influence on the risk. Given these considerations, the risk of falling asleep his said to be *conditionally independent* of the quality of sleep, *given* the lorry driver's fatigue.

To express it formally, let X_1, X_2 and X_3 be binary variables telling whether the lorry driver sleeps well the night before, whether he feels tired at the beginning of the trip, and whether he will fall asleep while driving. Then X_3 is independent of X_1, for any given value of X_2. In terms of probabilities, we have:

$$\mathbb{P}(X_3 \,|\, X_1 \text{ and } X_2) = \mathbb{P}(X_3 \,|\, X_2). \tag{1.7}$$

In such a case, knowing the values of X_1 and X_2 is not better than knowing only the value of X_2, and it is useless to describe the behavior of X_1, X_2, X_3 by a function of three variables; indeed, we may deduce from Equation (1.7):

$$\mathbb{P}(X_1, X_2, X_3) = \mathbb{P}(X_1)\,\mathbb{P}(X_2 \,|\, X_1)\,\mathbb{P}(X_3 \,|\, X_2), \tag{1.8}$$

which shows that the risk model can be constructed by successively studying the quality of sleep, then its influence on the state of fatigue, and then the influence of the state of fatigue on the risk of falling asleep.

1.3.2.2   The doped athlete example

Example 5. In a sports competition, each athlete undergoes two doping tests, aimed at detecting if he/she has taken a given prohibited substance: test A is a blood test and test B a urine test. The two tests are carried out in two different laboratories, without any form of consultation.

It is quite obvious in Example 5 that the results of the two tests are not independent variables. If test A is positive, then the participant is likely to have used the banned product; then test B will probably be also positive.

Now consider a participant who has taken the banned substance. Then tests A and B can be considered independent, since the two laboratories use different detection methods. Similarly, tests A and B can be considered independent when the participant has *not* taken the banned substance: the results of both tests are conditionally independent, given the status of the tested athlete. Formally, if X_1 is a binary variable telling whether the athlete is 'clean' or not, X_2 is the result of test A, and X_3 the result of test B, we can write:

$$\mathbb{P}(X_3 \,|\, X_1 \text{ and } X_2) = \mathbb{P}(X_3 \,|\, X_1). \tag{1.9}$$

Equation (1.9) can exactly be translated into 'knowing whether the athlete has taken the substance is enough information to estimate the chances of test B being positive'. A symmetrical equation holds regarding test A:

$$\mathbb{P}(X_2 \,|\, X_1 \text{ and } X_3) = \mathbb{P}(X_2 \,|\, X_1). \tag{1.10}$$

Here again, it is useless to describe the behavior of X_1, X_2, X_3 by a function of three variables. Equations (1.9) and (1.10) yield:

$$\mathbb{P}(X_1, X_2, X_3) = \mathbb{P}(X_1)\,\mathbb{P}(X_2 \,|\, X_1)\,\mathbb{P}(X_3 \,|\, X_1), \tag{1.11}$$

which means that considerations on the proportion of doped athletes $\mathbb{P}(X_1)$, and on the reliabilities of each tests are sufficient to construct the model.

1.4 Bayesian networks

1.4.1 Examples

In the lorry driver and doped athlete examples, we have identified the most direct and significant influences betweens the variables, and simplified the derivation of the joint probability distribution. By representing these influences in a graphical form, we now introduce the notion of Bayesian network.

In Example 4, our analysis has shown that there is an influence of variable X_1 on variable X_2, and another influence of variable X_2 on variable X_3; we have assumed that there is no direct relation between X_1 and X_3. The usual way of representing such influences is a diagram of nodes and arrows, connecting influencing variables (parent variables) to influenced variables (child variables). The structure corresponding to Example 4 is shown in Figure 1.3.

Similarly, the influences analyzed in Example 5 may be represented as shown in Figure 1.4.

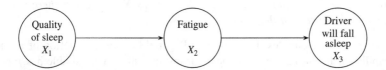

Figure 1.3   A representation of the influences between variables in Example 4. Variable X_3 is *conditionally independent* of X_1 given X_2.

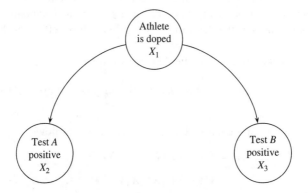

Figure 1.4   A representation of the influences between variables in Example 5. Variables X_2 and X_3 are *conditionally independent* given X_1.

Considering the graphical structures of Figures 1.3 and 1.4, and more precisely the parents of each variable, we observe that both Equations (1.8) and (1.11) can be written in the following form:

$$\mathbb{P}(X_1, X_2, X_3) = \mathbb{P}(X_1 \mid parents(X_1)) \, \mathbb{P}(X_2 \mid parents(X_2)) \, \mathbb{P}(X_3 \mid parents(X_3)).$$
(1.12)

Equation (1.12) is the formal definition of a Bayesian network, in the three-variable case: through a process of analyzing and sorting out the unconditional independences between the three variables, we have been able to convert $\mathbb{P}(X_1, X_2, X_3)$ into a product of three conditional probabilities. This definition is generalized in the next paragraph.

1.4.2 Definition

Definition 6 (Bayesian network) *Let us consider n random variables $X_1, X_2, \ldots, X_n$, a directed acyclic graph with n numbered nodes, and suppose node j ($1 \leq j \leq n$) of the graph is associated to the X_j variable. Then the graph is a* Bayesian network, *representing the variables $X_1, X_2, \ldots, X_n$, if:*

$$\mathbb{P}(X_1, X_2, \ldots, X_n) = \prod_{j=1}^{n} \mathbb{P}\left(X_j \mid parents(X_j)\right),$$
(1.13)

where: $parents(X_j)$ denotes the set of all variables X_i, such that there is an arc from node i to node j in the graph.

As shown in the examples, Equation (1.13) simplifies the calculation of the joint probability distribution. Let us suppose for instance that each variable has p possible values, and less than three parent variables. Then the number of probabilities in the model is lower than $n.p^4$, although the object can reside in p^n configurations. If $n = p = 10$, the reduction factor is greater than one hundred thousands.

A crucial point is that this simplification is based on an graphical, intuitive representation, and not on some highly technical considerations. A diagram of boxes and arrows can be easily interpreted, discussed and validated on a step-by-step basis by the stakeholders: there is no 'black box' effect in the modeling process.

Another important remark can be deduced from Definition 6.

Proposition 7. *Any joint probability distribution may be represented by a Bayesian network.*

Indeed, we may formally express $\mathbb{P}(X_1, X_2, \ldots, X_n)$ as follows:

$$\mathbb{P}(X_1, X_2, \ldots, X_n) = \mathbb{P}(X_1) \, \mathbb{P}(X_2, \ldots, X_n \mid X_1)$$

$$= \mathbb{P}(X_1) \, \mathbb{P}(X_2 \mid X_1) \cdots \mathbb{P}(X_3, \ldots, X_n \mid X_1, X_2)$$

$$= \cdots$$

$$= \mathbb{P}(X_1) \, \mathbb{P}(X_2 \mid X_1) \cdots \mathbb{P}(X_n \mid X_1, \ldots, X_{n-1}). \quad (1.14)$$

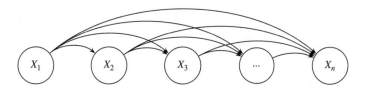

Figure 1.5 'Generic' structure of a Bayesian Network, suitable for any joint probability distribution of n random variables $X_1, \ldots, X_n$.

Equation (1.14) proves that the structure of Figure 1.5, with one arc from variable X_i to X_j, whenever $i < j$, is suitable to represent any joint probability distribution between n variables. In other words, there is no loss of generality in modeling a set of random variables with a Bayesian network.

Proposition 7 helps to answer a frequently asked question: *in Definition 6, why has the graph to be acyclic?* Besides the fact that Equation (1.13) would not make sense in the presence of loops, the hypothesis of graphical acyclicity is not at all restrictive: whatever the number and nature of the dependencies between the X_i variables, there is always at least one acyclic structure (i.e., that of Figure 1.5) that is suitable to represent the object.

Proposition 7 clearly shows the modeling power of Bayesian Networks. As mentioned above, any deterministic model is a particular case of probabilistic model; any probabilistic model may be represented as a Bayesian network.

2

Medical diagnosis

Agnieszka Oniśko

Faculty of Computer Science, Białystok Technical University, Białystok, 15–351, Poland

2.1 Bayesian networks in medicine

Medical decisions are hard. One of the reasons for this difficulty is that they often have to be made on the basis of insufficient and uncertain information. Furthermore, the outcome of the decision process has far-reaching implications on the well-being or even the very lives of patients. There is a substantial body of empirical evidence that shows that human decision-making performance is far from optimal. Furthermore, its quality decreases with the complexity of problems, time pressure, and high stakes. Therefore, given the increasing diversity of treatments and drugs, the accumulating body of knowledge about human health, the availability of diagnostic tests, including genetic tests, medical specialists often need to be assisted in their decision-making. At the same time, increasing costs of medical care and amounting risk of litigations following medical errors increase the pressures on the medical community to become cost-effective and to make fewer errors. Given all this, despite the fact that the medical community is rather conservative and resists technological assistance, computer support for medical decision-making is an inevitable fact. This is true in a number of aspects of medical care, including its initial phase when a physician has to come up with a preliminary prognosis or diagnosis and specify the possible directions for the patient's treatment.

Bayesian Networks: A Practical Guide to Applications Edited by O. Pourret, P. Naïm, B. Marcot
© 2008 John Wiley & Sons, Ltd

Computer-based tools have the potential to make a difference in medicine. Especially methods that are able to leverage on both available data and clinical experience and are at the same time based on sound foundations that offers some guarantees of accuracy, have a high potential. Bayesian networks are such tools and are especially suited for modeling uncertain knowledge. It is not surprising that some of their first applications are in medicine. These applications include first of all the management of a disease for individual patient: establishing the diagnosis, predicting the results of the treatment, selecting the optimal therapy. This chapter will discuss these three in depth. One should not forget, however, that other applications of Bayesian networks in medicine and related fields have been proposed and many others may be conceived. Chapter 3 of this book discusses a BN-based clinical decision support tool, and Chapter 4 is devoted to genetic models. Some other examples of applications are clinical epidemiology [183], disease surveillance [101], BARD (Bayesian Aerosol Release Detector) [68], or prediction of a secondary structure of a protein [399], or discovering causal pathways in gene expression data [508]. A special issue of the journal *Artificial Intelligence in Medicine* [288] is devoted to applications of Bayesian networks to biomedicine and health-care and lists several interesting applications.

2.1.1 Diagnosis

Medical diagnosis is often simplified to reasoning that involves building hypothesis for each disease given the set of observed findings that a patient is suffering from. This reasoning results from choosing the hypothesis that is the most probable for a set of observations. Formally, it may be expressed by Equation (2.1).

$$\text{diagnosis} = \max_i \mathbb{P}\left(D_i \mid E\right). \tag{2.1}$$

$\mathbb{P}\left(D_i \mid E\right)$ is the probability of the disease D_i given the evidence E that represents the set of observed findings such as symptoms, signs, and laboratory test results that a patient is presenting with.

Application of Bayesian networks to medical diagnosis was proposed more than a decade ago. The first expert system based on a Bayesian network model was CONVINCE [247]. Other early systems built based on a Bayesian network model were NESTOR, the system for diagnosis of endocrinological disorders [102], MUNIN, the system to diagnose neuromuscular disorders [16], or the classic, ALARM monitoring system [40]. Later systems are PATHFINDER IV, the system for diagnosis of the lymph system diseases [207] or the decision-theoretic version of QMR, the system for diagnosis in internal medicine that was based on the CPCS model (Computer-based Patient Case Simulation system) [428], [318]. Another diagnostic system based on the Bayesian network model was DIAVAL, the expert system for echocardiography [130].

2.1.2   Prognosis

Medical prognosis attempts to predict the future state of a patient presenting with a set of observed findings and assigned treatment. Formally, it may be expressed by Equation (2.2).

$$\text{prognosis} = \mathbb{P}\,(O\,|\,E,\,T)\,. \tag{2.2}$$

The variable E is the evidence, i.e., a set of observed findings such as symptoms, signs, and laboratory test results; T stands for a treatment prescribed for a patient, and the variable O is the outcome that may represent, for example, life expectancy, quality of life, or the spread of a disease.

There are fewer applications of Bayesian networks to medical prognosis than there are to medical diagnosis; for example, predicting survival in malignant skin melanoma [429], or NasoNet, a temporal Bayesian network model for nasopharyngeal cancer spread [172].

2.1.3   Treatment

An extension of Bayesian networks, influence diagrams, allows us to model decisions explicitly. This approach is capable of modeling different treatment decision within their uncertain variables and preferences that usually represent costs and benefits. Reasoning over possible treatment alternatives results in a selection of most optimal therapy. Examples of applications are: the system for assisting clinicians in diagnosing and treating patients with pneumonia in the intensive-care unit [286], the model for patient-specific therapy selection for oesophageal cancer [469], or the system for the management of primary gastric non-Hodgkin lymphoma [285].

2.2   Context and history

My experience in application of Bayesian networks in medicine covers diagnosis of liver disorders, diagnosis of prostate cancer and benign prostatatic hyperplasia, and a model for screening cervical cancer. The author of this chapter has been using Bayesian networks as a modeling tool for last eight years. In this chapter we will share with the reader our experience in building Bayesian network models based on HEPAR II, the model for the diagnosis of liver disorders.

2.2.1   The HEPAR II project

The HEPAR II project [350, 351] is a system for the diagnosis of liver disorders that is aimed at reducing the number of performed biopsies. The main component of HEPAR II is a Bayesian network model capturing over 70 variables. The model covers 11 different liver diseases and 61 medical findings, such as patient self-reported data, signs, symptoms, and laboratory tests results. HEPAR II is

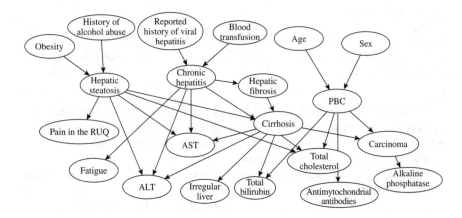

Figure 2.1   A simplified fragment of the HEPAR II network.

quantified by 1488 numerical parameters. Figure 2.1 shows a simplified fragment of the HEPAR II model.[1]

2.2.2   Software

An invaluable resource in building and testing Bayesian network models have been two software packages: GENIE, a development environment for reasoning in graphical probabilistic models, and SMILE, its inference engine; and Elvira, software offering innovative explanation capabilities. Both software packages are available free of charge for any use. GENIE has several thousand users, including universities, research centers and also industry.

GENIE offers several useful facilities such as a module for exploratory data analysis and learning Bayesian networks and their numerical parameters from data, dynamic Bayesian networks. It has a special module that addresses problems related to diagnosis. The program has a user-friendly graphical user interface that supports model building and analysis, and runs both under Windows and Linux. SMILE is its platform-independent inference engine that is available among others for Windows, Linux, Sun Solaris, and even Pocket PC. The software is fast, reliable, well documented, and well supported. GENIE and SMILE were both developed at the Decision Systems Laboratory, University of Pittsburgh, and are available for download.[2]

Elvira is a software package for building and reasoning in graphical probabilistic models [128, 260]. While Elvira is of somewhat academic appearance, speed, and reliability, it has implemented several great ideas in its user interface. Its explanation capabilities, for example, offered significant insight in refining the HEPAR II model [261]. Elvira allows for an explanation of the model (static explanation) and

---

[1]Full version of HEPAR II can be found at http://genie.sis.pitt.edu/networks.html
[2]http://genie.sis.pitt.edu/

explanation of inference (dynamic explanation). Elvira is written in Java and is, therefore, fully platform-independent. Elvira is the product of a joint project of several Spanish universities and is available for download.[3]

2.3   Model construction

There are two basic approaches to construct Bayesian network models: manual building based purely on human expert knowledge and automatic learning of the structure and the numerical parameters from data. HEPAR II was constructed based on a hybrid approach: the structure was built based on expert knowledge and available literature and the numerical parameters of the model, i.e., prior and conditional probability distributions for all the nodes, were learned from the HEPAR database. The HEPAR database was created in 1990 and thoroughly maintained since then by Dr. Wasyluk at the Gastroentorogical Clinic of the Institute of Food and Feeding in Warsaw. The current database contains over 800 patient records and its size is steadily growing. Each hepatological case is described by over 160 different medical findings, such as patient self-reported data, results of physical examination, laboratory tests, and finally a histopathologically verified diagnosis. The version of the HEPAR data set that was used to quantify HEPAR II consisted of 699 patient records.

2.3.1   Interactions with domain experts

Manual constructing a Bayesian network model typically involves interaction of the knowledge engineer, who is responsible for building the model, with domain experts. In case of the HEPAR II model, frequent short sessions with the expert worked well. In between these sessions, the knowledge engineer focused on refining the model and preparing questions for the expert. The refinement consisted of analyzing positive and negative influences[4] in the model when the model was fully quantified, i.e., when the numerical parameters of the network were already specified. It helps when the knowledge engineer understands the domain at least at a basic level. It is a good idea to read at least a relevant section of a medical textbook on the topic of the meeting with the expert, so the knowledge engineer is familiar with the terminology, the variables, and interaction among them. It is recommended, if the expert is comfortable with this, to record the sessions with the expert because it is often hard to process all the medical knowledge that is provided by a domain expert during a meeting. It is also recommended to organize brainstorming sessions with the participation of knowledge engineers and medical experts who are not directly involved in building the model. With respect

---

[3]http://www.ia.uned.es/~elvira.

[4]In causal models, most of the influences are positive, because usually the presence of the cause increases the probability of the effect's presence. Analogously, negative influences decreases the probability of the effect's presence.

to HEPAR II there were a few such sessions, and they addressed important issues and raised questions about the model.

2.3.2   Building the graphical structure of the model

The structure of the model, (i.e., the nodes of the graph along with arcs among them) was built based on medical literature and conversations with the domain experts, a hepatologist Dr. Hanna Wasyluk and two American experts, a pathologist, Dr. Daniel Schwartz, and a specialist in infectious diseases, Dr. John N. Dowling. The elicitation of the structure took approximately 50 hours of interviews with the experts, of which roughly 40 hours were spent with Dr. Wasyluk and roughly 10 hours spent with Drs. Schwartz and Dowling. This includes model refinement sessions, where previously elicited structure was reevaluated in a group setting. The structure of the model consists of 70 nodes and 121 arcs, and the average number of parents per node is equal to 1.73. There are on the average 2.24 states per variable.

2.3.2.1   Selecting variables and their domains

Building a model usually relates to the process of selecting the variables that represent certain features or events. In medical models, the variables are findings describing a patient condition, i.e., risk factors, symptoms, signs, results of physical examination and laboratory tests, and disorders that a patient can be potentially suffering from. Because the number of variables is typically large, it is often necessary to perform data reduction. This usually consists of feature selection and discretization of continuous variables.

2.3.2.2   Clarity test

One of the fundamental tools used in building decision-analytic models is the clarity test [220]. Essentially, for each element of the model, such as a variable and its states, clarity test probes whether it has been clearly defined. Vague and imprecise definitions of model elements will generally backfire at some stage of modeling, certainly when it comes to elicitation of numerical parameters.

When specifying states of a variable, the knowledge engineer should pay attention to whether the states represent mutually exclusive values. For example, in the initial version of the HEPAR II model, the possible disorders were modeled as one node. This representation did not reflect faithfully the domain of hepatology, because in practice liver disorders are not mutually exclusive. In case the possible states of a certain variable are not mutually exclusive, they should be split into several nodes.

A model is by definition a simplification of the real world. Most of the categorical variables in HEPAR II were modeled as binary nodes. In some cases, this simplification led to certain difficulties during the model building. The HEPAR data set that was available to me contained several simplifications that had impact on the model. For example, in case of the variable *Alcohol abuse*, the expert would have preferred to distinguish several states indicating different degrees of alcoholism depending on the amount, frequency, and duration of addiction. However, the variable was represented in the model as a binary node with two outcomes: *present* and *absent*. It is important to know that this kind of simplification often leads to difficulties in parameter elicitation from a human expert. There is no real need for this simplification from the point of view of the modeling tool, so one should carefully weight the resulting simplicity of the model against its expressiveness.

One of the purposes of modeling is documenting the problem. It is important for a software package to offer possibilities for entering comments into the model. GENIE offers the possibility of adding a comment to almost every part of the model. In addition to on-screen comments, the knowledge engineer can add comments to nodes, arc, node states, and even individual probabilities. It is very useful to record in the model how a particular variable was elicited. The knowledge engineer can then recall the source of information and the possible simplifying assumptions made in the process. For example, findings describing physical features of the liver, like its shape and density, can be elicited based on both ultrasound examinations and palpation. The knowledge engineer should also pay attention to variables that represent patient self-reported data. Some of these variables are not reliable enough, for example, *History of viral hepatitis* or *Alcohol abuse*. In case of *History of viral hepatitis*, it is possible that patients who suffer from an asymptomatic viral hepatitis seldom know about the presence of the virus. With respect to *Alcohol abuse*, it is a clinical rule of thumb that the patient usually consumes at least three times more alcohol than he or she reports to the doctor. Therefore, when using clinical data it is important to keep in mind that patient self-reported data may be unreliable. The knowledge engineer should know how to capture them correctly in the model.

2.3.2.3 Single-disorder vs. multiple-disorder diagnosis model

One of the advantages of Bayesian networks, compared to other modeling tools, is that they allow us to model the simultaneous presence of multiple disorders. Many approaches assume that for each diagnostic case only one disorder is possible, i.e., various disorders are mutually exclusive. This is often an unnecessarily restrictive assumption and it is not very realistic in medicine. The presence of a disorder often weakens a patient's immune system and as a result the patient may develop multiple disorders. It happens fairly often that a patient suffers from multiple disorders and a single disorder may not account for all observed symptoms. Worse even, a situation can arise that a single disorder offers a better explanation for all observations than any other single disorder, while the true diagnosis consists of, for example, two other disorders appearing simultaneously.

Initially, HEPAR II was a single-disorder diagnosis model (SDM). The model was extended later to a multiple-disorder diagnosis model (MDM) (see Figure 2.2). The model was transformed into a network that can perform multiple-disorder diagnosis with some benefits to the quality of numerical parameters learnt from the database. The transformation from single-disorder to multiple-disorder diagnosis

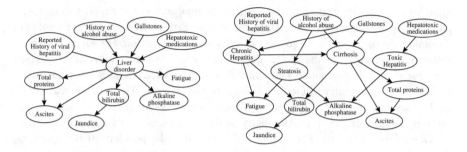

Figure 2.2 A simplified fragment of the HEPAR II model: Single-disorder diagnosis (left) and multiple-disorder diagnosis (right) version.

Table 2.1   Characteristics of SDM (single-disorder model) and MDM (multiple-disorder model) versions of HEPAR II.

Version	Size	Arcs	Connectivity	Parameters
SDM	67	78	1.16	3714
MDM	70	121	1.73	1488

model improved the diagnostic performance of HEPAR II [350]. The multiple-disorder diagnosis model is more realistic and has a higher value in clinical practice.

The structural changes of HEPAR II had an impact on the quantitative part of the model, i.e., they decreased the number of numerical parameters in the model. It often happens in learning conditional probability distributions from data that there are too few records corresponding to a given combination of parents of a node. Breaking the original disorder node, modeled in SDM, into several nodes representing individual disorders in MDM decreased the size of the conditional probability tables. Additionally, this transformation increased the average number of records for each combination of parents in a conditional probability distribution table. The multiple-disorder version of HEPAR II required only 1488 parameters ($\mu = 87.8$ data records per conditional probability distribution) compared to the 3714 ($\mu = 16.8$ data records per conditional probability distribution) parameters needed for the single-disorder version of the model. With an increase in the average number of records per conditional probability distribution, the quality of the model parameters improved. Table 2.1 presents characteristics of the two versions of the HEPAR II model.

2.3.2.4   Causal interpretation of the structure

During the initial modeling sessions of HEPAR II experts tended to produce graphs with arcs in the diagnostic direction. This was an initial tendency that disappeared after a few model-building sessions: the experts quickly got used to modeling the causal structure of the domain. Causal graphs facilitate interactions among multiple experts. Causal connections and physiological mechanisms that underlie them are a part of medical training and provide a common language among the experts participating in the session. The experts working on HEPAR II rarely disagreed about the model structure. A brief discussion of the pathophysiology of the disease usually led to a consensus.

During the model building, it is important for the expert to understand that each causal link in the graph represents a certain causal mechanism.[5] Hence, when the knowledge engineer poses a question to the expert, he or she should explicitly ask about the existence and operation of such mechanisms. However, sometimes following the causal structure is difficult because of lack of medical knowledge – then the model structure has to be simply based on the correlation between variables. For example, due to lack of causal knowledge, there was some difficulty with modeling the variable *Elevated triglicerides*. It is not known with certainty in medicine whether the variable *Elevated trigliceridies* is a cause or a symptom of *Steotosis* (fatty liver). Figure 2.3 presents two ways of modeling *Elevated triglicerides* that were considered in the HEPAR II model. The expert found it more intuitive to model *Elevated triglicerides* as an effect of *Steatosis*.

There were several links in the model that were refined from probabilistic to causal. For example, *Elevated serum urea* was modeled initially as an effect of *Cirrhosis* and as a cause of *Hepatic encephalopathy* (see the left-hand side network captured at Figure 2.4). Certainly, the link from *Elevated serum urea* to *Hepatic encephalopathy* did not represent a causal mechanism, therefore, this fragment of the network was refined into the right-hand side network in Figure 2.4.

Another example of refinement of causal structure is given in Figure 2.5. The figure captures a family of nodes related to the causes of *Hepatolmegaly* (enlarged liver size). It was not clear whether the link between *Hepatotoxic medications* and *Hepatomegaly* had been modeled correctly. The expert found it initially unnecessary to model this relationship since there was a path between these variables

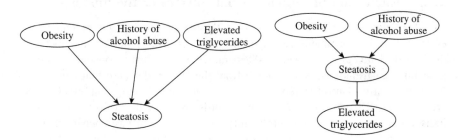

Figure 2.3 HEPAR II: Modeling the variable *Elevated triglicerides*.

---

[5]A causal mechanism is a mechanism describing the causal relations in the domain.

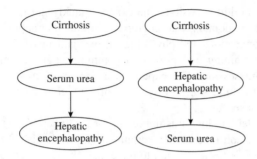

Figure 2.4 HEPAR II: Modeling causal mechanisms.

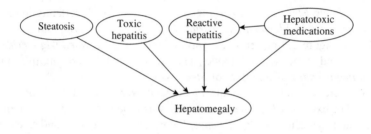

Figure 2.5 HEPAR II: Modeling the causes of the variable *Hepatomegaly*.

through *Reactive hepatitis* node. The knowledge engineer explained to the expert that if there is another causal mechanism that leads directly to enlarged liver size due to taking *Hepatotoxic medications*, then the relationship between the node *Hepatotoxic medications* and the node *Hepatomegaly* should remain.

2.3.3   Elicitation of numerical parameters of the model

Existing data sets of cases can significantly reduce the knowledge engineering effort required to parameterize Bayesian networks. The numerical parameters of the HEPAR II model (there were 1488 parameters), i.e., the prior and conditional probability distributions, were learned from the HEPAR database that consisted of 699 patient cases. Unfortunately, when a data set is small, many conditioning cases are represented by too few or no data records and they do not offer a sufficient basis for learning conditional probability distributions. In cases where there are several variables directly preceding a variable in question, individual combinations of their values may be very unlikely to the point of being absent from the data file. In such cases, the usual assumption made in learning the parameters is that the distribution is uniform, i.e., the combination is completely uninformative. The next subsection describes the approach that was applied to learn the numerical parameters of HEPAR II from sparse data sets such as HEPAR.

2.3.3.1   Enhancing parameters learned from sparse data sets

The proposed approach involved enhancing the process of learning the conditional probability tables (CPTs) from sparse data sets by combining the data with structural and numerical information obtained from an expert. The approach applied Noisy-OR gates [129, 210, 356].

A Noisy-OR gate is one of the canonical model which applies when there are several possible causes $X_1, X_2, \ldots, X_n$ of an effect variable Y, where (1) each of the causes X_i has a probability p_i of being sufficient to produce the effect in the absence of all other causes, and (2) the ability of each cause being sufficient is independent of the presence of other causes. The above two assumptions allow us to specify the entire conditional probability distribution with only n parameters $p_1, p_2, \ldots, p_n$. p_i represents the probability that the effect Y will be true if the cause X_i is present and all other causes X_j, $j \neq i$, are absent. In other words,

$$p_i = \mathbb{P}\left(y \mid \overline{x}_1, \overline{x}_2, \ldots, x_i, \ldots, \overline{x}_{n-1}, \overline{x}_n\right). \tag{2.3}$$

It is easy to verify that the probability of y given a subset X_p of the X_is that are present is given by the following formula:

$$\mathbb{P}\left(y \mid X_p\right) = 1 - \prod_{i:X_i \in X_p} (1 - p_i). \tag{2.4}$$

This formula is sufficient to derive the complete CPT of Y conditional on its predecessors $X_1, X_2, \ldots, X_n$.

Given an expert's indication that an interaction in the model can be approximated by a Noisy-OR gate [129, 210, 356], the Noisy-OR parameters for this gate were first estimated from data. Subsequently, in all cases of a small number of records for any given combination of parents of a node, the probabilities for that case as if the interaction was a Noisy-OR gate were generated. Effectively, the obtained conditional probability distribution had a higher number of parameters. At the same time, the learned distribution was smoothed out by the fact that in all those places where no data was available to learn it, it was reasonably approximated by a Noisy-OR gate.

This experiment has shown that diagnostic accuracy[6] of the model enhanced with the Noisy-OR parameters was 6.7% better than the accuracy of the plain multiple-disorder model and 14.3% better than the single-disorder diagnosis model. This increase in accuracy was obtained with very modest means – in addition to structuring the model so that it is suitable for Noisy-OR nodes, the only knowledge elicited from the expert and entered in the learning process was which interactions can be approximately viewed as Noisy-ORs. The study showed that whenever combining expert knowledge with data, and whenever working with experts in general, it pays off generously to build models that are causal and reflect reality as much as possible, even if there are no immediate gains in accuracy.

---

[6]The diagnostic accuracy of the model was defined as the percentage of correct diagnoses (true positive rate) on real patient cases from the HEPAR database.

2.3.3.2 Handling missing data

Missing values of attributes in data sets, also referred to as incomplete data, pose difficulties in learning Bayesian network structure and its numerical parameters.

Missing values in the HEPAR data set have been a major problem in the HEPAR II project. There were 7792 missing values (15.9% of all entries) in the learning data set. The HEPAR data set contained no records that are complete.

There was a study that investigated and compared the diagnostic accuracy of HEPAR II for several methods for obtaining numerical parameters of a Bayesian network model from incomplete data. The methods included the EM algorithm, Gibbs sampling, bound and collapse, replacement by 'normal' values, missing as an additional state, replacement by mean values, hot deck imputation, or replacement based on the k-NN method.

The diagnostic accuracy for most of the methods that were tested was similar, while 'replacement by 'normal' values' and the EM algorithm performed slightly better than other approaches. Even though the HEPAR data set contained many incomplete values and one would expect even small performance differences to be amplified, this did not happen.

The performance differences among methods virtually disappeared when a data set in which data were missing truly at random was generated. This is consistent with the observation made by Ramoni and Sebastiani [381], who found that performance differences among various methods for dealing with missing data were minimal. Their data set was real but the missing data were generated at random.

The advice to those knowledge engineers who encounter data sets with missing values is to reflect on the data and find out what the reasons are for missing values. In the case of medical data sets, the assumption postulated by Peot and Shachter[7] seems very reasonable [360]. Even in this case, however, the advise is to verify it with the expert.

2.4 Inference

The most crucial task of an expert system is to draw conclusions based on new observed evidence. The mechanism of drawing conclusions in a system that is based on a probabilistic graphical model is known as *propagation of evidence*. Propagation of evidence involves essentially updating probabilities given observed variables of a model (also known as *belief updating*).

For example, in case of a medical model without any observations, updating will allow us to derive the prevalence rate[8] of each of the disorders. Once

---

[7]Peot and Shachter argued convincingly that data in medical data sets are not missing at random and that there are two important factors influencing the probability of reporting a finding. The first factor is a preference for reporting symptoms that are present over symptoms that are absent. The second factor is a preference for reporting more severe symptoms before those that are less severe. In other words, if a symptom is absent, there is a high chance that it is not reported, i.e., it is missing from the patient record. And conversely, a missing value suggests that the symptom was absent.

[8]Prevalence rate is defined as the percentage of the total number of cases of a given disease in a specified population. In other words, it measures the commonality of a disease.

specific characteristics of a patient are entered, the model will produce the prevalence among the population group that the patient is coming from. Entering risk factors, symptoms, and test results in the course of the interaction with the system will further allow us to compute the probability distribution of each of the disorders for this particular patient case. This distribution can be directly used in assisting medical decisions, for example, by allowing the physician to focus on those disorders that are most likely. The inference in the HEPAR II model was performed with a use of SMILE and GENIE.

When a model is enhanced with measures expressing the benefits and costs of correct diagnosis and misdiagnosis, respectively, the system can further suggest the diagnosis that is optimal given the circumstances. The probabilistic approach allows for application of such methods as value of information (VOI) computation that essentially calculates the expected gain from performing various medical tests and allows for prioritizing various steps of the diagnostic procedure. This feature is implemented in the diagnostic module of GENIE that is described in Section 2.4.1.

2.4.1 Diagnostic GENIE

Figure 2.6 shows a screen shot of the user interface of diagnostic GENIE. The right-hand side of the window contains a complete list of all possible medical findings included in the HEPAR II model. The top right part of the window contains a list

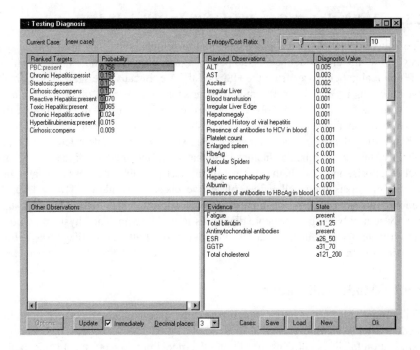

Figure 2.6 Screen shot of a diagnostic GENIE.

of those possible findings that have not yet been observed along with an indication of their diagnostic value for the pursued disorder (*Primary Biliary Cirrhosis (PBC)* in this case). Those findings that have been observed are brought over to the right bottom part of the window. Right-clicking on any of the finding brings up a pop-up menu that lists all possible values of the selected variable. By choosing one of these values the user can enter a finding. The top left column presents a ranked list of the possible diagnoses along with their associated probabilities, the latter being presented graphically. The probabilities are updated immediately after entering each finding. Updating the probabilities and presenting a newly ordered list of possible disorders takes in the HEPAR II model a fraction of a second and is from the point of view of the user instantaneous. This interface allows us further to save a patient case in a repository of cases and to return to it at a later time.

2.5   Model validation

Validation of the model justifies whether the model that was built is correct. Validation usually involves checking performance of the model and its further improvement. The performance of the model can be expressed by various measures, for example, sensitivity (true positive rate), specificity (true negative rate), false positive rate, or false negative rate.

2.5.1   Validation of the model with the clinical data

The HEPAR II model was validated based on the clinical data. The diagnostic accuracy of the model was defined as the percentage of correct diagnoses (true positive rate) on real patient cases from the HEPAR database. Because the same, fairly small database was used to learn the model parameters, the method of 'leave-one-out' was applied [323], which involved repeated learning from 698 records out of the 699 records available and subsequently testing it on the remaining 699th record. When testing the diagnostic accuracy of HEPAR II, we were interested in both (1) whether the most probable diagnosis indicated by the model is indeed the correct diagnosis, and (2) whether the set of w most probable diagnoses contains the correct diagnosis for small values of w (we chose a 'window' of $w = 1$, 2, 3, and 4). The latter focus is of interest in diagnostic settings, where a decision support system only suggest possible diagnoses to a physician. The physician, who is the ultimate decision-maker, may want to see several alternative diagnoses before focusing on one.

2.5.2   Model calibration

Calibration helps to identify inaccuracies and inconsistencies occurring in the model. Elvira software was used to calibrate the HEPAR II model [261]. Elvira's explanation facilities allow us to calibrate the model in two modes: (1) static explanation and (2) dynamic explanation.

2.5.2.1   Analysis of influences

Elvira's capabilities of static explanation involve coloring arcs of the model according to the sign of influence. This feature allows the user to observe the qualitative properties of the model. The positive influences are colored in red and negative in blue. Analysis of influences in HEPAR II concentrated on checking whether the expected positive and negative influences appeared to be really red and blue.

2.5.2.2   Analysis of patient cases

Elvira allows us to work with several evidence cases simultaneously and then to save them in files. Using this facility we performed the analysis of patient cases selected from the HEPAR data set. We were interested in observing the posterior probability distribution of the diseases for each patient case given different groups of evidence such as patient self-reported data, symptoms, results of physical examination and laboratory tests. This analysis has brought several corrections to the model. For example, when observing the variable *Enlarged spleen*, the expert noticed that the presence of the finding does not influence the probability distribution of *Toxic hepatitis* and *Reactive hepatitis*. Therefore, the arcs from *Toxic hepatitis* and *Reactive hepatitis* to *Enlarged spleen* were drawn.

One of Elvira's facilities allows us to analyze the paths from the evidence to the variable of interest. In other words, it helps to focus on one diagnostic hypothesis and check whether the observed evidence influences the selected hypothesis. Figure 2.7 shows an example of the explanation path that was generated for the

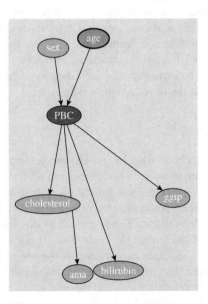

Figure 2.7   Elvira: Explanation paths for *PBC* (Primary Biliary Cirrhosis).

selected hypothesis (*Primary Biliary Cirrhosis (PBC)* in this case). The figure captures all the findings that were observed and that had possibly contributed to the hypothesis. Additionally, the nodes and links are colored, so it is easy to observe whether an observed finding has a positive or negative influence on the hypothesis.

2.5.3 Analyzing model sensitivity

There was an empirical study performed in which we systematically introduced noise in HEPAR II's probabilities and tested the diagnostic accuracy of the resulting model [349]. This was a replication of an experiment conducted by Pradhan et al. [373] and it showed that HEPAR II is more sensitive to noise in parameters than the CPCS network that they examined. The results showed that the diagnostic accuracy of the model deteriorates almost linearly with noise with a region of tolerance to noise much smaller than in CPCS. While the result is merely a single data point that sheds light on the hypothesis in question, this is an indication that Bayesian networks may be more sensitive to the quality of their numerical parameters than popularly believed.

There are two possible explanations of the results of this analysis. The first and foremost is that Pradhan et al. used a different criterion for model performance – the average posterior probability of the correct diagnosis. In the experiment with HEPAR II we focused on the diagnostic performance of the model. Another, although perhaps less influential factor, may be the differences between the models. The CPCS network used by Pradhan et al. consisted of only Noisy-OR gates, which may behave differently than general nodes. In HEPAR II only roughly 50% of all nodes could be approximated by Noisy-MAX [509]. The experiment also studied the influence of noise in each of the three major classes of variables: (1) medical history, (2) physical examination, (3) laboratory tests, and (4) diseases, on the diagnostic performance. It seemed that noise in the results of laboratory tests was most influential for the diagnostic performance of our model. This can be explained by the high diagnostic value of laboratory tests. The diagnostic performance decreases with the introduction of noise.

2.6 Model use

The HEPAR II model has been used by general medicine fellows participating in post-graduate training programs for physicians specializing in primary health-care in the Medical Center for Postgraduate Education in Warsaw.

There was performed an evaluation study of HEPAR II with a participation of its users, 23 general medicine fellows. The first result of this study was that diagnosis of liver disorders is far from trivial. On a subset of 10 cases randomly drawn from the database, HEPAR II was twice as accurate as the physicians (70% vs. 33.1% accuracy) and 40% better than the most accurate physician.

The second and quite important result of this study was that a diagnostic system like HEPAR II can be very beneficial to its users: interaction with the system almost doubled the medical doctors' accuracy (from 33.6% before to 65.8% after seeing the system's suggestion). In many test cases, a correct system's suggestion influenced the users to change their diagnosis. This improvement could still be potentially amplified by further work on the user interface, such as enhancing it with an explanation facility that would justify the results and help the user in judging whether the system advice is correct or not.

Primary care physicians face the daunting task of determining the source of discomfort based on patient-reported data and physical examination, possibly enhanced with the results of basic medical laboratory tests. In addition to the quantitative impact of the system on the quality of decisions, the experiment showed that the reaction of the users to the system was favorable and several of them said that the system had been understandable and that they had learned a lot from it. Therefore, HEPAR II may be a useful tool for general practitioners and could possibly improve the quality of their decisions.

2.7   Comparison to other approaches

In collaboration with Peter Lucas and Marek Druzdzel [353], we performed a study that has compared the Bayesian network approach with the rule-based approach. It was a comparison with HEPAR-RB,[9] the rule-based system for diagnosis of liver disorders [287]. The comparison of HEPAR-RB and HEPAR II showed that building the models in each of the two approaches has its advantages and disadvantages. For example, the rule-based approach allows us to test models by following the trace of the system's reasoning. This is in general possible with causal models, although automatic generation of explanations in Bayesian networks is not as advanced yet as it is in rule-based systems [258, 259]. An important property of the Bayesian network approach is that models can be learnt from existing data sets. Exploiting available statistics and patient data in a Bayesian network is fairly straightforward. Fine-tuning a rule-based system to a given data set is much more elaborate.

Rule-based systems capture heuristic knowledge from the experts and allow for a direct construction of a classification relation, while probabilistic systems are capable of capturing causal dependencies, based on knowledge of pathophysiology, and then enhance them with statistical relations. Hence, the modeling is more indirect, although in domains where capturing causal knowledge is easy, the resulting diagnostic performance may be good. Rule-based systems may be expected to perform well for problems that cannot be modeled using causality as a guiding principle, or when a problem is too complicated to be modeled as a causal graph.

The experiments confirmed that a rule-based system can have difficulty with dealing with missing values: around 35% of the patients remained unclassified by rule-based system, while in HEPAR II only 2% of patients remained unclassified. This behavior is due to the semantics of negation by absence, and is in fact a deliberate design choice in rule-based systems. In all cases, the true positive rate

---

[9]Actually, the system's name is HEPAR. *Hepar* is Greek for *liver* and this explains the similarity in names of the systems. To avoid confusion here the name HEPAR-RB will be used.

for HEPAR II was higher than for the rule-based system, although it went together with a lower true negative rate.

2.8   Conclusions and perspectives

Bayesian networks are recognized as a convenient tool for modeling processes of medical reasoning. There are several features of Bayesian networks that are specially useful in modeling in medicine. One of these features is that they allow us to combine expert knowledge with existing clinical data. However, the constructors of medical Bayesian network models should be aware of biases that can occur during combining different sources of knowledge [136]. Furthermore, Bayesian networks allow us to model multiple-disorder diagnosis, which is more realistic than some alternative modeling schemes. Other important advantage of Bayesian networks is flexible inference based on partial observations, which allows for reasoning over patient data during the entire course of the diagnostic procedure. A Bayesian network allows for modeling causality. From the point of view of knowledge engineering, graphs that reflect the causal structure of the domain are especially convenient – they normally reflect an expert's understanding of the domain, enhance interaction with a human expert at the model building stage and are readily extendible with new information.

On the other hand, Bayesian network models require many numerical parameters. If there are no data available or if the available data are incomplete it may lead to problems with parametrization of the model.

Time is an important factor of reasoning in medicine. Use of dynamic Bayesian networks and dynamic influence diagrams would bring modeling in medicine closer to the real world [151, 506].

Much work in this direction has been done by Tze-Yun Leong and collaborators, who, in addition to Bayesian networks and dynamic Bayesian networks, successfully use a combination of graphical models with Markov chains to address problems in different medical domains, including colorectal cancer management, neurosurgery ICU monitoring, and cleft lip and palate management [277].

Acknowledgment

The contents of this chapter are based in part on several publications resulting from the HEPAR II project [261, 350, 352, 353]. This work was supported by grant number 3T10C03-529. The author would like to thank her collaborators and co-authors, Marek Druzdzel, Hanna Wasyluk, Javier Díez and Carmen Lacave. Comments from Marek Druzdzel and Javier Díez improved the paper.

3

Decision support for clinical cardiovascular risk assessment

Ann E. Nicholson, Charles R. Twardy, Kevin B. Korb and Lucas R. Hope

Clayton School of Information Technology, Monash University, Australia

3.1 Introduction

In this chapter we describe a clinical decision support tool which uses Bayesian networks as the underlying reasoning engine. This tool, TakeHeartII, supports clinical assessment of risk for coronary heart disease (CHD). CHD comprises acute coronary events including myocardial infarction (heart attack) but excluding stroke. Improved predictions of cardiovascular deaths would allow for better allocation of health care resources and improved outcomes. Amid burgeoning healthcare costs (one billion per year is currently spent on anti-lipid and anti-hypertensive medication in Australia), cost effectiveness has become a dominating consideration for determining which preventive strategies are most appropriate. Improved assessment and better explanation and visualization of risk in a clinical setting may also help persuade the patient to adopt preventive lifestyle changes.

There have been many medical applications of BNs, including early networks such as the ALARM network for monitoring patients in intensive care [40], diagnosing oesophageal cancer [469], mammography [75] and diagnosing liver disorder [348]. Finally, the PROMEDAS medical decision support tool [155, 375] uses Bayesian networks, automatically compiling both the network and an interface from

Bayesian Networks: A Practical Guide to Applications   Edited by O. Pourret, P. Naïm, B. Marcot
© 2008 John Wiley & Sons, Ltd

the underlying medical database (which currently covers the areas of endocrinology and lymphoma diagnostics). Unfortunately too many of these applications appear to have been 'one-offs', with long development times using (mainly) expert elicitation and limited deployment. The lack of widespread adoption can be attributed to a wide range of factors, including (1) not embedding them in a more general decision support environment (rather than directly using BN software) that is more accessible for users who are domain experts but have little understanding of Bayesian networks and (2) the inability to easily adapt the parameterization of the network to populations other than those considered during the construction phase. In the development of TakeHeartII, we have attempted to address some of these issues.

To avoid the long development times required by expert elicitation, we have built BNs for predicting CHD in two other ways. First, we knowledge engineered BNs from the medical literature. Generally, 'knowledge engineering' means converting expert knowledge into a computer model. Here, rather than querying domain experts directly for information required to build a BN, we used published epidemiological models of CHD, supplemented by medical expertise when we were unsure of interpretation. The epidemiological models used were: (1) a regression model from the Australian Busselton study [251] and (2) a simplified 'points-based' model from the German PROCAM study [20]. Second, we applied a causal discovery program, CaMML [479, 480], to learn BNs from data. In [461] we briefly described the BNs knowledge engineered from the literature. Here, we provide more details on their construction and justify the knowledge engineering choices we made. This may assist others developing similar BN applications from the epidemiological literature for risk prediction of other medical conditions, as a case study in the steps that need to be undertaken. We used the Netica software [342] to implement the BNs; however, the approach described here should extend to any system allowing equations to specify the probability distributions for each node.

We have designed TakeHeartII for clinicians who know nothing about BNs. The clinical GUI is generated automatically from the underlying BN structure. Therefore it is simple to update the GUI when the BN is changed (for example, by adding a new node, or changing states). This separation also helps adapt the BN to a new dataset, or to new priors that reflect the population seen in a particular clinical environment.

3.2   Models and methodology

We begin with an overview of the two models we took from the epidemiological literature and our knowledge engineering methodology.

We selected two studies which presented regression models for predicting CHD. The first is the Busselton study [77], which collected baseline data every three years from 1966 to 1981 and has resulted in hundreds of research papers. We used the Cox proportional hazards model of CHD from [251]. The model has the form of a regression: each risk factor has a coefficient, and the risk factors are assumed to act independently. Therefore the structure of the model is a series of independent predictors leading to the target variable. The second study is the *Pro*spective *Ca*rdiovascular *M*ünster study, or PROCAM, which ran from 1979 to 1985, with followup questionnaires every two years. PROCAM has generated several CHD models. We use the 'simplified' model in [20]. Concerned that clinicians would not use a full logistic regression model, they converted it into a points-based system where the clinician need only add the points from each risk factor, and read the risk off a graph. Therefore the authors had already discretized the continuous variables, making it straightforward to translate to a BN. (However, this extra step designed to ease clinician's lives made for a more complicated model structure.) We refer the reader to [461] for further discussion of the Busselton and PROCAM studies.

There are at least four reasons to convert existing regression models to BNs. First, BNs provide a clear graphical structure with a natural causal interpretation that most people find intuitive to understand. Clinicians who are uncomfortable using or interpreting regression models should be much more amenable to working with the graphical flow of probability through BNs. Second, BNs provide good estimates even when some predictors are missing, implicitly providing a weighted average of the remaining possibilities. Third, BNs clearly separate prior distributions from other model parameters, allowing easy adaptation to new populations. Fourth, BNs can easily incorporate additional data, including (subjective) expert knowledge.

It is generally accepted that building a BN involves three tasks [256]: (1) identification of the important variables and their values; (2) identification and representation of the relationships between variables in the network structure; and (3) parameterization of the network, that is determining the CPTs associated with each network node. In our CHD application, step 3 is complicated by the fact that although many of the predictor variables are continuous, the BN software requires them to be discretized. We divided the process into: (3a) discretization of predictor variables; (3b) parameterization of the predictor variables; (3c) parameterization of the target variables.

3.3   The Busselton network

3.3.1   Structure

Knuiman et al. [251] described CHD risks separately for men and women. Rather than make two separate networks, we made Sex the sole root node in the network; this allows us to assess risk across the whole population. The other predictor variables in the study (such as Smoking) are conditional on Sex. This effectively gives separate priors for men and women (although those priors do not always differ much).

Figure 3.1 shows the full Busselton network. There is an arc from each predictor variable to the Score variable, indicating the predictors determine a risk score. That score is transformed into the 10-year risk of CHD event. Each state/value of this risk node represents a range of percentages; in the example, we can see that for a female patient with no other information provided, the BN computes a probability

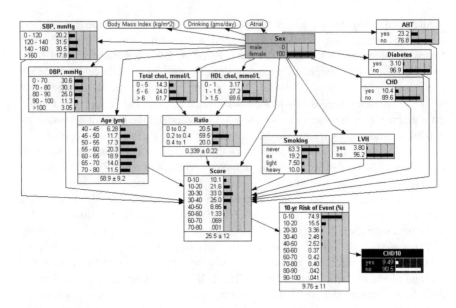

Figure 3.1 The Busselton network showing the prior distribution for females. The target node CHD10 is highlighted in the lower right of the figure. Reproduced with permission from the electronic Journal of Health Informatics.

0.749 of a 0–10% risk of a CHD event in the next 10 years, followed by 0.155 probability of a 10–20% risk, and so on. This risk assessment over percentiles is in turn reduced to a yes/no prediction of a coronary heart disease event, represented by the target node CHD10 (highlighted in the lower right of the figure); for this example, this risk is 9.43%.[1]

We note that the clinician should not care about the 'Score' and 'Ratio' nodes, as they are just calculating machinery. When embedding the BN in the clinical tool, we use a cleaner view of the Busselton network (depicted in [460]), where these have been absorbed away. This is a network transform (done using Netica's `Node Absorption`) that removes nodes from a network in such a way that the full joint probability of the remaining nodes remains unchanged. We left 10-year risk node in because clinicians may well prefer that to the binary CHD10, as the former gives a much better idea about the uncertainty of the estimate. Indeed, this is one of the benefits of using BNs.

3.3.2   Priors for predictor variables

Knuimuan et al. [251] reported summary statistics for their predictors. These become the priors for our population: one set for men and another for women. We generated parametric or multi-state priors from their summary statistics [251, Table 1]. The priors for the multi-state variables were entered as tables. The priors

---

[1]Note that this is high because the average age for the survey Busselton population is 58.9.

for continuous variables were specified using Netica's equation facility, assuming they were Gaussian distributions. To match their model, we also had to create a variable for the HDL/Total ratio. That variable is the child of HDL and Total Cholesterol.

3.3.3   Discretization of predictor variables

Seven predictors (including the computed BMI and Ratio) are continuous variables. Table 3.1 summarizes the discretization levels we chose. The rest of this section discusses our choices.[2]

Age: Knuiman et al. rejected those with baseline age <40 or ≥80, setting the bounds. To be practical, we chose five-year divisions within that range and then extended it to the limits of 0 and ∞.

Blood pressure: A general medical consensus discretization for blood pressure is shown in Table 3.2. However, since those divisions preclude the possibility of finding any harmful or helpful effects of *low* blood pressure, we added an extra range so we could distinguish low from normal.

Cholesterol: For cholesterol, the American Heart Association (AHA) and US National Cholesterol Evaluation Program (NCEP) websites gave guidelines for total, HDL, and LDL cholesterol in mg/dL. For total cholesterol they are 0–199,

Table 3.1   Discretization levels for each variable. See text for discussion.

Variable	Ranges
Age	0, 40, 45, 50, 55, 60, 65, 70, 75, 80, ∞
SBP	0, 100, 120, 140, 160, ∞
DBP	0, 70, 80, 90, 100, ∞
Chol	0, 5, 6, ∞
HDL	0, 1, 1.5, ∞
Ratio	0, 0.2, 0.4, 1
BMI	0, 20, 25, 30, ∞
Smoking	0, 0, 1, 15, ∞
Alcohol	0, 0, 0, 20, ∞

Table 3.2   Standard blood pressure classifications in the medical literature (units mmHg).

	Systolic	Diastolic
Normal	<120	<80
Pre-hypertension	120–140	80–90
Hypertension (Stage 1)	140–160	90–100
Hypertension (Stage 2)	>160	>100

---

[2]Additional discussion of discretization in Netica is given in [460].

200–239, and $\geq$ 240. For HDL they are <40, 40–59, and $\geq$60. Dividing by 39 gives mmol/L, and rounding gave the ranges above. Looking at the distribution, it may be argued that perhaps we should make finer divisions.

Ratio: The ratio HDL/Chol has to be between 0 and 1, and cannot realistically get near 1. The AHA said the 'optimal' Total/HDL is 3.5:1 (giving 0.28 for HDL/Tot) and the 'goal' was <5:1 (in our case >0.2). Another medical website said the average Tot/HDL was about 4.5 (0.22 for us), and a good ratio was 2 or 3 (for us, 0.5 or 0.33). From this, we decided to use the following divisions: anything <0.2 was bad, 0.2–0.4 were normal to good, and >0.4 was excellent. It seems unlikely that anyone could get much above 0.5. The nonlinearities introduced by dividing makes the use of ratios quite suspect. To prevent any strange errors, we forced values to lie between 0 and 1.

BMI: Body mass index was similar to cholesterol. According to [497], a BMI $\geq$25 is overweight, and $\geq$30 is obese. There is also long-term evidence that if BMI departs from about 23 in *either* direction, life expectancy goes down. So by symmetry we presumed that $\leq$20 was underweight. However, BMI was not used in the equation, as [251] did not find it to be a significant predictor.

Smoking and Alcohol: Discrete variables Smoking and Alcohol (Drinking) had quantities associated with them: number of cigarettes per day, and grams of alcohol per day. We modeled them as *continuous* variables to allow more flexibility. When defining discretising ranges, 'Never' and 'Ex' were 0 across the range. Light smoking was 1–14 cigarettes/day, and heavy smoking was $\geq$15. Light drinking was 0–20 grams/day, and heavy drinking was $\geq$20, as defined in the Busselton dataset and reported in [251]. Alcohol was excluded from the predictor variables because the authors could find no significant effect, even at hazardous drinking levels ($\geq$40 grams/day). The BN shows this by having no arc from Alcohol to CHD10.

Score and Risk: Discretising Score and Risk was straightforward. The multivariate score ranged from 0 on up, though by the time it reached 80, risk had saturated at 100%, so we just made eight divisions: 0–10, . . . , 70–80. Similarly, Risk was 0% to 100%, in 10 divisions.

3.3.4   Parameterizing the target variables

Score: Score is a *continuous* variable which is the weighted sum of all the predictor scores. The weights correspond to the Cox proportional hazards regression and are taken from [251, p. 750]. We have separate equations for men and women. Letting the binary variables (e.g., AHT, LVH, CHD, etc.) be 1 for 'yes' and 0 for 'no', these equations are:

Male: $0.53 \times$ Age $+ 0.055 \times$ SBP
$-56.26 \times$ Ratio $+ (0.43 \times$ AHT$)$
$+(3.83 \times$ LVH$) + (11.48 \times$ CHD$)$
$+(3.2$ if Smoke ≥ 15, or 3.01 if $15 >$ Smoke $\geq 1)$
$+10$

Female: $0.846 \times$ Age $+ 0.122 \times$ DBP
$- 33.25 \times$ Ratio $+ (0.5.86 \times$ AHT$)$
$+ (2.97 \times$ LVH$) + (7.85 \times$ DIABETES$)$
$+ (7.99 \times$ CHD$)$
$+ (6.16$ if Smoke ≥ 15, or 2.2 if $15 >$ Smoke $\geq 1)$
$+ 10$

Risk: Knuiman et al. did not provide the equation for their risk curve. We could just use their data points, but that would make the model fragile under a new discretization. So we fit a curve to the data. Not knowing the precise parametric form of the Cox model likelihood curve, we fit Weibull and logistic curves. As Figure 3.2 shows, the fit was good, but both curves underestimated the risk for low scores. We use the Weibull because it better accommodates the flat top, especially the (70, 99%) point. The general equation for the cumulative Weibull is:

$$F(x) = k \left(1 - e^{-\left(\frac{x-m}{a}\right)^g} \right).$$

Our curve starts at $x = 0$, and goes up to $y = 100\%$, so we know $m = 0$ and $k = 100$, and don't have to fit those. The fitted parameters are: $a = 54.5415$,

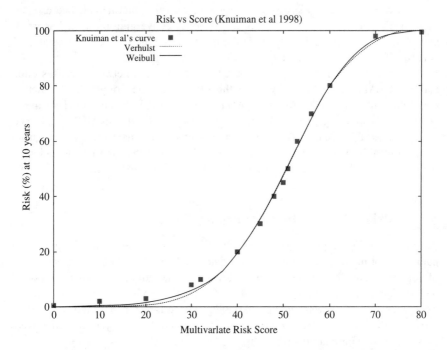

Figure 3.2 Risk versus Score. Least squares fit on data points are from Knuiman et al.'s Figure 3.

$g = 4.99288$. The resulting curve underestimates slightly the risk % for scores in the range 10–30, but this is a low risk category regardless. The actual curve may come from a more broad-tailed family like the Levy. Obviously, the model could be improved slightly by finding the correct family.

3.4   The PROCAM network

3.4.1   Structure

For PROCAM, CHD10 is a weighted sum of the eight risk factors. The full PRO-CAM BN is shown in Figure 3.3(a). There is one root node for each risk factor. Each of these root nodes has in turn one child, which is the associated scoring node. For example, there is an arc from the Age root node to its associated score node AgeScore. The eight score nodes are all parents of a combined score node, called here 'Procam Score', which combines the scores. This combined score node is then a parent of the 10 year risk node. However, although the scoring scheme was designed to be simpler than logistic regressions, the extra scoring nodes complicate the BN.

Figure 3.3(b) shows the essential structure, with the intermediate score nodes 'absorbed' away, as described earlier: eight nodes converging on a final score. It also shows a hypothetical case where the most likely outcome (0.929 probability) is a 10–20% risk of a CHD event.

Before moving on we note basic structural differences between the Busselton and PROCAM models. Most obviously, this PROCAM model omits Sex, predicting for males only. It also omits DBP, AHT, CHD, and LVH, but includes Family History. Instead of the ratio of HDL to Total cholesterol, it uses HDL, LDL, and Triglycerides individually. The discretization given in the next section came from Assman et al.'s paper [20] and is usually slightly finer than the one we adopted for the Busselton network. We discuss the details in the next section.

3.4.2   Priors for predictor variables

Assman et al. [20] reported summary statistics for their predictors. We generated parametric or multi-state priors as appropriate. Table 3.3 lists the predictors in order of importance in their Cox model, followed by the summary statistics reported in [20, Table 2]. For completeness, the BN equation follows. The priors for the multi-state variables were entered as tables. The priors for continuous variables were specified using Netica's equation facility. Except for Triglycerides, they were assumed to be Gaussian distributions.

Age: Assman et al. [20] excluded patients younger than 35 or older than 65 at baseline. We keep their exact range for now. In contrast, [251] used an age range of 40–80, which is significantly older, especially as CHD risk goes up dramatically with age.

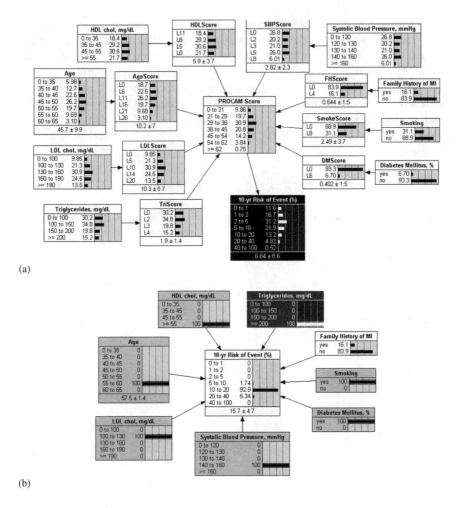

Figure 3.3 (a) The full PROCAM network; (b) with score nodes 'absorbed.' Reproduced with permission from the electronic Journal of Health Informatics [461].

Table 3.3   Variables and summary statistics from [20], Table 3.2, and corresponding BN equations.

	Summary Stats	BN Equation
Age	46.7 ± 7.5 years	$P(Age \mid) = NormalDist(Age,46.7,7.5)$
LDL cholesterol	148.5 ± 37.6 mg/dL	$P(LDL \mid) = NormalDist(LDL,148.5,37.6)$
Smoking	31.1% yes	
HDL cholesterol	45.7 ± 11.9 mg/dL	$P(HDL \mid) = NormalDist(HDL, 45.7, 11.9)$
Systolic BP	131.4 ± 18.4 mm Hg	$P(SBP \mid) = NormalDist(SBP, 131.4, 18.4)$
Family history of MI	16.1% yes	
Diabetes mellitus	6.7% yes	
Triglycerides	126.2* ± 65.9 mg/dL	$P(Tri \mid) = LognormalDist(Tri, \log(126.2), 0.45)$

(* = Geometric mean)

Triglycerides: Since the authors reported a geometric mean for Triglycerides, we infer the data must have a lognormal distribution, meaning that log(Triglycerides) would have a normal distribution. A look at the Busselton data confirmed the assumption, as does the common practice of having a LogTriglycerides variable. We then parameterize a lognormal distribution given a geometric mean m and a standard deviation s, as follows.[3] The NIST e-Handbook [333] says that a lognormal distribution uses $\mu = \log(m)$ as the scale parameter and has a shape parameter σ. It also gives an expression for the standard deviation s, which can then be inverted:

$$\sigma = \sqrt{\log\left(\frac{m \pm \sqrt{m^2 + 4s^2}}{2m}\right)} \qquad (3.1)$$

Therefore $m = 126.2$ and $\sigma = 0.4486$, which generates a curve with the appropriate parameters, shown in Figure 3.4. This provides the required Netica equation:

```
P(Tri|) = LognormalDist(Tri, log(126.2), .45).    (3.2)
```

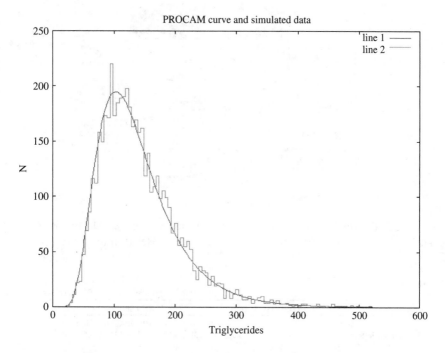

Figure 3.4   Lognormal curve for PROCAM m, σ, with simulated data superimposed. Standard deviation for the curve and data are 0.66.

---

[3]Note that (m, s) are the given geometric mean and sample standard deviation, while (μ, σ) are the scale and shape parameters that we are estimating.

3.4.3   Discretization of predictor variables

The discretization levels were given in [20]. We reproduce them here in Table 3.4. Age is in five-year bins, just like our Busselton network. LDL and HDL are measured in mmol/L instead of mg/dL and discretized more finely. The corresponding breaks for HDL would be $(0, 0.9, 1.15, 1.4, \infty)$, compared to the levels we adopted in the Busselton network: $(0, 1, 1.5, \infty)$. SBP inserts an extra level at 130, compared to Busselton's $(0, 120, 140, 160, \infty)$.

3.4.4   Points

Here we describe the straightforward translation of the discretized predictor variables into scores, by assigning point values to each of the levels of the predictors. In the BN, this means specifying the relationship between the Score node and its corresponding predictor variable, as in Table 3.5.

3.4.5   Target variables

The PROCAM score is a *continuous* variable which is a simple sum of the individual predictor scores. The work comes in translating this to probabilities. All these risk models have a sigmoid component to convert the risk score to a real risk. Therefore we can fit a logistic equation to 'Table 4, Risk of Acute Coronary Events Associated with Each PROCAM Score' [20]. We used the Verhulst equation

$$v(x) = \frac{N_0 K}{N_0 + (K - N_0)\exp(-r(x - N_1))}.$$

Table 3.4   Discretization levels for each variable. Values from [20], Table 3.3 and as implemented in Netica.

Variable	Levels (in paper)	Levels in network
Age	35, 40, 45, 50, 55, 60, 65	(35, 40, 45, 50, 55, 60, 65)
LDL	<100, 100, 130, 160, 190, ≥190	(0, 100, 130, 160, 190, INFINITY)
HDL	<35, 35, 45, 55, ≥55	(0, 35, 45, 55, INFINITY)
Tri	<100, 100, 150, 200, ≥200	(0, 100, 150, 200, INFINITY)
SBP	<120, 120, 130, 140, 160	(0, 120, 130, 140, 160, INFINITY)

Table 3.5   Equations defining scores for each predictor. The conditional 'if x then y else z' is written 'x ? y : z'. Terms 'L0', 'L16', etc. are state names assigned to have the corresponding numeric values 0, 16, etc. This is an artifact of the way Netica handles discrete variables that should have numeric values.

```
AgeScore     Age<40 ? L0 : Age<45 ? L6 : Age<50 ? L11 : Age<55 ? L16 :
             Age<60 ? L21 : L26
LDLScore     LDL<100 ? L0 : LDL<130 ? L5 : LDL<160 ? L10 : LDL<190 ? L14
             : L20
SmokeScore   Smoking==yes ? L8 : L0
HDLScore     HDL<35 ? L11 : HDL<45 ? L8 : HDL<55 ? L5 : L0
SBPScore     SBP < 120 ? L0 : SBP <130 ? L2 :
FHScore      FamHist==yes ? L4 : L0
DMScore      Diabetes == yes ? L6 : L0
TriScore     Tri<100 ? L0 : Tri<150 ? L2 : Tri<200 ? L3 : L4
SBP          SBP<140 ? L3 : SBP<160 ? L5 : L8
```

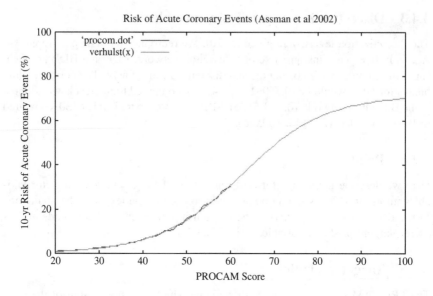

Figure 3.5 Logistic curve fitted to Table 4 in [20].

The corresponding graph is shown in Figure 3.5. The fit is good over the data range (20–60) and tops out at a 70% risk of an event in the next 10 years, no matter how high the PROCAM score goes. As there is always some unexplained variability, our domain experts consider this reasonable [283]. However, it conflicts with the Busselton study risk curve [251, p 750, Figure 3] shown in Figure 3.2, which goes up to 100%. [250] says his curve used the standard Cox formula involving baseline survival and defends the high maximum risk. He notes that age is a very strong determiner and the Busselton study included people who were up to 80 years old at baseline. This PROCAM study, by contrast, included only those up to 65 at baseline. For now we presume that our fit is reasonable, showing a difference between this PROCAM study and the Busselton study. To define the risk in Netica, we make 'Risk' a child of 'Score' and set its equation:

```
Risk (Score) =
(30 * 71.7) / (30 + (71.7 - 30) * exp(-0.104 * (Score - 59.5))).
```

Table 4 in [20] tops out at ≥30, and is much finer near the bottom, so we defined the following levels: (0, 1, 2, 5, 10, 20, 40, 100).

3.5   The PROCAM Busselton network

We wanted to know how well this simple PROCAM model predicts the Busselton data. To do this we must adapt both the data and the network so that all corresponding variables match. For example, some of the Busselton variables are measured on a different scale, or do not conform precisely to those in the PROCAM model.

Modifying the Busselton data   Knuiman et al. used only the 1978 cohort, for various reasons. So, first we selected the 1978 cohort from the original Busselton data. Survey and CHD events were then merged into a single datafile. A number of adaptations then had to be made to adapt Busselton variables to the PROCAM Bayesian network. Specifically:

1. Remove females, because the PROCAM study didn't include them.

2. Convert cholesterol from mmol/L to mg/DL. Multiply by 89 for triglycerides, or 39 for total and HDL.

3. Create the variable LDL from the other cholesterols (see below).

4. Move the target variable, CHD10 to the end.

5. Remove inapplicable variables.

Although we exclude females because we have no way to know the PROCAM model for females, we do not restrict the age to match the range used in the PROCAM study, because there is a natural extension, discussed below.

The conversion to HDL is done according to the Friedewald equation [159]. If the measures are in mmol/L, then the equation is: LDL = Total - HDL - TG/2.2. If the equation is in terms of mg/dL, then divide by 5 instead of 2.2. The Friedewald equation breaks down at high levels of Triglycerides. Above 200 mg/dL, 28% of the estimates were more than 10% inaccurate, and above 400 mg/dL, 61% were inaccurate [481]. We have accepted the inaccuracy, but another approach would be to make such values missing/unknown.

Modifying the PROCAM network   Having already transformed some of the Busselton variables, it remained to extend the variables in the PROCAM network to handle extended ranges, and optionally alter their priors to match our data.
Age: The Busselton baseline data includes people from 18 through 97 at baseline, although there are few above 80. Although Assman et al. [20] used only males aged 35–65 at baseline, we can see that age score is a linear function of age, allowing a natural extension to higher ages. We extended Age with new levels: 0, 20, 30 and 70, 80, INFINITY.
AgeScore: A quick plot shows that AgeScore $\approx \frac{26}{25}$ Age $- 41$, with AgeScore $= 0$ for Age < 40. But as we will now allow age to be much lower, we allow AgeScore < 0. We also get a slightly better fit to the PROCAM table by using a floor function, so we defined:

$$\text{AgeScore} = \text{floor}\left(\frac{26}{25}\text{Age}\right),$$

where floor supplies the largest integer less than or equal to its argument. Now AgeScore is automatically extended to any new ages. It is negative below 40, as younger people should be at lower risk. It is the natural extension of the risk function in the PROCAM model.

CHD10: We added a new binary variable CHD10, as a child of Risk. `P(CHD10|Risk) = Risk/100`.
Tri: We divided the highest range in two to better match the AHA categories found via websearch: 0–150 normal; 150–200 borderline; 200–500 high; ≥ 500 very high. So, `levels = (0, 100, 150, 200, 500, INFINITY)`. Recall that our calculated LDL values become suspect for high levels of Triglycerides.
TriScore: Was 0, 2, 3, 4, so we just gave '5' to the new upper category.
Score: Because AgeScore can be negative, we have to allow this to be negative as well. Because of sampling, the safest way is just to set the lower bound at $-\infty$, even though it should really be -41 or -42.

Modifying the priors   We now have an adapted network that will work on Busselton data, but still using the original PROCAM priors from their German population. However, even if the risk equation is exactly right for the Busselton population, this model may perform badly if the baseline population is different. For that reason, we may wish to modify the priors. Fortunately, in a Bayesian network, the priors are explicit and easy to modify. Therefore models learned for one population can be easily transported to another. Because we expect the prior distribution of risk factors to differ from Germany to Australia, we modified the priors to match the distribution in the 1978 cohort.

3.6   Evaluation

In previous sections we have described the construction of three BNs for CHD risk assessment, which we call Busselton, PROCAM-German (German priors) and PROCAM-adapted (Busselton priors). Evaluation of these networks, plus other BNs learnt from the epidemiological data, is provided in [461]; we summarize the evaluation here.

The evaluation used two metrics: ROC curves and Bayesian Information Reward (BIR).[4] ROC analysis applies to a set of binary (yes/no) predictions. One of the classes (here, a CHD event) is chosen as the positive class for analysis and the other is the negative (lack of CHD event). Each point on a ROC curve measures a machine learner's predictive performance (in terms of true positives) for a given false positive rate. Thus the point (0,0) corresponds to the policy of always predicting negative – no false positives, but no true positives either. Similarly, the point (1,1) corresponds to always reporting positive. (0,1) is the best possible result: correctly identifying all positive cases while completely avoiding incorrect classification of negatives. Intuitively, the closer to the top-left a curve is, the better the learner's performance. The area under the curve (AUC) averages the performance of the learner, with a perfect learner having AUC = 1 and a random predictor having AUC = 0.5 (a diagonal line). ROC analysis is meant to be used in lieu of a cost-sensitive predictive analysis, when explicit costs are unavailable. BIR [217] is

---

[4]We also looked at accuracy, log loss and quadratic loss, but results were not included for reasons of space. The results are much the same.

one of the log loss family of metrics, rewarding learners not just for right/wrong, but also for getting the *probability* of an event correct (i.e., calibration). It's a Bayesian metric, taking prior probabilities into account and maximally rewarding algorithms which best estimate the posterior probability distribution over predicted events.

In the first set of experiments, we compared the knowledge-engineered BNs – Busselton, PROCAM-German and PROCAM-Busselton as described above in Sections 3.3–3.5 – and a BN learned by the CaMML program [480].

In the second set of experiments, we compared CaMML against the standard machine-learning algorithms provided by Weka [501]: naïve Bayes, J48 (C4.5), AODE, logistic regression, and an artificial neural network (ANN) (all run with default settings). For comparison, we also included Perfect (always guesses the right answer with full confidence) and Prior (always guesses the training prior), which set the upper and lower bounds on reasonable performance, respectively.

We found that PROCAM-German does as well as a logistic regression model of the Busselton data, which is otherwise the best model. They had the same AUC, with about the same curve. This means they ranked cases in roughly the same order. We also found that PROCAM-German did just as well as the logistic regression on BIR and related metrics, which suggests that it *is* well-calibrated, regardless of the fact that the individual regression coefficients might be different.

3.7   The clinical support tool: TakeHeartII

3.7.1   The original Take Heart

Monash University's Epidemiological Modeling Unit developed the original Take Heart program [312] in conjunction with BP Australia's Occupation Health Services Department. Take Heart estimated CHD10 risk for approximately 900 BP Australia employees, starting from 1997 and extending for over two and a half years. Take Heart's epidemiological model used equations from the Multiple Risk Factor Intervention Trial (MRFIT) study [240], adjusted so that the background risk factor probabilities fit the Australian population. It used a Microsoft Access database to store cases, and Access macros for the calculations.

3.7.2   TakeHeartII architecture

Figure 3.6 shows the architecture for TakeHeartII, divided into the BN construction phase (on the left) and its use in a clinical setting (right). The construction phase depicts the general process described earlier in this chapter, with the BN built using a combination of models from the epidemiological literature, expert knowledge and data. It also includes an adaptation phase, where a BN (built using any method) can be adapted for a different population (such as re-parameterizing the PROCAM model from the Busselton dataset, described in Section 3.5).

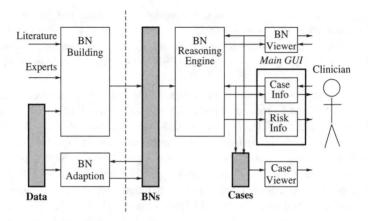

Figure 3.6   TakeHeartII Architecture: Construction and Adaptation (left) provide BNs used for risk assessment in a clinical setting (right).

3.7.3   TakeHeartII implementation

We wrote the scripting language *Modular Bayesian Network* (ModBayes) to implement TakeHeartII. ModBayes links a Bayesian network to graphical controls such as drop-down menus, buttons and check-boxes; as well as visualizations such as the chart and risk displays of Figure 3.7. It also manages a case file associated with the BN. ModBayes allows speedy organization of the controls and displays on screen: someone knowledgeable in the scripting language could redesign the layout in minutes. In addition, it automatically integrates with the powerful Bayesian Network viewer, *Causal Reckoner* [219, 255]. Causal Reckoner inputs BNs in the Netica .dne format, and uses Netica as the BN reasoning engine. It also provides additional functionality for network layout, causal manipulation and causal power. When evidence is entered in the Causal Reckoner, it appears on ModBayes's controls, and vice versa, which enables the BN display to change according to the needs of the user.

The scripting language itself is a dialect of Lisp, and thus is a full-featured programming language in its own right. This means that the scripting language can be extended (by an advanced user) to create arbitrary Bayesian network interfaces.

3.7.4   TakeHeartII interface

TakeHeartII's main interface, shown in Figure 3.7 is divided into two sections. The top section is for case information about the patient being assessed, provided in a survey form style. The clinician inputs information about the patient where known, which is then entered as evidence in BN by the reasoning engine.

The case title ('Cassy' in the figure) is an editable text field. This title becomes the case label on the risk assessment displays and on the case viewer (described below). In the survey form, answers to multiple choice questions are entered with drop-down boxes and yes–no answers are simple button clicks.

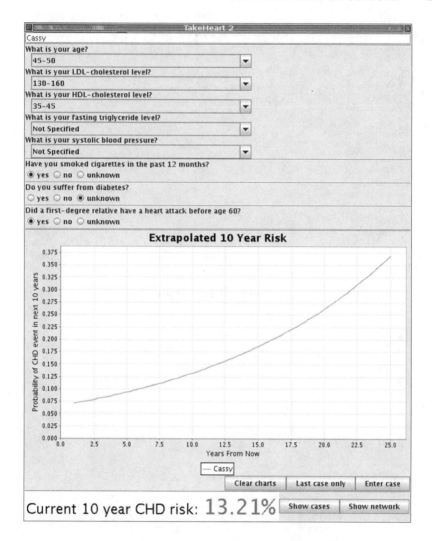

Figure 3.7   The main GUI for TakeHeartII.

At any stage, the clinician can ask TakeHeartII to make a risk assessment of a CHD in the next 10 years, by clicking the 'Enter case' button. If the clinician has entered age information, then TakeHeart displays the extrapolated 10 year risk for the current 'case', shown with label Cassy. The case is also saved (with its label) in the case database. The clinician can then modify the current case, for example to show the effect of a life-style change such as quitting smoking. The risk assessment then shows the risk for *both* cases (see Figure 3.8). The higher curve is her current projected risk, and the lower curve is her projected risk should she quit smoking; clearly, smoking has increased her risk of heart attack substantially. There is no formal limit to the number of cases that can be displayed together, however in

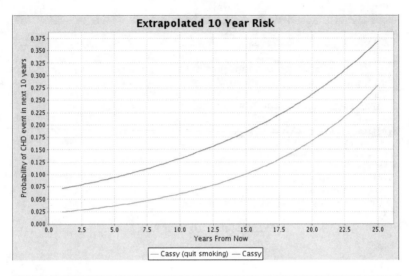

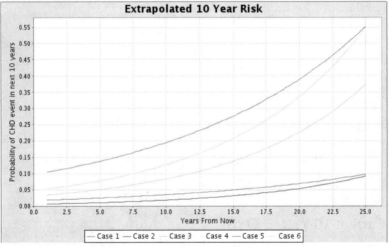

Figure 3.8   Extrapolated 10 year for (above) two cases and (below) six cases.

practice the size of the screen will dictate a limit of 6–8 cases. At any time, the clinician can clear the charts (to re-start) or show only risk given the current survey case information.

TakeHeart also provides risk assessment by *age*. Figure 3.9 shows a category plot of heart attack risk by age, again comparing smokers with nonsmokers (the form was filled out the same as for Cassy, but with age left blank). Again, this provides a clear visualization of the impact of smoking, almost doubling the risk for each age group.

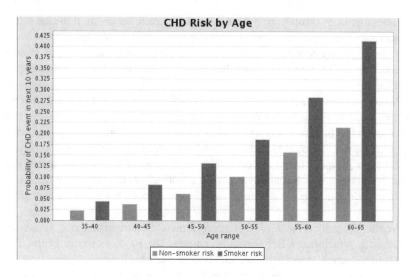

Figure 3.9   Heart attack risk by age for a smoker and a nonsmoker

How are the risk assessment values calculated? The category display is simple: the risk for each age group, given the other risk factors, is simply calculated in turn. The extrapolated risk chart is more complex. An exponential curve ($y = e^{mx+c}$), is fitted to two points: the current risk and the risk for someone five years older (or five years younger if the person is in the oldest age range). The exponential distribution is chosen because it matches observed data in how risk increases with age [312].

There are two additional displays provided with TakeHeartII. The first is Causal Reckoner's BN viewer, which is fully integrated into TakeHeartII: when the clinician enters evidence using the survey form, the BN display is also updated, and conversely, if the clinician chooses to enter evidence directly into the viewer (by clicking a value), the survey is updated. The BN viewer may also display non-input (i.e., 'intermediate') nodes, whose values are not included in the main survey form, such as the Score nodes in the full PROCAM BN (see Figure 3.3). Finally, the Case Viewer allows the clinician to look at the history of cases entered into TakeHeart.

Here we have only sketched TakeHeartII's main features. Full details, including the extrapolation algorithm/method, are provided in [218]. A prototype version of TakeHeartII is currently being evaluated by our medical research collaborators and a clinical evaluation is being planned.

3.8   Conclusion

In the first part of this chapter, we have provided a detailed description of the construction of two Bayesian networks for cardiovascular risk assessment. These networks were built using information available in the epidemiology literature, with only minor additional input from medical experts, avoiding much of the knowledge engineering 'bottleneck' associated with expert elicitation. We have also justified our knowledge engineering choices, providing a case study that may assist others develop similar BN applications in this way for risk prediction of other medical conditions.

We had to discretize our continuous variables. Often a tedious and uncertain step, we were fortunate in being able to use established diagnostic ranges, either from the models themselves or from other literature. Still, discretization introduces edge effects, and it would be better to perform continuous inference (perhaps via particle filter sampling), and discretize only at the end, if necessary to match diagnostic categories. Netica cannot yet perform 'native' continuous inference, but we should point out that this is a limit of our choice of tools, not the models or theory itself.

We then described our tool to support clinical cardiovascular risk assessment, TakeHeartII, which has a BN reasoning engine. The modular nature of the TakeHeartII architecture, and its implementation in the ModBayes scripting language, means that it can incorporate different BNs (for example changing to a BN that is a better predictor for a given population) without requiring major changes to the interface.

We are currently directing our efforts in two directions. First, at evaluating the TakeHeartII tool in a clinical setting, which may well lead to modifications of the interface. Second, at using data mining to obtain more informative and predictive BNs. For example, we have developed a method [346] for using expert knowledge to provide more direction to the causal discovery process and are currently applying that method to our CHD datasets.

Acknowledgment

Thanks to our experts Danny Liew and Sophie Rogers for their medical insights, and to David Albrecht for the inversion of the standard deviation of the lognormal distribution.

4

Complex genetic models

Paola Sebastiani
Department of Biostatistics, Boston University School of Public Health, Boston MA 02118, USA

and

Thomas T. Perls
Geriatric Section, Boston University School of Medicine, Boston, MA 02118, USA

4.1 Introduction

The map of our chromosomes produced by the Human Genome Project has documented similarities and differences between individuals and has shown that people share approximately the same DNA with the exception of about 0.1% nucleotide bases [452]. These variations are called *single nucleotide polymorphisms* (SNPs) and occur when a single nucleotide (A, T, C, or G) in the genome sequence is altered. Some of these variants may be the cause of 'monogenic diseases' in which a single gene is responsible for causing the disease. This happens when the gene is defective so that a mutation of the DNA sequence of the gene determines a change in the associated protein. Our cells contain two copies of each chromosome and monogenic diseases are classified as 'dominant' or 'recessive' based on the number of copies of the mutated variants that are necessary for the disease to manifest. A dominant disease needs only one mutated variant to manifest while a recessive disease needs two mutated variants to manifest and one mutated variant makes a subject only a carrier of the disease. A carrier can transmit the mutated variant to the offspring, but not the disease and a recessive disease can only be

Bayesian Networks: A Practical Guide to Applications Edited by O. Pourret, P. Naïm, B. Marcot
© 2008 John Wiley & Sons, Ltd

transmitted when both parents are carriers. This classification changes when the SNP is on chromosome X, so that one single variant can make an individual carrier or affected based on gender. Sickle cell anemia is a well known example of a recessive disease that is caused by a single mutation of the β-globin gene that determines a variant of the hemoglobin protein [442]. This was the first monogenic disease ever described and lead to Pauling's theory of molecular disease that opened a new chapter in the history of medicine [330].

Over the past decade, about 1200 disease-causing genes have been identified. This success was facilitated by studying well characterized phenotypes – the physical manifestation of the genotype or the combination of variants in the chromosome pair – and using specific techniques that are described in details for example in [54, 263]. However, monogenic diseases are typically rare while there are common diseases that are thought to have a genetic component but do not follow the rules of dominant or recessive disorders. Examples are many common age-related diseases such as diabetes, cardiovascular disease, and dementia that are presumed to be determined by the interaction of several genes (epistasis), and their interaction with environmental factors (gene environment interaction). These diseases are termed 'complex traits' and, as in general complex systems, they are characterized by a high level of unpredictability of the output (the disease or trait) given the same input (the genetic make up) [263].

While the discovery of monogenic disease-causing genes can be used to compute the risk for offspring to be affected given the parent genotypes in pre-natal diagnosis, discovering the genetic bases of common diseases and their interaction with the environment is the first step toward the development of prognostic models that can be used to compute the individual risk for disease and to suggest appropriate prophylactic treatments or lifestyle. The potential benefits of these discoveries on public health are immense, and the discovery of the genetic bases of common diseases is one of the priorities of medical research that is nowadays made possible by rapid innovations in biotechnology. Machine learning methods are providing an important contribution to this endeavor and Bayesian networks in particular have been shown to be ideal modeling tools to simultaneously discover the genetic basis of complex traits and to provide prognostic models [422].

4.2 Historical perspectives

The use of Bayesian networks in biomedical sciences can be traced as far back as the early decades of the 20th century, when Sewell Wright developed path analysis to aid the study of genetic inheritance [504, 505]. Neglected for many years, Bayesian Networks were reintroduced in the early 1980s as an analytic tool capable to encoding the information acquired from human experts [90, 356]. Compared to decision-rule based 'expert-systems' that were limited in their ability to reason under uncertainty, Bayesian networks were probabilistic expert systems that used probability theory to account for uncertainty in automated reasoning for diagnostic and prognostic tasks. This type of probabilistic reasoning was made possible by

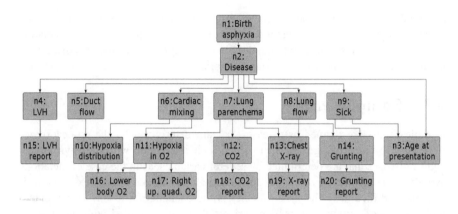

Figure 4.1   The CHILD network.

the development of algorithms to propagate probabilistic information through a network. The first algorithm was limited to networks with simple polytree structures [356], and was followed by more complex algorithms that could be applied to more general network topologies [86, 106, 273]. These algorithms placed Bayesian networks at the forefront of artificial intelligence research and medical informatics [44, 157, 434]. The toy network 'Asia' [106], the CHILD (or 'blue baby') system described in Spiegelhalter et al. [435] and the ALARM network [103] are probably the best known applications of Bayesian Networks in medical diagnoses. The CHILD network is the diagnostic system for telephone referrals of newborn babies with possible congenital heart disease displayed in Figure 4.1. The network consists of 20 nodes that include symptoms and clinical status of a newborn as well as laboratory results. The network is used to compute the probability that a newborn with sign of asphyxia has congenital heart disease by essentially using Bayes' theorem. Bayes' theorem facilitates the quantification of how two or more events effect each other in the production of one or more outcomes. To do this, the theorem updates the prior ('marginal') probability of each event or node, independent of any of the other events into the posterior ('conditional') probability of the event, which is influenced by other events.

The directed acyclic graph of the CHILD network has a very simple structure, with nodes that have only two parents or less so that at most tables of 4×2 conditional probabilities need to be elicited from experts. However, David Spiegelhalter acknowledged the difficulty in eliciting both the graphical structure and the conditional probability tables of the network from experts [435], and admitted that building the CHILD network was made possible by his very close relationship with the expert, who is actually his life partner. It is not until the development of intelligent statistical methods in the late 1980s, early 1990s that Bayesian networks became more easily available modeling tools [103, 185]. These statistical methods allow the machine learning from data of both the graphical structure and the set of conditional probabilities that define a Bayesian network. Popular algorithms such as the K2 [103], and some variations that make the use of coherent prior distributions [206] are implemented in commercial software, for example Discoverer and Hugin, and software that is freely available. The web site of the Association for Uncertainty in Artificial Intelligence[1] has a comprehensive list of software for learning and reasoning with Bayesian networks.

4.3   Complex traits

Complex traits are defined as diseases that are determined by the co-occurrence of several to numerous genetic factors and their interaction with the environment and/or diseases that are determined by several variables, as in the case of metabolic syndrome. Therefore the search for the genetic bases of many complex traits faces two major difficulties. The first difficulty is the challenge of discovering a potentially very large number of genetic variants that are associated with the disease, and their varying influences due to varying exposures to environmental conditions. The second difficulty is the definition of the correct phenotype to be used in the design of the study and the analysis of the data.

4.3.1   Genome-wide data

Until approximately 2004, genotyping costs limited the search for genetic variants responsible for disease to sets of candidate genes that were selected based on an educated guess. The ability to discover the genetic basis of complex disease was therefore limited to the set of candidates. The introduction of economically feasible high-throughput arrays for genome-wide genotyping in the last few years is now moving the field to genome wide association studies that can interrogate nearly the entire genome [193]. The technology allows the simultaneous genotyping of hundreds of thousands of SNPs that are a small proportion of the estimated 10 million existing SNPs but provide sufficient coverage of the variations. For example, either the Sentrix HumanHap550 Genotyping Beadchip (Illumina, San Diego, CA) or the Affymetrix GeneChip © Human Mapping 500K Set consists of more than 500 000 SNPs providing comprehensive genomic coverage across multiple populations. This coverage based on a relatively small number of variants is made possible by the block structure of the human genome that is due to the patterns of linkage disequilibrium (LD) [110, 170]. The latter is the effect of nonrandom association of SNPs that results in the inheritance of blocks or 'haplotypes' of nearby SNPs in the same chromosome. This linkage of genetic variants implies that SNPs in the same haplotypes are mutually informative and that a subset of them is sufficient to tag blocks of the human genome with a certain precision [228, 421].

4.3.2   Complex phenotypes

The difficulty in defining the phenotypes is apparent in the study of those diseases that are determined by multiple variables contributing to the disease condition or syndrome, for example diabetes, metabolic syndromes, neurological degenerative

---

[1]http://www.auai.org

disorders such as Parkinson's or Alzheimer's disease, hypertension or various forms of cardiovascular disease. Dissecting the genetic basis of these diseases requires a phenotype to be defined as the response to different genetic characteristics. In the case of diabetes, one could measure blood sugar level after a night fasting, or base the diagnosis on the glucose tolerance test, and have the glucose level define the phenotype as a quantitative trait. Alternatively, a blood sugar level higher than 140 mg/dl can be translated into a diagnosis of diabetes that is then used as a binary phenotype in the genetic analysis. More complicated is the definition of those phenotypes in which the diagnosis has to be based on the compilation of different symptoms, like the diagnosis of Alzheimer's disease. Such symptoms or measurements are not independent because they are the result of the same disease, so an alternative approach is to leverage on the mutual information that these measurements bring about the same phenotype and use a multivariate model as phenotype of a genetic study. The availability of large amounts of data available from detailed cohort studies like the Framingham Heart Study[2] would make this idea feasible for a variety of complex diseases.

4.3.3   Genetic dissection: The state of the art

Quantitative strategies for the discovery of the genetic variants associated with complex traits are still in their infancy and it is widely acknowledged that the modeling of genotype/environmental factors/phenotype(s) data requires alternative solutions to those currently available to genetic epidemiologists [512]. Reviews by Risch [54], Cardon and Bell [82] and Hoh and Ott [213] survey the major approaches to the statistical analyses of multiple genotypes in either case-control or family based studies and highlight the limitations of the 'one SNP at a time' procedure that is predominantly used to analyze these data. The procedure examines one genetic variant at a time in relation to a presumably well-defined phenotype [234]. This reductionistic approach risks not capturing the multigenic nature of complex traits and typically identifies too many associations because of dependencies between SNPs in contiguous regions as a result of LD, and the evolutionarily induced dependency between SNPs on different chromosomes [170]. A further limitation of the one-at-time variant approach is the inability to discover associations that are due to interdependent multiple genotypes. Hoh and Ott point out the case in which the simultaneous presence of three genotypes at different loci leads to a disease. The three genotypes themselves have the same marginal penetrance and would not be found associated with the disease in a one-at-a-time search. This situation is an example of the well known Simpson's paradox [492]: the fact that marginal independence of two variables does not necessarily imply their conditional independence when other variables are taken into account. These two situations are described in Figure 4.2, which shows a possible dependency structure between a phenotype represented by the node P and the four SNPs represented by the nodes S1–S4 in the graph. The two SNPs S2 and S3 are associated with the phenotype P and their association is represented by the edges between P and S2, as well as P

---

[2]http://www.framingham.com/heart

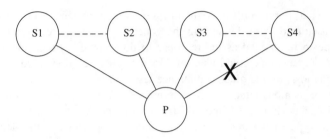

Figure 4.2 Example of redundant and missed association induced by a one-at-a-time search. The SNP S1 is an older mutation in linkage disequilibrium with the more recent mutation S2 and an association between S1 and the phenotype P would be redundant and could potentially be misleading. The association between S4 and P is only found conditional on S3.

and S3. The node S1 represents an older SNP that is associated with the SNP S2 through evolution. The two SNPs do not need to be necessarily on the same genes, or on the same region of linkage disequilibrium. The SNP S4 is not associated with the phenotype when considered individually, but only in conjunction with the SNP S3. A simple one-at-a-time search would probably lead us to identify S1 as associated with P and hence introduce redundancy, if S1 and S2 are in linkage disequilibrium, or introduce a false association if S1 and S2 are not on the same gene or region of linkage disequilibrium. Another limitation of the one-at-a-time search is the likely loss of the association between S4 and P.

Multivariate statistical models, such as logistic regression, can circumvent these limitations by examining the overall dependency structure between genotypes, phenotype, and environmental/clinical variables. However, traditional statistical models require large sample sizes and/or experimental and control samples that are distinctly different enough in terms of the phenotype of interest to confer significant power. Solutions proposed include efficient parameterizations of genotype probabilities so that a two loci interaction models that would require eight parameters can be estimated using two or three parameters [299], and multifactor dimensionality reduction (MDR). MDR reduces dimensionality by pooling combinations of genotypes from multiple loci into a smaller number of groups so as to improve the accuracy in the identification of SNPs that are associated with disease risk [324, 397]. A software package for implementing MDR with up to 15 polymorphisms is freely available,[3] and strategies are currently being explored to expand the algorithm to genome wide association studies in which several hundreds thousands of loci are investigated and data from related individuals can be analyzed [304].

Although these methods can help reduce the complexity of gene–gene and gene–environment interaction models, the amount of data produced by the new genotyping technology requires novel techniques that go beyond 'traditional statistical thinking' in order to accommodate the complexity of genetic models.

---

[3]http://www.epistasis.org/software.html

Machine learning methods used in data mining have the ability to extract information in large databases [198] and, for example, classification and regression tree (CART) [376] and Random Forests [61] have been proposed to model complex gene–environment interaction when the phenotype is a well defined variable. CART is a multivariate statistical technique that creates a set of if-then-rules linking combinations of genotypes and environmental exposures to the phenotypes. The if-then-rules are created with a recursive procedure that groups data in sets in order to maximize the utility of all of the information [198]. Random forests have shown particular promise for the analysis of genotype/phenotype correlations, taking into account gene–gene interactions. The intuition is to use permutation methods and bootstrapping techniques to create thousands of CART models. From the analysis of these models structures, one can produce a summary importance measure for each SNP that takes into account interactions with other SNPs influencing the phenotype [291]. It has been shown that when unknown interactions among SNPs and genetic heterogeneity exist, random forest analysis can be significantly more efficient than standard univariate screening methods in pointing to the true disease-associated SNPs.

Contrary to all these machine learning methods that require a well defined phenotype to be applied, Bayesian networks can model multiple random variables and accommodate the complex structure of gene environment interactions together with phenotypes defined by multiple variables. By using these models, the probability distribution of the many variables describing a complex system is factorized into smaller conditional distributions, and this factorization permits an understanding of the entire complex system by examining its components [106, 271]. Compared to standard regression models in which the correlation between the variables leads to multicollinearity and lack of robustness of model fitting, Bayesian networks leverage on the mutual correlation between variables to define the conditional probability distributions. These conditional distributions become the 'bricks' used to build a complex system in a modular way. Specifically, the modularity provides the critical advantage of breaking down the discovery process into the search for the specific components of a complex model.

4.4   Bayesian networks to dissect complex traits

4.4.1   Bayesian learning

The modularity is one of the critical features to derive or learn Bayesian networks from large data sets with many variables. There are several strategies for learning both the graphical structure and the tables of conditional probabilities, and well-established techniques exist to induce BNs from data in an almost automated manner. These strategies include standard association tests [111, 185, 271] and Bayesian procedures that were proposed in [436] for the estimation of the conditional probability tables, and subsequently model selection in [103] and [106, 202]. The underlying premise of a Bayesian model selection strategy is to assess each dependency model by its posterior probability and select the model with maximum posterior probability [242]. The posterior probability is therefore the scoring metric that is used to rank models. Its value is computed by using Bayes' theorem to update the prior probability with the marginal likelihood and, because it is common to assume uniform probabilities on the space of models, the posterior probability is proportional to the marginal likelihood that becomes the scoring metric for model selection.

The marginal likelihood is the expected value of the likelihood function when the parameters follow the prior distribution. The likelihood function is the joint density of the data, for fixed parameter values, and it is determined by the nature of the network variables and the assumption of marginal and conditional independencies that are represented by the network graph. The prior distribution of the model parameters, on the other hand, represents the prior knowledge about the parameter distribution so that the marginal likelihood is an average value of the likelihood functions determined by different parameters values. Compared to the maximum likelihood that returns the likelihood function evaluated in the estimate of the parameters, the marginal likelihood incorporates the uncertainty about the parameters by averaging the likelihood functions determined by different parameter values. This conceptual difference is fundamental to understanding the different approach to model selection: in the classical framework, model selection is based on the maximized likelihood and its sampling distribution to take into account sampling variability for fixed parameter values. In the Bayesian framework, model selection is based on the marginal likelihood which takes into account the parameter variability for fixed sample values. Therefore, no significance testing is performed when using this approach to model selection. Our experience with the Bayesian procedure to model selection is that it is usually more robust to false associations. This is due to the use of proper prior distributions that can tune the level of evidence needed to accept an association.

As an example, suppose the network variables are all categorical, and there are not missing values in the data set. We denote by y_{ik} the kth category of the variable Y_i in the network, $k = 1, \ldots, k_i$ and $i = 1, \ldots, q$, and by $pa(y_i)_j$ the jth configuration of the parents $pa(y_i)$ of the variable Y_i, $j = 1, \ldots, j_i$. Given a network structure, the model parameters are the conditional probabilities $\theta_{ijk} = \mathbb{P}\left(Y_i = y_{ijk} | pa(y_i)_j\right)$, and the hyper-Dirichlet distribution is the conjugate prior [116, 179]. The marginal likelihood can be calculated in closed form and it is given by the formula:

$$\prod_i \left\{ \prod_j \frac{\Gamma(\alpha_{ij})}{\Gamma(\alpha_{ij} + n_{ij})} \prod_k \frac{\Gamma(\alpha_{ijk} + n_{ijk})}{\Gamma(\alpha_{ijk})} \right\} \tag{4.1}$$

where n_{ijk} is the frequency of $(y_{ik}, pa(y_i)_j)$ in the sample, $n_{ij} = \sum_k n_{ijk}$ is the marginal frequency of $p(y_i)_j$, the values α_{ijk} are the hyper-parameters of the hyper-Dirichlet distribution, and $\alpha_{ij} = \sum_k \alpha_{ijk}$. With large networks it is convenient to use symmetrical hyper-Dirichlet prior distributions that depend on one hyper-parameter value α. By symmetry, this value is used to specify all the other

	Y_2 (parameters)		Y_2 (frequencies)		Y_2 (hyper-parameters)	
	True	False	True	False	True	False
	θ_{21}	θ_{22}	$n_{321}+n_{322}$ $+n_{341}+n_{342}$	$n_{311}+n_{312}$ $+n_{331}+n_{332}$	$\alpha/2$	$\alpha/2$

(Network diagram at left: Y_2 with an arrow to Y_3; $Y_1 \rightarrow Y_3$.)

Parents	Parents configuration		Y_2 (parameters)		Y_2 (frequencies)		Y_2 (hyper-parameters)	
$pa(y_3)$	Y_1	Y_2	True	False	True	False	True	False
$pa(y_3)_1$	False	False	θ_{311}	θ_{312}	n_{311}	n_{312}	$\alpha/8$	$\alpha/8$
$pa(y_3)_2$	False	True	θ_{321}	θ_{322}	n_{321}	n_{322}	$\alpha/8$	$\alpha/8$
$pa(y_3)_3$	True	False	θ_{331}	θ_{332}	n_{331}	n_{332}	$\alpha/8$	$\alpha/8$
$pa(y_3)_4$	True	True	θ_{341}	θ_{342}	n_{341}	n_{342}	$\alpha/8$	$\alpha/8$

Figure 4.3 Example of a network with 3 variables and parameters θ_{ijk}, frequencies n_{ijk} and hyper-parameters α_{ijk}. In this network, only the node Y_3 has parents $pa(y_3)$ defined by the joint distribution of Y_1 and Y_2. The table at the top shows the marginal distribution of Y_2, the frequencies n_{2k}, and the hyper-parameters that are determined by distributing the overall precision α in the 2 cells of the probability table. The table at the bottom shows the parameters of the 4 conditional distributions of Y_3 given the parents. The last four columns show the frequencies and hyper-parameters that are determined by distributing the overall precision α in the eight cells of the table.

hyper-parameters as $\alpha_{ij} = \alpha/j_i$ and $\alpha_{ijk} = \alpha_{ij}/k_i = \alpha/(j_i \times k_i)$. The main intuition is to use the value α – the prior precision – as an overall sample size, and then distribute the counts in each conditional probability table as if the probabilities were all uniform. Figure 4.3 provides an example. Once the best network is identified, the conditional probabilities are estimated as $\hat{\theta}_{ijk} = (n_{ijk} + \alpha_{ijk})/(n_{ij} + \alpha_{ij})$.

In many studies, environmental, nongenetic factors, and phenotype may be continuous variables and to fully extract information from epidemiological studies, we need to induce BNs from data that consist of categorical variables (genotypes, haplotypes), continuous or discrete phenotypes (survival time, or clinical measurements), and continuous and/or discrete environmental variables. The challenge posed by using a mix of continuous and categorical variables is that the calculation of the posterior probability of each model may not be possible in closed form for some choice of probability distributions, and functional dependencies between parents and children nodes. When the number of variables is small, we can resort to stochastic methods to estimate the marginal likelihood, or some penalized forms such as the Deviance information criterion (DIC) [433], or the Akaike information criterion (AIC) or the Bayesian information criterion (BIC) [242]. The AIC of a network M is given by the formula:

$$\text{AIC} = -2\log(l(\hat{\theta}|M)) + 2n_p$$

where $l(\hat{\theta}|M)$ is the likelihood function $l(\theta|M)$ evaluated in the estimate $\hat{\theta}$ of the parameters, n_p is the number of parameters, and log is the natural logarithm. The BIC is the modification of AIC given by the formula:

$$\text{BIC} = -2\log(l(\hat{\theta}|M)) + \log(N) \times n_p$$

where N is the sample size. Compared to the AIC, the BIC penalizes models with the extra term $(\log(N) - 2) \times n_p$ that is positive as long as $\log(N) > 2$ and hence the sample size is greater than $\exp(2) = 7.4$. The DIC was introduced recently as a way to penalize model complexity when the number of parameters is not clearly defined. This criterion is defined as

$$\text{DIC} = -2\log(l(\hat{\theta}|M)) + 2\hat{n}_p = -2E(\log(l(\hat{\theta}|M)) + \hat{n}_p$$

where $\log(l(\hat{\theta}|M))$ is defined as above and $E(\log(l(\hat{\theta}|M))$ is the expected log-likelihood. Both quantities can be computed using stochastic methods [433].

4.4.2   Model search

In building genetic models, the 'common disease, common allele variants' assumption described in [374] would support that the risk for common diseases is modulated by particular combinations of alleles that are common in the population. Individually, each allele would not be sufficient to predict the disease outcome, which can be predicted only by their simultaneous presence. Therefore, in deciphering the genetic basis of complex traits, we expect to find a large number of SNPs involved in modulating disease severity, each having a small effect on the phenotype. This potential complexity makes building standard regression, or prognostic models unfeasible because the associations between SNPs and possible environmental variables with the phenotype are competitive and the presence of an association between a SNP and the phenotype affects the association between the phenotype and other SNPs. The consequence is a limit on the number of SNPs that can be detected as determinant or parents of the phenotype and a potentially serious limitation on the predictive accuracy of the model. This limitation is removed with the use of diagnostic models, in which we model the dependency of SNPs and environmental variables on the phenotype (see Figure 4.4) and the association of each SNP with the phenotype does not affect the association of other SNPs with the phenotype. This structure has the additional advantage of representing the correct data generating mechanism of a cross-sectional study via the 'retrospective likelihood' that models the distribution of the genetic causes conditional on the phenotypic effect rather than the physical/causal process underlying the biological system that relates genotype to phenotype. The prediction of the phenotype given a particular genetic profile is not explicitly represented by this model structure, but needs to be computed by using Bayes' theorem using probabilistic algorithms [106]. A particular advantage of this 'inverted' dependency structure is the ability to represent the association of independent as well as interacting SNPs with the phenotype and, as shown in Figure 4.4, the structure allows to remove redundancy by conditional

Prognostic models

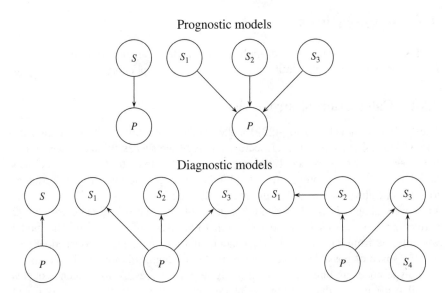

Diagnostic models

Figure 4.4 Network structure to build prognostic models (top) and diagnostic models (bottom). The node P represents the phenotype, while the nodes $S, S_1, \ldots, S_4$ represent SNPs and/or environmental variables. In the prognostic model, the phenotype node is modeled as a child of the genotype nodes. In the diagnostic model, the phenotype node can only be a parent of all the other nodes.

independence, but enriches the Markov blanket of the phenotypes by those nodes that are only conditionally dependent on the phenotype given other nodes.

To search for prognostic models, we assume that the node representing the phenotype in the network can only be a parent of any other nodes, while to search for the dependency structure among the remaining variables we exploit the network modularity that is induced by the marginal and conditional independencies to build local models and join them. For this step we use the K2 algorithm introduced by Greg Cooper [103] that requires an initial ordering on the variables to avoid the introduction of loops. In practice we have observed that ordering nodes by their marginal information content leads, on average, to better networks. The prediction of the phenotype given a particular genetic profile is not explicitly represented as in a traditional regression model but needs to be computed by Bayes' theorem using probabilistic algorithms [106].

This search strategy for the special case of categorical variables is implemented in the program Discoverer. Continuous variables are automatically discretized using either quartiles, or by dividing the range of definition into intervals of equal length. The program has one algorithm for probabilistic reasoning that can be used to infer the distribution of any variable in the network given information on other variables. This reasoning tool is used to validate the predictive accuracy of networks induced from data in independent data sets or by standard cross-validation.

4.5   Applications

In this section we report some applications of Bayesian network to model the genetic basis of a complex trait, and to model complex phenotypes.

4.5.1   Calculation of stroke risk

The BN model developed in [422] for the genetic dissection of stroke in subjects with sickle cell anemia is a small scale example of the successful and powerful use of Bayesian networks to model the genetic basis of complex traits. This model was based on a BN that captures the interplay between genetic and clinical variables to modulate the risk for stroke in sickle cell anemia subjects. Available for this study were about 250 SNPs in candidate genes of different functional classes in 92 patients with nonhemorrhagic stroke and in 1306 disease-free controls. The model was inferred from data using the Bayesian model search strategy described earlier, and the space of possible models was restricted to diagnostic models in which the phenotype node can only be a parent of all the other nodes. A consequence of this model restriction is that, during the model search, the phenotype node is tested as a parent of each SNPs. To further reduce the complexity of the search, SNPs were ordered by entropy so that less variable SNPs were only tested as parents of more variable SNPs. This search favors more recent mutations as possibly implicated with the phenotype. Intuitively, lethal older mutations are eliminated because they lead to early death, therefore in our model search we privileged more recent mutations as responsible for disease.

The network identified by this search is depicted in Figure 4.5 and describes the interplay between 31 SNPs in 12 genes that, together with fetal hemoglobin (a protein that is present in adult subjects with sickle cell anemia) modulate the risk for stroke. The strength of the associations of each SNP in the Markov blanket and the phenotype was measured by the Bayes factor of the model with the association, relative to the model without the association. For further details refer to the original manuscript [422]. This network of interactions included three genes, BMP6, TGFBR2, TGFBR3 with a functional role in the TGF-beta pathway and also SELP. These genes and klotho (KL), a longevity-associated gene in animals and humans [18], are also associated with stroke in the general population.

We initially validated the network by five-fold cross-validation and the network reached 98.5% accuracy. However, the lack of reproducibility of genetic associations in different studies has made the genetic community very cautious about associations found in one single data set. Therefore, we validated our model in a different population of 114 subjects including seven stroke cases and 107 disease free subjects. The model reached a 100% true positive rate, 98.14% true negative rate and an overall predictive accuracy of 98.2%. This accuracy confirms the predictive value of the model that can be used for risk prediction of new sickle cell anemia subjects based upon the genetic profile. The model can also be used as a diagnostic tool, to identify the genetic profiles that modulate increase the risk for stroke. This use suggested that genes in the TGFBeta signaling pathway could

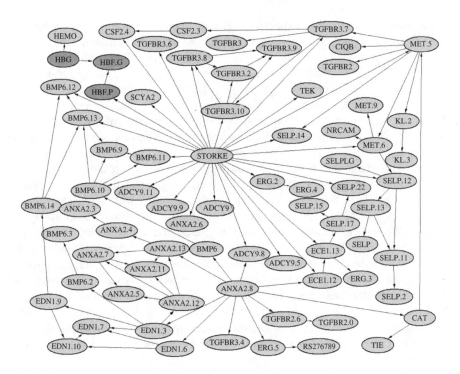

Figure 4.5   The Bayesian network describing the joint association of 69 SNPs with stroke. Nodes represent SNPs or clinical factors, and the numbers after each gene name distinguishes different SNPs on the same gene. SNP are represented as blue nodes and clinical variables (HbF.G: fetal hemoglobin g/dL; HbF.P: fetal hemoglobin percent; HbG: total hemoglobin concentration; THALASSEMIA: heterozygosity or homozygosity for α-thalassemia deletion) are represented by green nodes. Twenty six nodes (HbF.P and 25 SNPs on the genes ADCY9, ANXA2, BMP6, CCL2, CSF2, ECE1, ERG, MET, SELP, TGFBR3 and TEK) are children of the phenotype, and seven more nodes are parents of the children node (BMP6, BMP6.14, ANXA2.8, SELP.17, CAT, CSF2.3). These 32 nodes are the Markov blanket and are sufficient to predict the risk for stroke.

play a crucial role in stroke and other vaso-occlusive complications of sickle cell disease and triggered follow up work [33, 420].

4.5.2   Modeling exceptional longevity with Bayesian networks

Evidence is rapidly accumulating to support the existence of significant familial and heritable components to the ability to survive to extreme old age [362, 363, 364]. This exceptional longevity (EL) can be defined as survival beyond a certain age or percentile of survival but this definition does not take into account the physical status of subjects reaching exceptional ages. Many different paths depending upon secular trends in life expectancy, gender, social, environmental and behavioral factors and their interactions with a spectrum of genetic variations lead to EL. These paths generate exceptional survival (ES) phenotypes [361]. Definitions of ES might include, for example, disability-free survival past a specified age, or disease-free survival past a specified age. Figure 4.6 illustrates, albeit in a reductionist manner, the complex interaction of disease predisposing and longevity enabling exposures and genetic variations that lead to different patterns of ES and EL. The need for a well characterized phenotype that goes beyond simple survival is widely acknowledged. For example, based on the occurrence of the most common diseases of aging, Evert et al. noted that centenarians in the New England Centenarian Study (NECS) fit into one of three categories: 'survivors', 'delayers' or 'escapers' [147]. Survivors were individuals diagnosed with age-related illness prior to age 80 (24% of the male and 43% of the female centenarians). Delayers were individuals who delayed the onset of age-related diseases until at least age 80 (44% of the male and 42% of female centenarians). Escapers were individuals who attained their 100th year of life without the diagnosis of an age-related disease (32% of the male and 15% of the female centenarians). In the case of cardiovascular disease (CVD) and stroke, which are typically associated with relatively high mortality risk compared with more chronic diseases such as osteoporosis, only 12% of subjects were survivors and the remaining 88% of centenarians delayed or escaped these diseases. In Figure 4.6, these categories are examples of ES phenotypes.

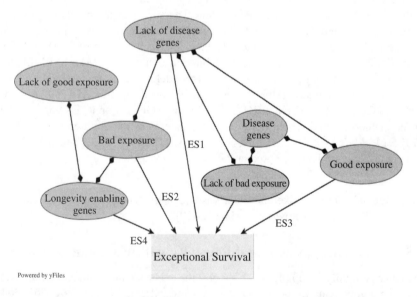

Figure 4.6 There are multiple paths (exceptional survival subphenotypes, ES1, ..., ES?) composed of different combinations of deleterious, neutral and beneficial exposures and genetic variations leading to exceptional longevity (EL).

This classification of centenarians into the three morbidity groups survivors, delayers and escapers was based on an 'educated' selection of disease and their age of onset to help distinguish between different patterns of ES. Additional phenotypic data suggest that there are several ES phenotypes that are associated with gender, age, and other determinants of physical and cognitive functional status which could provide more informative phenotypes in genome wide association studies. We are currently using Bayesian networks to discover these patterns of ES from data in two ways, using either unsupervised or supervised methods. In the unsupervised procedure, patterns of ES are represented by sets of probability distributions of the phenotypic data, and the analysis aims at grouping those subjects whose phenotypic data follow the same probability distribution. In the supervised procedure, we use survival to a certain age as the variable to be predicted, using information about the other phenotypic data. Small scale examples of these two approaches follow.

4.5.2.1 Unsupervised approach

We use Bayesian network modeling to discover patterns of ES using data about hypertension (HTN), congestive heart failure (CHF), stroke, and gender in 629 subjects from the NECS. We focus on a class of diagnostic models in which the root node is a hidden variable that labels groups of subjects with the same phenotypic patterns and has to be discovered from data. This model is known as a mixture model with a hidden mixing variable [156], and it is commonly used for clustering. Compared to traditional clustering methods that are mainly descriptive techniques, this approach is based on a probabilistic definition of clusters and takes into account sampling variability by computing the probability that each subject in the sample belongs to one of the clusters. The mixture model can be represented by the Bayesian network displayed in Figure 4.7 (left panel), and consists of the hidden variable that represents the cluster membership (this is the root node in the figure) and the other observable variables (the children nodes Gender, Stroke, HTN and CHF) that are assumed to be independent, conditional on the hidden variable. The states of the hidden variable define the cluster profiles as the set of conditional distributions of the observable variables. By using this model to represent groups of subjects, the objectives of the analysis become to estimate:

1. the number of clusters in the data that is represented by the number of states of the hidden varable;

2. the cluster membership of each subject that is computed as the most likely state of the hidden variable for each subject;

3. the cluster profiles that are characterized by the set of conditional distributions of the variables in each cluster.

Consistent with other approaches [91], we model the hidden variable with a multinomial distribution with h categories, and the prevalence of disease and gender using binomial distributions with parameters that change in each cluster. Because

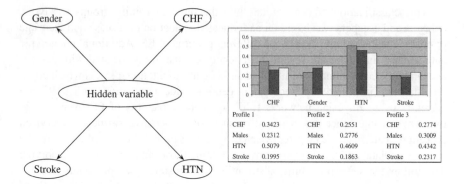

Figure 4.7   Left: Graphical description of a mixture model to cluster subjects based on information about gender, history of stroke, hypertension (HTN) and congestive heart failure (CHF). Each state of the hidden variable represents a cluster that is characterized by the set of probability distributions of the four variables, conditional on the hidden variable. Right: Profiles of the three clusters identified by the Bayesian network modeling.

of the hidden variable, the estimation cannot be done in closed form and we used the Markov Chain Monte Carlo methods implemented in the program Win-Bugs 1.14 [453]. Given the data, the algorithm produces a sample of values from the posterior distribution of the model parameters and the states of the hidden variable. These quantities are used to estimate the states of the hidden variable and the parameters of the conditional distributions of the phenotypic data that define the cluster profiles. To identify the optimal number of clusters, we can repeat the analysis for different number h of states of the hidden variable, score each model using the DIC, and choose the model with the minimum DIC as suggested in [433]. In our analysis, a model with three clusters gives the best fit and it is characterized by the profiles summarized in Figure 4.7: the barplots in light gray depict the profile of Cluster 1 that is characterized by the probability distributions of CHF, Gender, HTN and Stroke described in the figure. The barplots in dark gray and white depict the profiles of cluster 2 and 3. Centenarians in cluster 1 have the highest risk for CHF, HTN and are more likely to be females, compared to clusters 2 and 3. Centenarians in cluster 3 are at highest risk for stroke, but at lowest risk for HTN. The three clusters suggest three different patterns of ES. The first two clusters are characterized by a higher risk of cardiovascular disease (CHF and HTN), while subjects in cluster 3 have a higher risk of cerebral-vascular accidents (stroke). The analysis also provides the probability of cluster membership for each subject in the study: for example the first subject in the data set has a 40% chance of being a member of cluster 1, while the second subject has a 40% chance of being a member of cluster 3. This score can be used as the phenotype of some genetic association study to search for modulators of different patterns of ES rather than focusing on a specific disease.

4.5.2.2   Supervized approach

We can use Bayesian network modeling to describe inter-relationships between being a centenarian, socio-demographic characteristics, sex, education, and the health history of diseases in the same set of NECS subjects. Compared to the previous analysis, we now consider the binary variable 'age at death > 100' (node Centenarian) as the variable to be predicted. Figure 4.8 displays the network that we generated using the computer program Bayesware Discoverer. To generate the network, we limited attention to diagnostic models in which the age at death can only be a parent node of all the other variables. We ordered the other variables by entropy, so that more entropic variables can only be parents of less entropic variables, and we used a hyper-Dirichlet prior distribution on the model parameters, with $\alpha = 12$. Our experience is that increasing the overall precision α makes the network more robust to sampling variability, and in practice we use a value α between 1 and 16, and assess the sensitivity of the network structure to different prior distributions. This network was built using the program Bayesware Discoverer that provides validation tools to assess the goodness of fit of the network learned from data. These tools include standard cross-validation techniques that can assess the robustness of the network to sampling variability. The program has

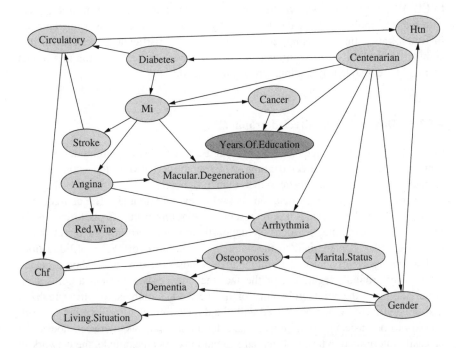

Figure 4.8   The Bayesian network that summarizes the mutual relations between socio-economic, demographic and health variables in a samples of 600 subjects of the NECS.

different algorithms for probabilistic reasoning that can be used to infer the distribution of any variable in the network given information on other variables. Using one of these algorithms, we assessed the accuracy of the network to predict EL given particular patterns of ES. The network reached 81% accuracy using five-fold cross-validation in identifying those centenarians who lived past 100 years.

The network describes expected associations such as those between cancer, years of education and age at death, or the association between HTN and cardio-vascular and cerebro-vascular accidents (nodes stroke and circulatory disease). Inspection of the probability distributions associated with these nodes shows that, for example, centenarians with both circulatory problems and diabetes have 2.4 times the odds for stroke compared to centenarians with diabetes alone. These associations change when we consider gender and history of HTN and show that female centenarians with a history of HTN have a higher prevalence of stroke than male centenarians, while no history of HTN makes the prevalence of stroke in male centenarians higher compared to female centenarians. When we consider profiles of centenarians by gender, male centenarians present with a healthier profile characterized by a smaller prevalence of hypertension, congestive heart failure, and a much smaller prevalence of dementia (21% compared to 34% in females) and osteoporosis (8% compared to 34%). The complexity of these combinations of disease or syndromes confirms that there are different patterns of survival leading to EL. However, it also shows the challenge in summarizing them in prototypical patterns. The multivariate model described by the BN provides a powerful approach in which, rather than 'labeling' centenarians by some particular pattern of ES that summarizes the information in many variables, we use this information in a model that can be expanded by adding genetic information.

4.5.2.3   Directed or undirected graphs?

One of the major challenges we face in using Bayesian networks to model complex traits is to explain the directed graph of associations between the variables. We typically work with diagnostic rather than prognostic models because they are better able to incorporate associations between several variables. The consequence is that most of the directed arcs represent associations from effects to causes. In Figure 4.8, for example, the focus on diagnostic model determines the directed arcs from the node 'Centenarian' to the node 'Marital.Status' or 'Years.of.Education' and would inevitably raise the comment that those arc directions should be the 'other way around', regardless of the fact that they only represent a convenient factorization of the joint probability of the network variables. We have realized that the undirected graph associated with the directed graph of a Bayesian network appears to be easier to communicate, and the undirected links between nodes can be easily interpreted in terms of mutual associations. As an example, the network in Figure 4.9 is the undirected graph that can be obtained from the Bayesian network in Figure 4.8 using the moralization procedure [106]. This procedure allows to drop the arc directions when all parents of each node are linked to each other, and

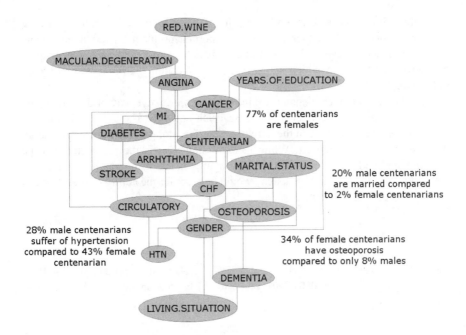

Figure 4.9 The undirected graph obtained by the Bayesian network in Figure 4.8 using the moralization procedure, in which the arc directions are dropped once all the parents of each node are linked to each other. The graphical display was generated using the yEd graph editor.

this guarantees that the undirected graph shows the same Markov properties of the directed graph.

The graphical display was generated using the yEd graph editor.[4] This program provides tools for annotating links between nodes and we are experimenting the addition of labels that quantify some of the association displayed by the network. For example, the graph in Figure 4.9 provides information about the relation of gender and living to 100 years, as well as information about some disease prevalence by gender. This representation is more informative than a simple graph that does not quantify the effects of the associations.

4.6 Future challenges

Bayesian networks provide a flexible modeling framework to describe complex systems in a modular way. The modularity makes them the ideal tool for modeling the genetic basis of common diseases and their interaction with the environment but there are computational issues to address. With the availability of genome wide data, we face the challenge to model hundreds of thousands of variables, and we

---

[4]http://www.yWorks.com

need algorithms able to efficiently search for good models. Extending learning algorithms to mixed variables is another important problem to be solved. In the context of genetic analysis, the major challenge is to extend the learning algorithms to family based data in which subjects in the same family cannot be considered independent.

Another important challenge is to find a more informative way of displaying the amount of information condensed in a Bayesian network. The graphical structure displays the relations between the variables, but does not provide information about the effects of these relations. While in regression models the parameters can be interpreted directly in terms of effects, this synthesis is not so straightforward with networks. A more informative display would help confer the ability of these models to describe complex systems.

Acknowledgment

Research supported by the National Institutes of Health: National Heart Lung Blood Institute (NIH/NHLBI grant R21 HL080463-01) and National Institute on Aging (NIH/NIA 5K24-AG025727-02).

5

Crime risk factors analysis

Roongrasamee Boondao

Faculty of Management Science, Ubon Rajathanee University
Warinchumrab, Ubon Ratchathani 34190, Thailand

5.1 Introduction

Throughout recorded history crime has been a persistent societal problem. With increasing urbanization the problem has increased in both magnitude and complexity. To provide effective crime control, with the limited resources available to them, law enforcement agencies need to be proactive in their approach. Crime pattern analysis can guide planners in the allocation of resources. Since both cities and crime patterns are dynamic in nature, a system is required which can develop a model that is not only able to make accurate predictions but can also be constantly updated to accommodate changes in various parameters over time. This chapter describes a model, using a BN, which was used in a pilot scheme to analyze the factors affecting crime risk in the Bangkok metropolitan area, Thailand. In the pilot project both spatial and nonspatial data describing populations of areas, locations and types of crimes, traffic densities and environmental factors were collected.

BNs have become increasingly popular as decision support tools among those researching the use of artificial intelligence, probability and uncertainty. The Bayesian paradigm has been adopted as a most useful approach for inference and prediction in the crime problem domain [345]. A key feature of this approach is the use of existing data to set up the initial model and the continuing enhancement of the model's predictive capabilities as new data is added. This is particularly relevant to police work as there is always a large amount of data on hand to set up the initial model and a continuous flow of fresh case information to constantly

Bayesian Networks: A Practical Guide to Applications Edited by O. Pourret, P. Naïm, B. Marcot
© 2008 John Wiley & Sons, Ltd

refine the model. The system can provide a model for crime risk analysis that constantly adapts in response to changes in the pattern of crimes in the area under consideration. The ability to update the initial distribution on receipt of new data makes the Bayesian approach a natural choice for the analysis of crime risk factors in an area.

5.2   Analysis of the factors affecting crime risk

The series of steps used in the analysis of the factors affecting crime risk by means of a BN are illustrated in Figure 5.1.

1. *Identify crime pattern characteristics.* The first step in the process was the identification of crime pattern characteristics. This was carried out by means of an intensive study of crime theory and a series of discussions with a number of police officers with considerable experience in the fields of crime investigation and suppression. The purpose of this step was to identify the range of factors that were most likely to have an influence on the crime risk in each district of the area under study.

2. *Establish the relationships between various factors.* After developing an outline of the crime pattern, it was necessary to establish the relationships between the various crime factors that were identified in the previous step. Consideration was given to the various factors identified and the linkages

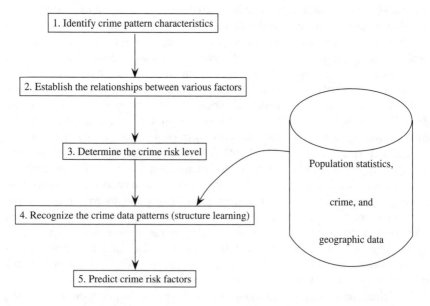

Figure 5.1 The process for analysis of the factors affecting crime risk. Reproduced by permission of WSEAS.

between them to determine which factors were likely to influence the crime rate and the types of crimes likely to be committed in particular regions of the study area.

3. *Determine the crime risk level.* The crime risk level must be determined. After the establishment of the relationships, the crime risk level of each factor needs to be decided. The results of steps 2 and 3 were derived from expert elicitation in the field of crime control.

4. *Recognize the crime data patterns (structure learning).* In this step the crime data patterns were recognized using structure learning. Hugin Researcher 6.3 software was used to recognize the crime data patterns.

5. *Predict crime risk factors.* Finally, the results of predicting crime risk factors were obtained. The details of the model are explained in section 5.5.

5.3   Expert probabilities elicitation

To elicit probabilities from the experts involved some difficulties. In practice, the police officers found it difficult to express their knowledge and beliefs in probabilistic form, so it was necessary to apply the process for eliciting expert judgements of Keeney and von Winterfeldt [244]. The formal process to elicit probabilities from experts consists of seven components:

- Step 1: Identification and selection of the issues was achieved through a broad review that included the opinions of police officers in the field of crime control and stakeholders involved in the problem.

- Step 2: Identification and selection of experts; Clemen and Winkler [97] suggests that five specialists are usually sufficient. The selection was based on the need for diversity resulting in the selection of two crime control experts, one expert in crime prevention, another in crime suppression and one involved in crime control policy setting.

- Step 3: Discussion and refinement of the issues was conducted in the first meeting of the experts. The aim of the meeting was to clearly define the events and quantities that were to be elicited. After this meeting, the experts were given some time to study the issue, think about the elicitation events and variables. This process took two months and involved another meeting among the experts to better define the events and quantities.

- Step 4: Training for elicitation, after the issues were clarified, involved a one-day seminar in which the experts were trained in concepts of probability, elicitation methods, and biases in probability assessment.

- Step 5: Elicitation was carried out at a second meeting of the experts. During this process, the experts were interviewed individually by an analyst. The elicitation sessions lasted three hours.

- Steps 6 (Analysis, aggregation, and resolution of disagreements) and 7 (Documentation and communication). After, the elicitation, individual decompositions were analyzed and re-aggregated to obtain comparable probability distributions over the relevant event or quantity for each expert. Subsequently, the experts' probability distributions were aggregated by taking the average of the individual distributions. Finally, the results were documented by the researcher.

5.4   Data preprocessing

The purpose of the data preprocessing is to prepare the raw data for the subsequent process of analysis. The three steps which were involved are described below.

Data consolidation is the process of gathering all the data from multiple sources into one centralized location. In this research, the data were collected from the National Statistical Office of Thailand, the Royal Thai Police, the Bangkok Metropolitan Administration, and the Ministry of Transportation.

Data selection is the process of selecting, from the wide range of data collected, a data collection suitable to be used in the next step. In this research, data from January 2000 to December 2003 was selected.

Data transformation is the process of organizing data into the same format. Since the data were gathered from multiple sources there was a need to transform wide-ranging data into a set of variables with suitable formats that would be useful for meaningful analysis.

As a result of the above steps of data preprocessing, a set of data consisting of 1000 records was obtained. This set of data was used in the next step: the recognition of the crime data pattern. These data consisted of 20 variables which were classified into five groups of factors: population, crime locations, types of crimes, traffic, and environment (see Figure 5.2).

The 20 variables were classified into a range of states as shown in Table 5.1.

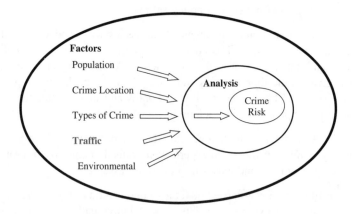

Figure 5.2   The set of variables for crime risk factors analysis.

Table 5.1   The names and states of the variables used for the analysis.

Type of data	Names and states of variables	Explanation of variable name
Population factor		
– pop_factor	0 = not sensitive, 1 = sensitive	A group of population factors
– pop_density	0–4	Population density
– pop_gender	0 = not sensitive, 1 = sensitive	Population gender
– pop_male	0–4	Male population
– pop_female	0–4	Female population
– pop_income	0–4	Population income level
– pop_primary	0–4	Population employed in primary sector
– pop_secondary	0–4	Population employed in secondary sector
– pop_tertiary	0–4	Population employed in tertiary sector
– pop_age	0 = not sensitive, 1 = sensitive	Population age
– infant	0–4	Infant population group
– young	0–4	Young population group
– working	0–4	Working population group
– old	0–4	Elderly population group
Traffic factor		
– traffic_volume	0–4	Traffic volume
– low_standard_housing	0–4	Number of low standard housing areas
Environment factor		
– drugs_sale	0–4	Number of drug–sale areas
– lighting	0 = not enough, 1 = enough	Lighting degree
– hiding	0 = cannot hide, 1 = can be hide	
Crime factor		
– rape	0–4	Number of rapes cases
– robbery	0–4	Number of robbery cases
– murder	0 = no, 1 = yes	Status of murder, no = no cases yes = one or more cases
Location factor		
– crime_location_factor	0 = not sensitive, 1 = sensitive	Crime location factor
– nightclubs	0–4	Number of nightclubs
– shopping_center	0–4	Number of shopping centers
– movie_theatre	0–4	Number of movie theatres
– bank	0–4	Number of banks
– hotel	0–4	Number of hotels

0–4 means: 0 = very low; 1 = low; 2 = medium; 3 = high; 4 = very high

5.5   A Bayesian network model

The BN model developed for use in this particular research project was based on the crime pattern analysis work carried out by Brantingham and Brantingham [60] and the theory of crime control through environmental design [231, 393]. Pattern theory focuses attention and research on the environment and crime, and maintains that crime locations, characteristics of such locations, the movement paths that bring offenders and victims together at such locations, and people's perceptions of crime locations are significant objects for study. Pattern theory synthesizes its attempt to explain how changing spatial and temporal ecological structures influence crime trends and patterns. The model was constructed and tested using Hugin software. This software was also used to analyze the relationships within the data. Figure 5.3 illustrates the BN model for analysis of the factors affecting crime risk.

5.5.1   Conditional probability tables

The conditional probability tables (CPTs) and prior probabilities in the crime risk factors analysis model's BN were defined using the best estimates of a number of experienced officers of the Royal Thai Police.

The model was constructed, and the probabilistic values were calculated and stored in the Conditional Probability Table (CPT).

5.5.2   Sample of conditional probability table

The probability for each input node was calculated using the data contained in the training examples for each state and is shown in Table 5.2. The variables have five

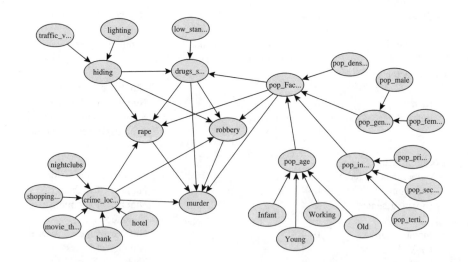

Figure 5.3 A BN model for crime risk factors analysis. Reproduced by permission of WSEAS.

Table 5.2   The partial conditional probability table (CPT) of Robbery given Drug_sales, Pop_factor, Crime_location, Robbery and Rape.

Drug_sales						0				
Pop_factor						0				
Crime_location						0				
Murder			0					1		
Rape	0	1	2	3	4	0	1	2	3	4
0	1	0.5	0.2	0.2	0.2	0.75	1	1	0.2	0.2
1	0	0.5	0.2	0.2	0.2	0.25	0	0	0.2	0.2
2	0	0	0.2	0.2	0.2	0	0	0	0.2	0.2
3	0	0	0.2	0.2	0.2	0	0	0	0.2	0.2
4	0	0	0.2	0.2	0.2	0	0	0	0.2	0.2

states: 0 = very low, 1 = low, 2 = medium, 3 = high, and 4 = very high. For instance, from Table 5.2, the probability of 'robbery = low' given 'drug sales = very low' is 0.2.

5.5.3   Learning Bayesian network

The process of learning by a Bayesian network [86, 162, 232] is divided into three parts as shown below in Figure 5.4.

1. *Input.* The input dataset consists of prior crime data and background knowledge that reflects the human experts' opinions as to which graphical and/or probabilistic structure is more likely than others to represent the relationships among the variables.

2. *Output.* The output is a revised Bayesian network that gives a much better representation of the probability distribution, independence statements and causality as a result of adjustments to the network to make it fit the data more closely.

3. *Learning algorithm.* The learning algorithm involves two tasks. The first is to determine which links are to be included in the DAG, and the second is to determine the parameters.

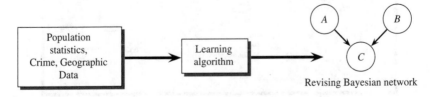

Figure 5.4 Process of learning Bayesian network.

Figure 5.5 The iteration steps of the Expectation Maximization algorithm.

5.5.4   The Expectation Maximization algorithm

The Expectation Maximization (EM) algorithm [86, 125] was used within the Hugin software for learning the data. The EM algorithm is used to learn incomplete data, which makes it particularly suitable for crime data which, by its very nature, tends to be incomplete. In this study, since we have a complete set of data, the EM algorithm is used to count the frequency of probabilities. The EM algorithm consists of two iteration steps:

- Expectation step (E-step) calculates the expectation of the missing statistic.

- Maximization step (M-step) maximizes a certain function.

The two steps are performed until convergence (Figure 5.5).

5.6   Results

In Figure 5.6 the final model with initial probabilities is shown, and Figure 5.7 gives the posterior probabilities, given 'Murder' is 'Yes'.

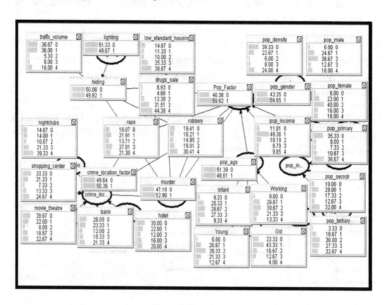

Figure 5.6 Initial probabilities.

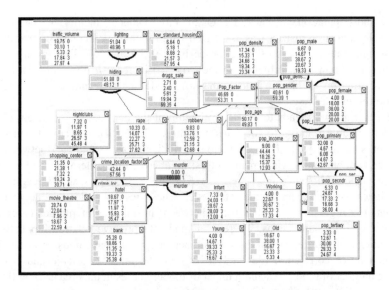

Figure 5.7   Crime risk probabilities given that murder is 'yes'.

Table 5.3   The model probabilities given that Murder is Yes.

Factor	Name of variable	Posterior probabilities	
		State 'very low'	State 'very high'
Environment	drugs_sale	0.2710	0.6935
	low_standard_housing	0.0664	0.5795
Crime	rape	0.1033	0.2762
	robbery	0.0983	0.4268
Location	nightclubs	0.0732	0.4548
	shopping_center	0.2135	0.3071
	movie_theatre	0.2874	0.2259
	bank	0.2528	0.2538
	hotel	0.1867	0.3547
Traffic	traffic_volume	0.1875	0.2797
Population	pop_density	0.1734	0.2334
	pop_male	0.0667	0.1933
	pop_female	0.0400	0.2000
	pop_income	0.0900	0.1293
	infant	0.0733	0.1200
	young	0.0400	0.1667
	working	0.0400	0.1733
	old	0.1667	0.0533

According to Table 5.3, the factors that considerably affected crime risk in the Bangkok Metropolitan Area were environment, types of crimes, crime location, traffic and population, in that order. Of the environmental factors, the number of drug-sale areas in a district was associated with the highest increase in probability (very high state, 0.4438 to 0.6935) – see Figures 5.6 and 5.7. This means that it has the most powerful influence on the expected increase in the murder rate. By concentrating on the elimination of the drug trade, the government could greatly reduce the murder rate.

Further analysis leads to the determination of the correlation coefficient of the murder variable relative to the other variables (Table 5.4). It was calculated by Pearson's method [67]. The correlation coefficients (see Table 5.4) between the murder variable and environment factors; number of drug-sale areas and number of low standard housing areas are $r = 0.617$, and $r = 0.589$ respectively and the correlation is significant at the 0.05 level. These results indicated that there is a relatively high positive relationship between the number of murder cases and number of drug sales areas. There is a moderate positive relationship between the number of murder cases and the number of low standard housing areas. The coefficients of correlation (denoted r) between the murder variable and type of crime; number of rape cases and number of robbery cases in the area are $r = 0.519$, and $r = 0.470$ respectively. The coefficients of correlation (r) between the murder variable and location factors: number of nightclubs, number of shopping centers, number of movie theatres, number of banks, and number of hotels in the area are

Table 5.4   Coefficient of correlation of the murder variable relative to other variables.

Factor	Name and state of Variables	Coefficient of correlation (r)
Environment factor	drugs_sale	0.617
	low_standard_housing	0.589
Crime factor	rape	0.519
	robbery	0.470
Location factor	nightclubs	0.503
	shopping_center	0.438
	movie_theatre	0.529
	bank	0.601
	hotel	0.528
Traffic factor	traffic_volume	0.250
Population factor	pop_density	0.030
	pop_male	0.341
	pop_female	0.331
	pop_income	0.296
	infant	0.215
	young	0.286
	working	0.304
	old	0.328

Correlation is significant at the 0.05 level.

$r = 0.503$, $r = 0.438$, $r = 0.529$, $r = 0.601$, and $r = 0.528$ respectively. These results mean that there is a moderately positive relationship between the number of murder cases and number of rape cases, number of robbery cases, number of nightclubs, number of shopping centers, number of movie theatres, number of banks, and number of hotels in the area. If the values of these variables are high, the number of murder cases will also tend to be high. There is a weak relationship between the number of murder cases and the traffic factor and population factor.

From the posterior probability of murder rate and the coefficient of correlation results, we can see that the environment factors have the greatest effect on the expected increase in murder rate and have a relatively high correlation to the murder variable.

5.7 Accuracy assessment

The prediction accuracy of the model was evaluated using Rockit software, Beta Version [400] to construct and analyze the Receiver Operating Characteristic (ROC) curve. Rockit employs the LABROC5 algorithm, a quasi-maximum likelihood approach, to analyze multiple value data. ROC analysis [491] comes from statistical decision theory and was originally used during World War II for the analysis of radar images. From the computer science point of view, ROC analysis has been increasingly used as a tool to evaluate discriminate effects among different methods.

The ROC curve relies heavily on notions of sensitivity and specificity and these values depend on the specific data set. It has the sensitivity plotted vertically

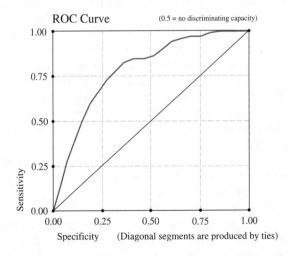

Figure 5.8 Receiver Operating Characteristic curve of 'murder' variable. Reproduced by permission of WSEAS.

and the reversed scale of the specificity on the horizontal axis. The scale of the horizontal axis is also called the false positive rate (Figure 5.8). The sensitivity and specificity, and therefore the performance of the model, vary with the cut-off. It simply looks at the area under the ROC curve. The pattern of the preferred ROC curve is a curve that climbs rapidly towards the upper left-hand corner of the graph. This means that the probability rate to reject bad items is high and the probability rate to reject good items is low. A value of the area under the ROC curve close to 1 indicates an excellent performance in terms of the predictive accuracy.

It should be noted that for forecasting purposes using ROC, the accuracy required should be greater than 0.5. Therefore, the value of 0.77 (see Figure 5.8) obtained for the area under the ROC curve for the murder variable indicated a good performance of the model in terms of its predictive accuracy. This accuracy suggests that this machine learning technique can be used to analyze crime data and help in crime control planning.

5.8   Conclusions

The BN can provide useful information for crime risk factors analysis. The Bayesian technology has proved very flexible and suitable to the requirements of this research project. The Hugin software used with the EM algorithm has made the creation of a sophisticated network relatively straightforward. In this research, some factors that affect the crime risk in the Bangkok Metropolitan Area, were analyzed. The factors considered were classified into five groups. These groups were population, crime location, crime type, traffic and environment. From the output of the model, the factors that considerably affected crime risk in the Bangkok Metropolitan Area were, in order of descending influence, environment, types of crimes, crime location, traffic and population. The Receiver Operating Characteristic (ROC) analysis was used to test the accuracy of the model. The area under the ROC curve for the model is 0.77 which indicates that the model is performing well.

The result from this analysis can be used to help in crime control planning and environmental design to prevent crime. Based on data from the study, the environmental factor, the number of drug-sale areas in a district, had the most powerful influence on the expected murder rate. By concentrating on the elimination of the drug trade the government could greatly reduce the murder rate. The government can apply the model to a specific area to analyse the crime problems. For instance, if we know that an area in the district has high crime rate as a result of environment factors, the government can solve the problem by increasing the number of patrol officers or improving the lighting on the streets in that area.

For future work, the BN for crime risk prediction, which was tested only for the crime of murder in the prototype, may, with further research, be extended to cover the full range of criminal activities, and further refined to improve its performance. With some modification it could also be applied to the study of road accidents which are a major cause of death and injury in Thailand, eventually enabling the police to adopt a proactive approach to accident prevention.

Acknowledgment

The author would like to thank Hugin Expert for providing Hugin software for Bayesian Network Analysis. Sincere thanks to T.J. King for his assistance in preparing the manuscript. Section 5.2, Figures 5.1, 5.3 and 5.8 are reproduced from [53] with permission of WSEAS.

6

Spatial dynamics in the coastal region of South-Eastern France

Giovanni Fusco

Associate Professor of Geography and Planning, University of Nice-Sophia Antipolis, UMR 6012 ESPACE, France

6.1 Introduction

Modeling spatial dynamics through Bayesian networks is a challenging task, over which research has been developed at the University of Nice-Sophia Antipolis since 2003. In this chapter the example of modeling socio-demographic and urban dynamics during the 1990s in the coastal region of South-Eastern France will be shown. This application illustrates how spatial databases and a modeler's previous knowledge can converge to produce synthetic models shedding light on the complex dynamics characterizing European metropolitan areas in the last decades. The study area encompasses three French departments (the Alpes-Maritimes, the Bouches-du-Rhône and the Var) as well as the Principality of Monaco (Figure 6.1). Two metropolitan systems have emerged during the last decades: the Metropolitan Area of Provence, including Marseilles, Aix-en-Provence and Toulon, and the Metropolitan Area of the French Riviera around Nice, Monaco and Cannes. The basis of the formation of metropolitan areas resides in phenomena taking place at the international level [474]: globalization, competition among international cities, the development of inter-metropolitan networks, the concentration of rare urban functions in the hubs of these networks, etc. Modeling these phenomena was not the aim of this particular application. The goal was to analyze spatial dynamics

Bayesian Networks: A Practical Guide to Applications   Edited by O. Pourret, P. Naïm, B. Marcot
© 2008 John Wiley & Sons, Ltd

Figure 6.1   The study area: Coastal South-Eastern France.

characterizing the emergence of metropolitan systems at a local level. The model had to be able to reproduce the forms and the impacts of the territorial transformation of the case study. Gaining knowledge over the study area and offering support to decision makers was the ultimate objective of this research. Several phenomena contribute to this territorial transformation. Infrastructure offer, accessibility, mobility behavior, socio-economic dynamics, urban sprawl, the development of tourism and environmental impacts are all capital issues for the two metropolitan areas of the region. Geographers and planners have analyzed several of these issues individually, but a comprehensive model of their interaction within a regional space is still missing. From this point of view, the metropolitan areas of coastal South-Eastern France present peculiar situations which need attentive consideration. The above mentioned interaction of phenomena is particularly problematic as it takes place in a region marked by strong physical constraints, increased competition among different land-uses and contrasted population patterns between the costal area and the hinterland. In this application, we will focus on a limited number of elements measured at the municipal level: population dynamics, housing, land-use dynamics, transportation and population mobility. The reference period will be limited by the last two French national censuses (1990 and 1999). Modeling will be the key tool in dealing with the complexity of spatial dynamics. The complexity will also be reduced through the use of indicator-based models. Indicators are synthetic parameters describing the average situation of each spatial unit (here the municipalities) for a more complex phenomenon. Indicators can only give an approximate knowledge of real phenomena. Consequently, possible models have to integrate uncertainty in the most direct and explicit way. The availability of spatial databases at the municipal level, the opportunity to explore complex causal links among indicators and the necessity to integrate uncertainty, all make Bayesian networks an extremely appropriate technique for the development of models.

6.2   An indicator-based analysis

6.2.1   The elements of the model

Overall, the study area is exerting a strong attraction for population and businesses at the national level. Resident population and urban areas are growing steadily, from 3577 000 inhabitants in 1990 to 3778 000 in 1999. The growth rate between the censuses is thus +5.62%, to be compared with a national average of +3.36%. The positive migration balance accounts for three quarters of the demographic growth. The total urbanized area has grown 70% over the same period, as a result of strong urban sprawl. Population and more recently jobs, are increasingly attracted by suburban and ex-rural locations. This scattering of urban functions over most of the study area does not contradict the concentration process benefiting the two metropolitan systems at the national level. The impacts of metropolization differ according to the level of observation. At a global level, metropolization appears as a process concentrating population and activities within the metropolitan areas (centripetal forces). Nevertheless, at a local level, the trend is towards an increasing diffusion from the metropolitan core to new suburbs and ex-rural areas (centrifugal forces).

In order to reduce the complexity and the breadth of the analysis, we described the socio-demographic and spatial phenomena through a limited number of key indicators. An indicator is a parameter capable of apprehending more complex phenomena in a synthetic manner. Indicators can give us an approximate knowledge of the functioning of real systems [313, 367]. Calculating indicators for spatial units always presents the problem of the 'spatial resolution' of information. In this study, we have always opted for indicators calculated at the communal level, the commune being the smallest scale for which information was available homogeneously.

Considering the characteristics of the study area and the databases available, 10 themes were chosen: the site of the commune, its location in relation to infrastructures, urban structures, spatial interaction, employment, migratory fluxes, characteristics of the population, housing, daily mobility and the dynamics of the land use. For each theme, a very limited number of indicators were selected (from two to six). The 39 indicators retained are enumerated in Table 6.1. The indicators were thus calculated for 436 spatial units: the 435 communes of the three French departments and the Principality of Monaco. Two kinds of variables were chosen as indicators: those describing the dynamics between 1990 and 1999 (dynamic variables) and those describing the situation of the study area in 1990 (static variables). Our goal was to apprehend both the changes induced by the metropolisation process during the 1990s and the situation at the beginning of this period.

Several sources and tools were used in order to create our database. The indicators of employment, migratory fluxes, population, housing and daily mobility

Table 6.1   The 39 indicators of the model.

Category and Label (*)	Definition
Site:	
(S): dist_littoral	Average distance of the commune from the shoreline (km)
(S): Contrainte_topo	Proportion of surfaces having a slope $\geq 20\%$ within the municipal boundaries (%)
Situation:	
(S): dist_echangeur	Average distance of the commune from the closest highway interchange (km)
(S): temps_aeroport	Road access time to the closest international airport (min)
Urban Structure:	
(S): Pop_AU	Population of the Metropolitan Area in 1990 (for rural communes outside of metropolitan areas, municipal population)
(S): poss_exp_urb	Proportion of the municipal territory free of strong topographical constraints and not builtt up in 1988 (%)
(S): dens_net	Net population density in 1990 (inhabitants/urbanized ha)
(D): Var_dens_net	Variation of the net population density between 1990 and 1999 (inhabitants/ha)
Spatial Interaction:	
(S): acc_emplois	Accessible jobs in less than 30 minutes through the road network in 1990 (thousands)
(D): taux_croiss_acc_emp	Growth rate of the accessibity to jobs between 1990 and 1999 (%)
(S): pot_emplois	Urban job potential in 1990 (ad defined by INSEE PACA)
(S): Pot_habitat	Urban housing potential in 1990 (ad defined by INSEE PACA)
Employment:	
(S): mix_empl_pop	Jobs/resident population ratio in 1990 (%)
(D): taux_croiss_empl	Job growth rate between 1990 and 1999 (%)
(D): var_mix_empl_pop	Variation of the jobs/population ratio between 1990 and 1999
Migration:	
(D): part_pop_meme_com	Share of the population residing in the same dwelling in 1990 and 1999 (%)
(D): part_pop_meme_log	Share of the population residing in the same commune in 1990 and 1999 (%)
(D): taux_var_mig	Migratory population growth rate between 1990 and 1999 (%)
Population:	
(D): taux_var_pop	Total resident population growth rate between 1990 and 1999 (%)

Table 6.1   (*continued*)

Category and Label (*)	Definition
(S): Vieillesse	Old age index of the resident population (60 years old and more/0–14 years old)
(D): var_vieillesse	Variation of the old age index of the resident population between 1990 and 1999
(S): taille_menage	Average household size in 1990
(D): taux_var_taille_men	Variation rate of the average household size between 1990 and 1999 (%)
Housing:	
(D): part_log_neufs	Proportion of dwellings built between 1990 and 1999 (%)
(S): part_res_princ	Share of main homes in the total housing stock in 1990 (%)
(D): var_part_res_princ	Variation of the share of main homes between 1990 and 1999
(S): part_res_sec	Share of secondary homes in the total housing stock in 1990 (%)
(D): var_part_res_sec	Variation of the share of secondary homes between 1990 and 1999
(S): part_log_vac	Share of unoccupied dwellings in the total housing stock in 1990 (%)
(D): var_part_log_vac	Variation of the share of unoccupied dwellings between 1990 and 1999
Mobility:	
(S): Part_navetteurs	Share of commuters in the active population in 1990 (%)
(D): Var_part_navetteurs	Variation of the share of commuters in the active population between 1990 and 1999
(S): Part_men_plurimot	Share of multi-motorized (having more than one motor vehicle) households in 1990 (%)
(D): Var_part_men_plurimot	Variation of the share of multi-motorized households between 1990 and 1999
(S): Part_men_no_auto	Proportion of households without a car in 1990 (%)
(D): Var_part_men_no_auto	Variation of the proportion of households without a car between 1990 and 1999
Land Use:	
(D): poids_exp_urb	Relative size of new urbanization between 1988 and 1999 (% of municipal surface)
(D): poids_depr	Relative size of agricultural shrinkage between 1988 and 1999 (% of municipal surface)
(D): poids_stab	Relative size of land use stability between 1988 and 1999 (% of municipal surface)

(*) (S) and (D) respectively denote *static* and *dynamic* variables

were calculated from the data of the French national censuses. The indicators concerning the other themes were produced through GIS applications and simulations on graphs, using INSEE data [225], land use data [131, 182], IGN's Carto Database [227] and a digital terrain model of the study area. In general, indicators were selected according to the characteristics of the study area. As far as the site is concerned, topographical constraints and the attraction of the coastal area clearly play a fundamental role in the ongoing formation of metropolitan areas [29] and characterize differently the 436 communes.

The development of the highway network during the 1970s and the 1980s was the only infrastructural support to the metropolitan expansion of Provence and of the French Riviera for a long time. The present configuration of the network (Figure 6.1) was completed at the end of the 1980s. This network has been a powerful vector of urban sprawl, following both a North–South and East–West axis. It also allowed a first, spontaneous phase of networking of the two metropolitan areas among them and with metropolitan areas outside the study area.

The connection of the two metropolitan areas to world networks has been traditionally assured by the two international airports of Marseilles and Nice, often through connections to the Parisian hub. The strong dependence of the study area on highway and airport infrastructures guided our choice of the two indicators of geographical situation, whose objective is to evaluate the position of every commune in relation to these two structuring features. In 1999, the Mediterranean High-Speed Train (TGV Méditerranée) was put in service in the western part of the study area. By 2020 a new high speed line should link the metropolitan areas among them and within the rest of the French and European networks [392]. New forms of metropolitan development could then arise. They could be less car dependent and generate less urban sprawl, but perhaps more selective in the networking of hierarchical metropolitan centres.

The indicators of urban structure describe the two main phenomena operating in the study area: the integration of cities and villages in vast metropolitan areas polarized by metropolitan centers and urban sprawl. As for the latter, the parameter of net population density (that is population as referred to urbanized land only) has often been pointed out as being the most pertinent indicator [169, 337]. This indicator can only be calculated through the integration of census and land use data.

Road accessibility and the reciprocal attraction among the communes of the study area are the main indicators of spatial interaction. Among them, the two potential parameters were drawn from previous work [226]. The urban job potential measures the ability of a city, through the jobs it offers, to attract active population residing in distant communes. It is a function of the mass of the commune (through its total number of jobs), of the commuter distances of people working there, and of regression coefficients which were calculated using the census data of 1990. Similarly, an urban housing potential can be defined. The latter measures the ability of a city, through the dwellings it offers, to attract the active population working in distant communes.

The indicators of employment, migration, population, housing and daily mobility are classical parameters drawn from census data. Their interest in the analysis of regional dynamics of metropolitan development has already been pointed out [29]. As for land use, we will focus on the ongoing dynamics during the 1990s [182]. The analysis has been carried out through three broad land use classes:

- The first class encompasses all urban land uses (including industrial areas and transportation infrastructure within rural areas);

- the second class covers all agricultural land;

- the third class corresponds to natural surfaces (forests, prairies, marshes, rocks, inner water bodies, etc.).

The land use dynamics affecting our study area can be summarized by the scheme in Table 6.2. Three main phenomena can be pointed out:

- stability within the three classes of land use (light gray in the scheme): urban, agricultural or natural land in 1999 was already characterized by the same land use in 1988;

- new urban development (medium gray): agricultural or natural land in 1988 became urban in 1999;

- agricultural shrinkage (dark gray): agricultural areas in 1988 have been abandoned becoming natural in 1999.

To conclude this section, a few remarks will be made on the advantages and the shortcomings of the choice of the municipal level for calculating data.

In choosing to calculate indicators for every commune, a spatial viewpoint is privileged. We can analyze how the metropolitan development affects each territorial component independently from its population. A sociological study focusing on the way the metropolisation process affects the population would have preferred spatial units with comparable populations. The most populated cities would then be divided in smaller districts, whereas rural communes would be aggregated to form vaster territories. This method is currently used in mobility surveys within metropolitan areas. In these works, carried out through sampling, suburban and rural areas are always apprehended through a very limited number of spatial units, producing what we could call a compression of geographical information in the

Table 6.2   Scheme of land use dynamics in the study area between 1988 and 1990.

Land use in 1999 / Land use in 1988	Urban	Agricultural	Natural
Urban	Stability		
Agricultural	Urban Development	Stability	Agricultural Shrinkage
Natural	Urban Development		Stability

areas undergoing the most radical changes as a consequence of the metropolization process.

Nevertheless, the choice of the municipal level poses a few problems. When indicators describe phenomena linked to the population of the spatial unit (such as the unemployment rate, the population growth rate, etc.), the values of the smallest communes (25 communes in the study area have less then 100 inhabitants) can be statistical outliers as they depend on a very small number of individuals.

6.2.2   Recent trends and the situation of coastal South-Eastern France in 1990

6.2.2.1   Patterns of differentiated evolutions

As we could see in the previous section, metropolitan growth, urban sprawl and the attraction of migratory fluxes are the general trend of the study area during the 1990s. Nevertheless, when the municipal level is considered, the picture is highly contrasted.

Figure 6.2 shows the variation rate of resident population for the 436 communes of the study area. Big cities at the core of the metropolitan process (Marseilles, Nice, Toulon, Cannes) stagnate or lose population. Only Aix-en-Provence shows a clearly positive evolution. Other areas with declining population are to be found in the mountainous regions of the Alpes-Maritimes and in the Rhone delta. Areas with steady demographic increases emerge around metropolitan areas: the Gardanne district North of Toulon (around the metropolitan area of Provence), the close

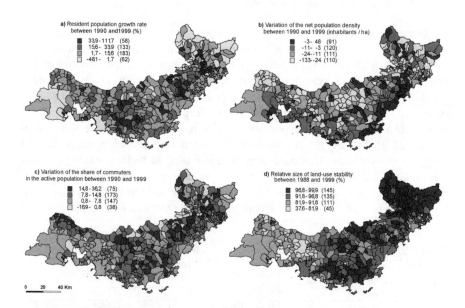

Figure 6.2   Spatial dynamics in the study area between 1990 and 1999.

hinterland North of Frejus, Cannes and Nice (around the metropolitan area of the French Riviera).

As for urban development, city structures sprawl and become less dense in the entire study area. Differences can nevertheless be found at the communal level. Part (b) of Figure 6.2 shows the variation of net population density between 1990 and 1999. The loss of density is extreme in the areas which are the furthest from the coast. The coastal area experienced a less pronounces loss of density, with a few communes actually increasing the net density of their urban tissue.

The growing integration of rural areas in the daily functioning of metropolitan areas is a main feature of the metropolitan process. Patterns of commuting trips to jobs are a first symptom of this integration. The number of commuters (i.e., of people having their job outside the commune of residence) grows almost everywhere in the study area (Figure 6.2c). Only a few communes at the outer margins of the mountainous areas show a stagnating or declining number of commuters. Commuter growth is the strongest around metropolitan areas: a large front of metropolitan integration through commuting is taking shape East and North of the metropolitan area of Provence and West and North of the French Riviera.

As for land use dynamics, Figure 6.2(d) shows the relative size of land use stability in the communes of the study area. Different trends have emerged in the area during the 1990s: stability in almost all the hinterland of the Var and Alpes-Maritimes departments, moderate change (mainly urban development) around Nice and Toulon, intense change (a combined effect of urban development and agricultural shrinkage) almost everywhere in the Bouches-du-Rhone department.

6.2.2.2   The situation in 1990: A region with strong internal contrasts

In order to understand the changes observed during the 1990s, it is useful to consider the situation of the study area at the beginning of the decade. Deep differences were already shaping the region at the time, as shown in Figure 6.3 for a limited number of elements.

The possibilities of urban development, for example, were not the same in the East and in the West of the study area (Figure 6.3a). Besides urban planning and environmental regulations, strong topographical constraints and previous urban developments were already limiting future urban developments in the Alpes-Maritimes in 1990: only the Western section of the French Riviera still possessed a significant amount of undeveloped land.

On the other hand, the entire Bouches-du-Rhône department (except for the highly urbanized area of Marseilles and its topographically constrained northern hilly suburbs), as well as the inner section of the Var department, could offer to future urban development vast agricultural and natural surfaces. As far as spatial interaction is concerned, the communes of the study area entered the decade with completely different situations as well. Using the FRED software [124], we calculated the accessibility level of every commune to the regional jobs through the road network (Figure 6.3b). The cores of the two metropolitan areas appear clearly as high accessibility areas, allowing significant agglomeration economies

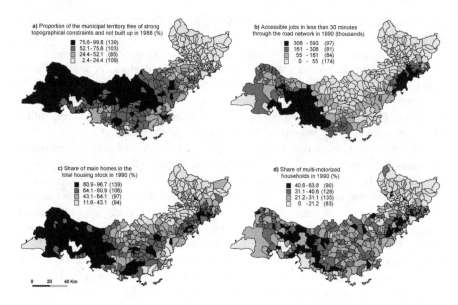

a) Proportion of the municipal territory free of strong
topographical constraints and not built up in 1988 (%)

■ 75.6 - 99.8 (139)
■ 52.1 - 75.6 (103)
□ 24.4 - 52.1 (85)
□ 2.4 - 24.4 (109)

b) Accessible jobs in less than 30 minutes
through the road network in 1990 (thousands)

■ 308 - 593 (97)
■ 161 - 308 (81)
□ 55 - 161 (84)
□ 0 - 55 (174)

c) Share of main homes in the
total housing stock in 1990 (%)

■ 80.9 - 96.7 (139)
■ 64.1 - 80.9 (106)
□ 43.1 - 64.1 (97)
□ 11.6 - 43.1 (94)

d) Share of multi-motorized
households in 1990 (%)

■ 40.6 - 63.8 (90)
■ 31.1 - 40.6 (128)
□ 21.2 - 31.1 (135)
□ 0 - 21.2 (83)

0   20   40 Km

Figure 6.3   The situation of the study area in 1990.

to firms and household. From this point of view, the Marseilles/Aix-en-Provence couple emerges as the real core of the metropolitan area of Provence, whereas the Toulon area shows a relative accessibility deficit. The far hinterland of the Var and Alpes-Maritimes departments, as well as the Maures massif and the outer area of the Camargue are, in 1990, spaces suffering from a severe accessibility handicap. Other remarkable differences within the study area emerge from 1990 census data. The space of main homes, for example, encompasses the Bouches-du-Rhone department, the Toulon area and the close outskirt of the city of Nice (Figure 6.3c). Several communes of the French Riviera show significant levels of secondary homes and unoccupied dwellings: Antibes, Cannes, Menton, Frejus, Saint-Raphael. In the far hinterland of the Var and Alpes-Maritimes departments, as well as on the Var coast, main homes only account for 11–43% of all dwellings.

In a region where the networking of metropolitan components was mainly done through the highway network, and where urban sprawl and commuting affect large areas around the metropolitan cores, mass multi-motorization of households is an essential element of the metropolization process (Figure 6.3d). In 1990, in the suburban areas around Aix and Marseilles, in the West, and around Nice, Antibes and Cannes, in the East, more than every other household already possessed several motor vehicles. On the other hand, multi-motorization was marginal in the far mountainous hinterland, as well as in the main coastal cities (Marseilles, Toulon, Nice, Cannes, Antibes).

6.3   The Bayesian network model

In the previous section, we analyzed a limited number of territorial variables through thematic maps. The spatial distribution of these variables is clearly not casual. Maps suggest the existence of links among the different phenomena. It is the combination of these links that makes up the complex metropolization process of coastal South-Eastern France. But how can these interrelations be apprehended passing to all the 39 indicators?

Increasing the number of maps and diagrams would make a global view of the process impossible. Only global models can handle the complexity of the interrelated phenomena we are studying. The approximate knowledge given by indicators compels models to deal with uncertainty in the relationships among phenomena. The availability of a large spatial database, the opportunity to explore possible causal links and the necessity of integrating uncertainty, make Bayesian networks an extremely appropriate technique for the development of models.

The capabilities of Bayesian networks in modeling are just beginning to be discovered in geography. Bayesian networks' probabilistic approach to causality is particularly suited for representing links among territorial variables [169]. Relations among socio-demographic, economic and urban indicators are rarely strictly deterministic. They are much more often fuzzy relations: variable A somehow influences variable B, but it is always possible that the value of the dependent variable contradicts the general sense of the relationship linking it to the parent variable.

The first applications of Bayesian networks in geography concerned image processing in GIS. Stassopoulou et al. [440] developed a Bayesian network to assess the risk of desertification of areas subject to fire in the Mediterranean. The network was not generated from data, as the modelers formalized directly their knowledge of functional relations among variables. In France, Cavarroc and Jeansoulin [87], within the works of the Cassini research group, evaluated the contribution of Bayesian networks in the search of spatio-temporal causal links in remote sensing images. Interesting applications have been developed at the Department of Geography of the Pennsylvania State University: the group of C. Flint [152] in electoral geography, W. Pike [366] in the analysis of water quality violations. In Italy, F. Scarlatti and G. Rabino, at the Milan Politecnico, proposed the use of Bayesian networks for landscape analysis [415]. These pioneer studies showed the great potential of Bayesian networks in modeling systemic links among space variables. As for the territory-transportation interaction, the interest of Bayesian networks has been pointed out by P. Waddell [476] without producing any operational model. The models developed at the University of Nice [167, 168, 169] are the first accomplishments in this direction.

6.3.1   Creating the network through structure and parameters learning

One of the most interesting applications of Bayesian networks is causal knowledge discovering, that is automatic research of possible causal links among the variables of a data base. With a set of observed variables at hand, even in the absence of temporal information, powerful algorithms can determine the most probable causal links among variables. The network thus produced can be a possible causal model for the phenomena described in the database (nevertheless, the modeler must always evaluate the pertinence of the network). Pearl and Russell [358] point out that patterns of dependency in data, which can be totally void of temporal information, are the result of certain causal directionalities and not others. Put together systematically, such patterns can be used to infer causal structures from data and to guarantee that any alternative structure compatible with the data is less probable than the one(s) inferred. This task can be accomplished, probabilistically, by Bayesian learning algorithms. The most complete algorithms can learn both network structure and parameters from data.

The search for the most probable network given the database of 39 indicators was accomplished through the software Discoverer [36]. Discoverer uses only discrete variables. Every indicator was thus discretized in four classes according to the Jenks algorithm (minimizing the sum of absolute deviations from class means). In order to generate the structure and the parameters of the network, Discoverer uses the K2 algorithm, developed by G. Cooper [103]. K2 selects the a posteriori most probable network within a subset of all possible networks. This subset is defined by the modeller who has to define a hierarchy among variables, thus constraining the network search. The position of each variable in the hierarchy determined the number of variables which will be tested as possible parents. Let X be the last variable in the hierarchy. The implementation of the algorithm begins by this variable and first calculates the log-likelihood of the model without any link towards X. The next step is the calculation of the log-likelihood of all the models having one link towards X. If none of these models has a higher log-likelihood than the model without links, the latter is assumed as being the most probable in the explanation of X, and the algorithm goes on evaluating the second lowest variable in the hierarchy. If at least one of these models has a higher log-likelihood than the model without links toward X, the corresponding link is kept and the search continues by trying to add a second link towards X, and this until the log-likelihood stops growing. Once a model has been retained for a variable, the algorithm searches the most probable links explaining the next variable in the hierarchy using the same approach.

K2 can be implemented through two different search strategies, mixing the modeler's knowledge and the knowledge acquired from data: Greedy and ArcInversion. In the Greedy strategy, the variables tested as possible parents of a given variable are only those who have a higher position in the causal hierarchy. On the other hand, the ArcInversion strategy allows variables who have a lower position to be tested as possible parents as well, as long as this doesn't produce a closed cycle in the network structure. Contrary to other applications of Bayesian Network modeling in geography [167, 168, 169], in this research, the ArcInversion strategy was used. More probable links could thus be explored, even if they contradicted the modeler's starting hypotheses.

Table 6.3   The hierarchy among the variables used to generate the network.

Indicator (S/D)	Theme	Indicator (S/D)	Theme
1 dist_littoral (S)	Site	20 var_vieillesse (D)	Population
2 contrainte_topo (S)	Site	21 taille_menage (S)	Population
3 dist_echangeur (S)	Situation	22 taux_var_taille_men (D)	Population
4 temps_aeroport (S)	Situation	23 part_log_neufs (D)	Housing
5 pop_AU (S)	Urb. Struct.	24 part_res_princ (S)	Housing
6 poss_exp_urb (S)	Urb. Struct.	25 var_part_res_princ (D)	Housing
7 dens_net (S)	Urb. Struct.	26 part_res_sec (S)	Housing
8 acc_emplois (S)	Sp. Interact.	27 var_part_res_sec (D)	Housing
9 taux_croiss_acc_emp (D)	Sp. Interact.	28 part_log_vac (S)	Housing
10 pot_emplois (S)	Sp. Interact.	29 var_part_log_vac (D)	Mobility
11 pot_habitat (S)	Sp. Interact.	30 Part_navetteurs (S)	Mobility
12 mix_empl_pop (S)	Employment	31 Var_part_navetteurs (D)	Mobility
13 taux_croiss_empl (D)	Employment	32 Part_men_plurimot (S)	Mobility
14 var_mix_empl_pop (D)	Employment	33 Var_part_men_plurimot (D)	Mobility
15 part_pop_meme_com (D)	Migrations	34 Part_men_no_auto (S)	Mobility
16 part_pop_meme_log (D)	Migrations	35 Var_part_men_no_auto (D)	Mobility
17 taux_var_mig (D)	Migrations	36 var_dens_net Dyn.	Urb. Struct.
18 taux_var_pop (D)	Population	37 poids_exp_urb (D)	Land Use
19 vieillesse (S)	Population	38 poids_depr (D)	Land Use
		39 poids_stab (D)	Land Use

Table 6.3 shows the causal hierarchy among the 39 indicators which was used by K2 in order to generate the network. Indicators of the site (the distance from the coast and the topographical constraints) were considered as primordial elements, followed by indicators of the situation of the commune and then by those of urban structure and spatial interaction. Indicators of employment, of migration, of population, of housing and of daily mobility are the next in the hierarchy. The parameters of land use are considered to be the result of all the phenomena taking place in the study area (only dynamic variables were considered for the land use, describing its evolution over the decade). For every theme, static variables precede dynamic variables in the causal hierarchy. According to the ArcInversion strategy, modeling hypotheses are not binding for the network search: K2 can produce causal links refuting the proposed hierarchy, as long as they don't create cycles.

Several hypotheses were tested, resulting in different causal hierarchies. Here, we focused on the hierarchy which produced the best Bayesian network in terms of causal explanation of the phenomena which were to be modeled.

6.3.2   A well-connected network

The Bayesian network generated using the variable hierarchy of Table 6.3 is shown in Figure 6.4. It is a completely connected network and it constitutes a convincing causal model of the spatial dynamics within coastal South-Eastern France in the last decade. No indicator of the dynamics between 1990 and 1999 ended up being

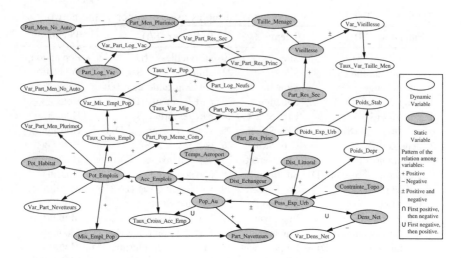

Figure 6.4   The Bayesian network modeling spatial dynamics in coastal South-Eastern France.

a parent of indicators of the 1990 situation, which would have clearly been a modeling mistake of the learning algorithm. On the other hand, several dynamic variables depend directly on the situation in 1990, as the analysis of the network will show.

As the understanding of spatial processes is the main goal of exploratory Bayesian network modeling within our research, a detailed analysis of the network will be of capital importance. In order to facilitate the reading of the network, different colors were used to represent static and dynamic variables. We also indicated the pattern of the causal relation modeled by an edge in the graph. This pattern is a summary of the probabilistic information of the associated CPTs. A positive relation, for example, means that it is very probable that low (high) values of the child variable correspond to low (high) values of the parent variable. Similarly, negative relations can be defined, as well as U shaped relations (which are negative for low values of the parent variable and positive for higher values) and reverse U shaped (positive for low values of the parent variables and negative for higher values). More complex patterns can finally be detected (they are represented in Figure 6.4 with the ± symbol).

6.3.2.1   The effects of site and transportation networks on employment and demography

The starting point in the reading of the network are the two indicators of the site, which are the only independent variables of the model (right lower corner of Figure 6.4). Topographic constraints limit the possibility of urban expansion. On the other hand, the distance from the coast has a positive effect on the possibilities of future urban expansion (areas further inland were not exposed to the strong real estate pressure of coastal municipalities), apart from areas where topographical constraints are extreme (as in the mountainous hinterland of the Alpes-Maritimes). The distance from the coast also affects the distance from highway interchanges: public authorities gave priority to coastal areas in the development of highway infrastructures. In 1990 the distance of the commune from the coast is thus still the first explicatory factor of its distance from the highway network. The distance from highway interchanges also depends on the possibilities of urban expansion. With the exception of the coastal areas (which were first and foremost connected to the highway network), the probability of finding a highway interchange in the hinterland is higher in areas lacking constraints from topography and previous urban development.

The relations among the distance to highway interchanges, access time to airports and job accessibility constitute the core of the impact of transportation infrastructures over the functioning of the regional metropolitan systems. Access time to airports grows for communes far away from interchanges, confirming the strong dependence of the performance of the regional road network on highway infrastructures. Accessibility to jobs decreases both with distance from highway interchanges and with access time to airports, as the latter polarized much of job localizations in the study area.

Job accessibility affects, in its turn, the population size of the metropolitan area (big integrated metropolitan areas develop in areas characterized by high accessibility levels, allowing consistent agglomeration economies), which also depends on the possibilities of urban expansion. In areas with high accessibility levels, vast opportunities of urban expansion favor the emergence of integrated metropolitan areas (as is the case for the Aix-Marseille region). On the contrary, in areas with low accessibility levels, vast spaces presenting few constraints from topography and previous urban developments become an obstacle to territorial integration within a metropolitan area (as is the case for the communes of the Rhone delta and central Var).

The growth rate of job accessibility between 1990 and 1999 is the first dynamic variable we find in the reading of the network. Significantly enough, it depends negatively on the 1990 level accessibility. This means that the metropolisation process in the region implies a catching up on accessibility level: areas with lower accessibility in 1990 show the most significant relative gains during the decade. Accessibility growth rate is also a consequence of the population size of the metropolitan area: with an equal accessibility level in 1990, the growth rate is higher in rural communes (outside metropolitan areas) and in those within big metropolitan areas, whereas it is lower in communes within minor urban areas. Another phenomenon of catching up concerns net population densities (in the lower right corner of Figure 6.4). Communes with denser urban structures experience the most severe density loss during the decade. Population density in 1990 was in its turn a consequence of the possibilities of urban expansion. Density decreases where land for urban development is abundant, but the relation inverts its sign for the communes having the greatest land availability. These areas (the Rhone delta and central Var) had not yet been affected by the metropolization process at the beginning of the decade. Lacking strong real estate pressure, urban sprawl was still limited.

Going on with the analysis of the network, jobs accessibility is the main explanatory variable of the urban jobs potential. The latter plays a fundamental role in the functioning of the model, as it directly determines six other phenomena:

- the urban housing potential, through a positive relation;

- the jobs/population ratio, through a positive relation as well;

- the variation of the proportion of commuters between 1990 and 1999 (the increase of commuting is one of the main aspects of the metropolization process, as shown in Figure 6.2(c); nevertheless, the communes with the highest job potential show a lower increase);

- the variation of the proportion of multi-motorized households (once again through a negative relation, as multi-motorization grows the most in the new residential areas where job potential is the lowest);

- the share of population residing in the same commune in the previous census (this is a positive relation because residential migration occur seldom towards areas with important job concentrations; they increasingly concern communes with low job potential in the suburbs or even in the rural areas);

- job growth rate, whose relation is first positive (the communes with the lowest job potential, either specialize in accommodating new residents or get deserted, either way their jobs stagnate or diminish) but later becomes negative (the communes with the highest job potential are already saturated and can rarely increase their jobs significantly).

Not surprisingly, the share of the population residing in the same commune in the last two censuses explains the migratory population growth rate (through a negative relation) and the share of the population residing in the same dwelling (through a positive relation). The latter relation implies that households seldom move within the same commune and that their residential cycle increasingly concerns the whole metropolitan region. Finally, the migratory population growth rate is the main cause of inter-census population growth, confirming a demographic trend that, for every commune as well as for the whole region, is mainly driven by the arrival of new inhabitants.

6.3.2.2   Dwelling characteristics and land use dynamics

The Bayesian network gives an unexpected explanation of the dwelling characteristics of the communes within the study area. The share of main homes depends, through a negative relation, on the distance from highway interchanges. Households residing in the study area prefer to live close to the nodes of the highway network, in order to benefit from the multiple opportunities of jobs, commerce and services offered within the region. The highway network is thus the first agent of the metropolization process within coastal South-Eastern France. The share of secondary homes decreases as the share of main homes increases (the third category, unoccupied dwellings is seldom significant). This means that secondary homes are relatively further away from highways, preferring the proximity to environmental amenities to the access to transportation networks.

The explanation of land use dynamics is another interesting contribution of the Bayesian network. The communes having the most important share of main homes in 1990 are also those where new urban developments are the most important. A specific trait of the metropolisation process within the study area is thus urban sprawl alimented by new main homes and concentrated where main homes were already predominant. Secondary homes drive urban sprawl only in a few exceptional areas (as in the Eastern section of the French Riviera and in the Gulf of Saint-Tropez). In both cases the result is an increased specialization of vast territories as residential areas.

The perspectives of future urban expansion nourish the expectations of land owners (agricultural shrinkage is directly linked to the possibilities of future urbanization). It is thus in the plain, where future capital gains linked to new developments are possible, that most agricultural activities shut down, and not in the mountain, where the mechanization of agriculture is objectively most difficult. The main exception to this rule comes from the Rhone delta, where agricultural shrinkage takes place even without strong speculative pressure on the land.

6.3.2.3   Demography and changes of dwelling characteristics

Dwelling characteristics are also the basis of important socio-demographic phenomena. The old age index of resident population depends on the share of second homes through a positive relationship. Evidently, the residential and tourist municipalities which offer plenty of second homes attract seniors and keep away households with children, who cannot find affordable dwellings of an appropriate size. The negative relationship between the old age ratio and the average household size confirms this explanation. As far as demographic characteristics are concerned, some catching up has taken place between 1990 and 1999: the Vieillesse→Var_Vieillesse relationship shows that the old age ratio grows in the youngest municipalities and diminishes (or is stable) in the older ones. At the same time, a spatial fracture emerges: some municipalities, characterized by a high old age index in 1990, keep on aging more than the average of the study area, contradicting the general trend of catching up. It is the case for a small number of villages in the mountain areas of the Var and the Alpes-Maritimes, who are still untouched by the dynamics of metropolitan integration.

The variation of the average household size follows the variation of the old age index through a negative relationship. It also contributes to the homogenization of population characteristics in the study area. This relationship is nevertheless more fuzzy and shows a higher number of exceptions. The Bayesian network shows a positive link between the average household size and the share of households having more than one motor vehicle, as well as a negative link between the latter and the share of non-motorized households. These relationships describe the situation in 1990 and illustrate an important aspect of an automobile dependent metropolization process [141, 337, 494]. Going back up the causal links within the Bayesian network, bigger households settled in municipalities served well by the highway network, in order to take advantage of the most jobs, services and leisure facilities offered within the metropolitan area. They also show the most significant and complex mobility demand (with a lot of commuting), due to the presence of several active members and children. Multi-motorization and an almost exclusively automobile mobility are the logical consequence for most of these households living in peripheral residential areas where alternatives to private cars are hardly present.

Municipalities burdened by a higher proportion of unoccupied dwellings show the higher share of households without a car. It is the case of the villages of the mountainous areas which, in 1990, were still left out of the metropolization integration process. To a lesser degree, it is also the case of big cities (Marseilles, Nice, Toulon, etc.). Here, public transit offer and high urban densities make the choice of not having a car possible for some households. At the same time, complex urban cycles produce a considerable number of unoccupied dwellings (these phenomena have to be explained at an intra-urban level).

Finally, some catching up has taken place during the 1990s. The reduction of the share of households without a car (which was observed everywhere in the study area) was the strongest in the less motorized municipalities. Having a car becomes a necessary condition for metropolitan life in the coastal region of South-Eastern France. At the same time, the share of unoccupied dwellings diminishes where it was the highest and increases where it was the lowest, showing a communicating vessels effect within a real estate market becoming increasingly integrated at a regional level.

The description of the changes of dwelling characteristics will be concluded by taking into account the role of the population growth rate. Population growth leads to the construction of new dwellings (positive link Taux_Var_Pop→Part_Log_Neufs) and increases the share of main dwellings (positive link Taux_Var_Pop→Var_Part_Res_Princ). The strongest population growth rates, observed at the edge of metropolitan areas, ignite a construction led economy seeking to accommodate a new active population. The latter settle in these fast developing municipalities, keeping their jobs in the core of the metropolitan areas (this reading also takes into account the causal chain

Acc_Emploi → Pot_Emploi → Part_Pop_Meme_Com → Taux_Var_Mig).

In order to sum up the above mentioned phenomena, dwelling characteristics are influenced by transportation networks and are at the basis of the main socio-demographic dynamics. The latter feed back in the dwelling characteristics as they produce, directly or through the automobile system, the main changes of the dwelling characteristics.

6.3.2.4   An overall vision

The Bayesian network generated from our geographic data-base models and summarizes the relationship among spatial variables describing the municipalities of coastal South-Eastern France. It is the synthesis of the metropolization dynamics which marked the 1990s and the situation at the beginning of the decade. More specifically, the links among static variables represent the heritage of systemic relationships established within the study area before the 1990s. They could be thought of as the structure of the regional system. The links among static variables and dynamic variables represent the effect of this inherited structure on current spatial dynamics.

There are, finally, a few links among dynamic variables. Some of them are logic truisms, and put together they don't give any understandable reading of the spatial dynamics. This indicates that the spatial dynamics of the study area cannot be explained without reference to the situation in 1990. The inertia of past conditions is very strong. Observed dynamics often result in reducing the gaps among territories, with respect to the initial conditions in 1990. From this point of view, the metropolization process in the 1990s consists in the diffusion and generalization of situations which would formerly distinguish cities, towards new suburban and ex-rural areas within the region. The metropolisation process is thus homogenizing the study area, countering strong inherited contrasts and geographical situations (distance from the sea, topographical constraints) which cannot, even so, disappear.

6.3.3   Evaluating the model

In this section the produced Bayesian network will be evaluated in three different ways: through its likelihood given the data, through its predictive strength and through its ability to model every spatial unit of the study area.

6.3.3.1   A globally robust model

Bayesian networks are normally evaluated through their likelihood, i.e., the a posteriori probability knowing the data [39]. As this probability is extremely small, its logarithm will be preferred, the log-likelihood. The Bayesian network of Figure 6.4, though being the most probable given the data and the variable order, has a log-likelihood of -18185.4 (Table 6.4). This corresponds to an average per variable log-likelihood of -466.3. In order to assess these values, we will compare them to those of the model of stochastic independence, where the 39 variables are completely unconnected and the co-occurrences are simply fortuitous (it is the zero hypothesis of all significance tests of probabilistic models).

The log-likelihood of such a model is -22215.3, showing a difference of 4029.9 with the Bayesian network. This means that the produced model is 1.46×10^{1750} times more probable than the model of stochastic independence, given our data.

In order to appreciate the robustness of the model better with respect to the initial hypotheses, two more Bayesian networks were produced. The first uses only

Table 6.4   The log-likelihood evaluation of the Bayesian network.

	Complete BN	St. Ind. Mod.	BN 1990 State	St. Ind. Mod.	BN 1990– 1999	St. Ind. Mod.
Number of variables	39	39	19	19	20	20
Log-likelihood	−18185.4	−22215.3	−8838.6	−11115.5	−9757.1	−11100.8
Log-likelihood per variable	−466.3	−569.6	−465.2	−585.0	−487.9	−555.0
Log-likelihood gain with respect to the St. Ind. Mod.	4029.9		2276.9		1343.7	

St. Ind. Mod.: Stochastic Independence Model

the 19 indicators describing the situation of the study area in 1990, the second uses the 20 indicators of the dynamics between 1990 and 1999. The first Bayesian network contains exactly the edges linking the same 19 variables within the complete Bayesian network of Figure 6.4. The per variable log-likelihood of this network is nearly the same as the one of the complete Bayesian network (Table 6.4).

The second Bayesian network, on the other hand, has a per variable log-likelihood considerably lower than the one of the complete Bayesian network (−487.9 instead of −466.3). Albeit more probable than the model of stochastic independence, this Bayesian network presents relatively fuzzy CPTs (representing weaker relationships among variables than those of the complete Bayesian network). This confirms the remark that it is impossible to explain the spatial dynamics between 1990 and 1999 in coastal South-Eastern France, without considering the condition of the spatial units in 1990.

Through the cross-validation test, Discoverer can assess the predictive force of Bayesian networks using one database only. The test first divides the database in k folds. Then, for every fold, it predicts the values of a subset of variables using the CPTs estimated from the rest of the database. Our database was thus divided in 10 folds and 39 cross-validation tests were performed in order to predict every single variable given the other 38. By comparing the predicted value of every record with its known value, a prediction accuracy rate could be calculated for every variable (Table 6.5). Globally, the network's accuracy rate is 64.84% (but the accuracy varies between 45 and 84% according to the variable). This value is sufficiently high for a large-scale geographical model of interaction among land use, transportation, socio-demographic and housing indicators (the model of the stochastic independence would have an average accuracy rate of 25%). The perspective of using Bayesian networks for strategic simulation of spatial dynamics is thus open.

Table 6.5   The cross-validation test of the Bayesian network.

Cross validation results on 1 test

	Corr.	Incorr.	Acc.	σ		Corr.	Incorr.	Acc.	σ
Var_Dens_Net	194	242	45%	2.38	Var_Vieillesse	289	147	66%	2.26
Var_Part_Navetteurs	199	237	46%	2.39	Taux_Croiss_Empl	298	138	68%	2.23
Part_Log_Vac	208	228	48%	2.39	Poids_Exp_Urb	309	127	71%	2.18
Part_Navetteurs	211	225	48%	2.39	Var_Mix_Empl_Pop	309	127	71%	2.18
Var_Part_Men_Plurimot	222	214	51%	2.39	Pop_Au	310	126	71%	2.17
Var_Part_Men_No_Auto	223	213	51%	2.39	Var_Part_Res_Princ	311	125	71%	2.17
Part_Log_Neufs	225	211	52%	2.39	Taille_Menage	313	123	72%	2.16
Dens_Net	239	197	55%	2.38	Var_Part_Res_Sec	320	116	73%	2.12
Pot_Habitat	240	196	55%	2.38	Vieillesse	323	113	74%	2.10
Part_Pop_Meme_Log	240	196	55%	2.38	Pot_Emplois	331	105	76%	2.05
Taux_Croiss_Acc_Emp	243	193	56%	2.38	Poss_Exp_Urb	334	102	77%	2.03
Mix_Empl_Pop	246	190	56%	2.38	Contrainte_Topo	337	99	77%	2.01
Part_Pop_Meme_Com	248	188	57%	2.37	Taux_Var_Mig	338	98	78%	2.00
Temps_Aeroport	252	184	58%	2.37	Poids_Stab	339	97	78%	1.99
Taux_Var_Taille_Men	252	184	58%	2.37	Acc_Emplois	343	93	79%	1.96
Dist_Littoral	260	176	60%	2.35	Dist_Echangeur	345	91	79%	1.94
Poids_Depr	260	176	60%	2.35	Taux_Var_Pop	354	82	81%	1.87
Part_Men_Plurimot	266	170	61%	2.34	Part_Res_Princ	361	75	83%	1.81
Var_Part_Log_Vac	279	157	64%	2.30	Part_Res_Sec	367	69	84%	1.75
Part_Men_No_Auto	287	149	66%	2.27	Total	11025	5979	65%	2.29

6.3.3.2   General trends and particular behaviours

The Bayesian network model can finally be evaluated through its capacity to explain every single record of the database. In our case this would mean evaluating how the Bayesian network models every spatial unit of the study area following a model-residual approach. The residual analysis is a classical practice in geographical modeling, as residuals reveal the particular behavior of the different spatial units with respect to the overall behavior described by the model [66, 409]. This approach changes somehow using a Bayesian network model. The residuals with respect to the predicted value of a given variable are not the most interesting quantities for a global evaluation. A much more significant quantity is the overall probability that the set of values characterizing every spatial unit can be produced by the Bayesian network. This probability takes into consideration the prediction of all the variables at the same time. The map of the probabilities of spatial units according to the Bayesian network is shown in Figure 6.5(a). Some municipalities in the Northern Var, the Eastern Alpes-Maritimes and in the West of the Bouches-du-Rhône seem much less likely than the others (log-likelihood between −58.5 and −41.2). On the other hand, the municipalities of the Aix-Marseilles metropolitan area are the closest to the general behavior of the Bayesian network model.

Actually, we should evaluate the probability of producing the set of values of every spatial unit through the Bayesian network, with respect to the intrinsic

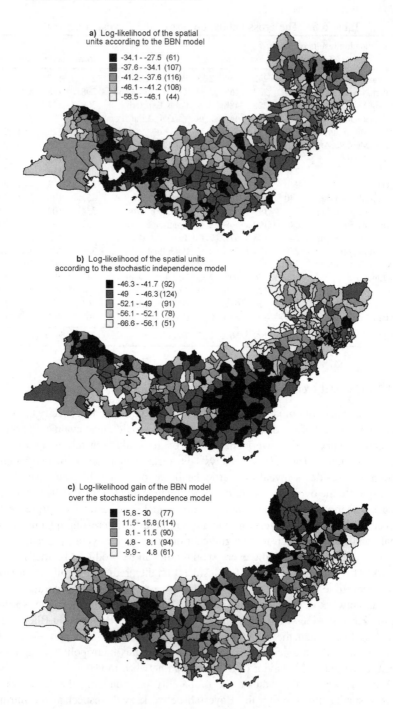

Figure 6.5 The probabilistic evaluation of spatial units.

probability of every set of values. Some municipalities present combinations of particularly rare values: the four classes of every variable do not have the same marginal probabilities. Particular combinations of values could thus be much rarer than the most common ones. The map in part (b) of Figure 6.5 shows the probabilities of spatial units in the case of perfect stochastic independence among variables. The villages of the mountainous hinterland of Var and Alpes-Maritimes have particularly rare combinations of values of the 39 indicators. Some incorrect data, due to the extremely small populations of these spatial units could partly explain this situation.

The ratio of the intrinsic probability of the spatial unit and the probability according to the Bayesian network (it is a difference in logarithms) gives the likelihood gain of the Bayesian network model with respect to stochastic independence (part (c) of Figure 6.5). The municipalities which are better described by the CPTs of the Bayesian network are those of the Provence metropolitan area, those of the mountainous hinterland of the Var and the Alpes-Maritimes and (with a few exceptions) those in the Western part of the French Riviera.

The Bayesian network is a weaker model for some municipalities of the central Var, of the Westernmost part of the Bouches-du-Rhône and, above all, of the Eastern part of the French Riviera. In these areas, spatial dynamics are the farthest from the overall behavior of the Bayesian network model. They are very peculiar territories over which the metropolization process shows different outcomes. The real estate market of the East of the French Riviera, for example, is deeply marked by important cross-border interactions with the Principality of Monaco and the metropolitan areas of North-Western Italy.

6.4   Conclusions

6.4.1   Results and comparison with other modeling approaches

The research presented in this chapter shows several innovations in the modeling of spatial dynamics within a regional space. Spatial indicators derived from census data (population, migration, jobs, housing, and mobility) have been coupled with indicators derived from GIS applications and from modeling on transportation graphs. An overall model of interaction among indicators was then produced using the Bayesian network technique. The model was derived from data but could integrate soft hypotheses from the modeler (in the form of a hierarchy among variables) and uncertainty in the relationships among indicators.

The Bayesian network model was used as an explanatory tool in geographical analysis. It simplifies, sometimes considerably, the complexity of causal links among variables. Highlighting the most statistically significant links, it allows for a clearer view of the main spatial phenomena. The metropolization process within coastal South-Eastern France is thus reduced to the essentials and becomes much more understandable.

More particularly, the Bayesian network model highlights homogenization dynamics related to the spatial diffusion of the metropolization process. It is the case of urban densities, socio-demographic characteristics, dwelling characteristics and mobility behavior.

Another result of the Bayesian network model is the importance of the 1990 situation in order to explain spatial dynamics in the study area between 1990 and 1999. This suggests that spatial dynamics are deeply influenced by the historic heritage of the region.

Besides these general trends, specific areas show particular dynamics which are not explained by the Bayesian network model. It is namely the case of the Eastern section of the French Riviera. These dynamics are the consequence of peculiarities in the socio-economic structure and in the geographical position of these areas. Finally, transportation offer and accessibility indicators play a key role within the Bayesian network model. This confirms that the metropolitan integration process in the region is strongly driven by highway and airport facilities. A marked aspect of the metropolization of coastal South-Eastern France is thus intense urban sprawl, resulting from a spatial redistribution of inhabitants and economic activities within a wider urban time-space, made possible by the automobile technology [494].

In what follows, we will compare our modeling experience using Bayesian networks with other techniques currently employed by modelers on wide geographic databases.

Bayesian networks are much more powerful tools of knowledge discovery than classical multiple regression models. Regressors can be discovered through appropriate step-wise procedures, but the model is calibrated for the prediction of one or several pre-established dependent variables. Bayesian networks don't need pre-established target variables and can later be used to predict any variable from any subset of other variables. Moreover, the interaction between the modeler and the search algorithms allows for the construction of complex models taking into consideration all possible links among variables (instead of needing independent regressors).

Bayesian networks should also be compared to multivariate statistical models (hierarchical clustering, factor analyses, etc.), which are powerful tools of knowledge discovery [408]. Bayesian networks have the advantage of producing an explicit model of relations among variables which can later be used for probabilistic simulations. On the other hand, Bayesian network packages should further develop interfaces for batch evaluation of the database records. Record evaluation is an important aspect of modeling in geography and the ability to discover relations among both variables and statistical units is a strong point of classic multivariate models.

Bayesian network could finally be compared to neural networks applications in geography [127]. Both Bayesian networks and NNs give versatile models derived from data which are suitable for probabilistic prediction. The advantage of Bayesian networks resides in the possibility of having an explicit model of relations among domain variables. In this respect NNs almost constitute black boxes. It would be impossible to conceive an application as the one carried out in this chapter using NNs, as the model was mainly used for domain analysis, and not for simulation.

6.4.2   Perspectives of future developments

The Bayesian network model proposed in this chapter is the result of the first phase of a wider research program on modeling spatial dynamics in a regional space. The main goal of this application was to validate the new modeling approach through Bayesian networks. This will be further developed in order to be applied to wider spatial databases. Our work could thus highlight a certain number of desirable improvements.

First of all, the geographic database should be developed to include the spatial structure of daily and residential mobility (indicators should be calculated from Origin/Destination matrixes). Indicators on economic activities, professional skills, revenues and real estate market should also be included. These elements would permit a better understanding of mobility phenomena and economic specialization of certain geographic areas within the metropolization process. The geographic database should also encompass a wider time span: spatial dynamics evaluated over several decades (within the limits of available data) would be a much more powerful tool for understanding today's phenomena.

As far as the modeling technique is concerned, Bayesian networks provided a first model of the interaction among the selected spatial variables. The model could describe (and explain) important spatial dynamics in the coastal region of South-Eastern France in a synthetic way.

More specifically, the ArcInversion search strategy made the modeler's hypotheses on variable order less constraining in the structure generation, when compared to faster greedy strategies. The produced Bayesian network is statistically more robust and can integrate unforeseen explanations of the phenomena taken into consideration. Search strategies even less dependent on the modeller's hypothesis could be tested, as the generation of the essential causal graph based only on conditional independence and dependence statements (it is an option available in several Bayesian network packages like Hugin and BayesiaLab).

Integrating in the same Bayesian network geographic variables evaluated at different dates allows for the detection of feedbacks among variables. This would otherwise be impossible within the structure limitations of directed acyclic graphs. In our model, a feedback was detected among population and housing variables. These results open the perspective of using dynamic Bayesian networks to produce dynamic regional models derived from longitudinal geographic databases. Such models could eventually be used for strategic prospective simulation of regional development.

7

Inference problems
in forensic science

Franco Taroni and Alex Biedermann

*The University of Lausanne – Faculté de Droit et des Sciences
Criminelles – École des Sciences Criminelles, Institut de Police
Scientifique, 1015 Lausanne-Dorigny, Switzerland*

7.1 Introduction

7.1.1 Forensic science, uncertainty and probability

Forensic science can be considered as a discipline that seeks to provide expertise that should assist both investigative and legal proceedings in drawing inferences about happenings in the past. Such events are unique, unreplicable and remain unknown to at least some degree, essentially because of mankind's limited spatial and temporal capacities. However, past occurrences may leave one with distinct remains, in the context also referred to as scientific evidence, that may be discovered and examined. For example, DNA profiling analyses may be applied to a blood stain recovered on a crime scene and the results compared to those obtained from a sample provided by a suspect. Or, fibers may be collected on a dead body and compared to sample fibers originating from the trunk of a vehicle in which the victim may have been transported.

Although forensic scientists can benefit from a wide range of sophisticated methods and analytical techniques for examining various items of evidence, there usually is a series of factors that restrain the inferences that may subsequently be drawn from results. For instance, there may be uncertainty about whether the blood stain on the scene has been left by the offender. On other occasions, the blood stain

Bayesian Networks: A Practical Guide to Applications   Edited by O. Pourret, P. Naïm, B. Marcot
© 2008 John Wiley & Sons, Ltd

may be degraded or only of minute quantity. In a fibers case, the control samples may not be representative or the characteristics of the fibers may not allow for much discrimination. Thus, there currently is a great practical necessity to advise customers of forensic expertise (lawyers, jurors or decision makers at large) about the significance of findings in a case at hand. Forensic scientists need to qualify and, where possible, quantify the states of available knowledge while relying upon appropriate means in dealing with the sources of uncertainty that bear on a coherent evaluation of scientific evidence.

According to a viewpoint maintained in a predominant part of both literature and practice within forensic and legal areas, the management of such sources of uncertainty should be approached through probability theory. There is a particularity with that point of view which is worth mentioning at this stage. It is the idea that forensic scientists assist the evaluation of scientific evidence through the assessment of a likelihood ratio (V), which is a consideration of the probability of the evidence (E) given both, competing propositions that are of interest in a particular case – typically forwarded by the prosecution and the defense (for instance, '$H, \bar{H}$: the crime stain comes (does not come) from the suspect') – and auxiliary circumstantial information (habitually denoted I, but often omitted for brevity):

$$V = \frac{\mathbb{P}(E \mid H, I)}{\mathbb{P}(E \mid \bar{H}, I)}. \tag{7.1}$$

The term 'proposition' is interpreted here as a statement or assertion that such-and-such is the case (e.g., an outcome or a state of nature). It is assumed that personal degress of belief can be assigned to it.

A Bayesian framework is accepted so that the provision of prior probabilities for the target propositions, as well as the calculation of posterior probabilities given the evidence, is reserved to the recipient of expert information (e.g., a juror). The main reason for this is that such calculations require assumptions to be made about other (nonscientific) evidence pertaining to the case [229]. It is thought that this segregation helps to clarify the respective roles of scientists and members of the court. In addition, it is often argued that actors within judicial contexts should be aware of probability because it is a preliminary to decision making and it can guard one against fallacies and demonstrably absurd behaviour [389].

Thoroughly developed collections of probabilistic approaches are currently available to assist scientists in assessing the evidential value of various kinds of scientific evidence [9], but the practical implementation of probabilistic analyses may not always be straightforward. Difficulties may arise, for instance, when one needs to cope with multiple sources of uncertainty (with possibly complicated interrelated structure), several items of evidence or situations involving acute lack of data.

Such issues are discomforting scientists and other actors of the criminal justice system, which illustrates the continuing need for research in this area. One direction of research that has gained increased interest in recent years focuses on a formalization of probability theory that consists of its operation within a graphical environment, that is, Bayesian networks.

7.1.2   Bayesian networks in judicial contexts

The study of representational schemes for assisting reasoning about evidence in legal settings has a remarkably long history. In the context, the charting method developed by Wigmore [495] is a frequently referenced – however essentially non-probabilistic – predecessor of modern network approaches to inference and decision analyses that can be traced back to the beginning of the twentieth century. But it is only about two decades ago that researches have begun to show interest in graphical approaches with genuine incorporation of probability theory. Examples include decision trees and a modified, more compact version of these, called 'route diagrams' [163, 164]. Since the early 1990s, however, it is Bayesian networks that have advanced to a preferred formalism among researchers and practicioners engaged in the joint study of probability and evidence in judicial contexts, notably because of that method's efficient representational capacities and thorough computational architecture. It thus seems interesting to note that – compared to other domains of applications – researchers in law were among the pioneers who realized the practical potential of Bayesian networks.

In judicial contexts, one can generally distinguish different ways in which Bayesian networks are used as a modeling technique. Legal scholars focus on Bayesian networks as a means for structuring cases as a whole whereas forensic scientists concentrate primarily on the evaluation of selected issues that pertain to scientific evidence.

Many studies with an emphasis on legal applications thus rely on Bayesian networks as a method for the retrospective analysis of complex and historically important *causes célèbres*, such as the Collins case [144], the Sacco and Vanzetti case [237, 418], the Omar Raddad case [281] or the O.J. Simpson case [451]. Other contributions use Bayesian networks to clarify fallacious arguments [150].

Forensic applications of Bayesian networks range from offender profiling [6, 8], single [7, 174] and complex [115, 449] configurations of different kinds of trace evidence as well as inference problems involving the results of DNA analyses. The latter is an imporant category that covers studies focusing on small quantities of DNA [148], cross-(reciprocal-)transfer evidence [10], relatedness testing with or without missing data on relevant individuals [117] or mixed stains [326].

These conceptual and practical studies represent a relatively rich variety of inferential topics. The reported works jointly support the idea that graphical probability models can substantially improve the evaluation of likelihood ratios used for the assessment of scientific evidence. In particular, they allow their user to engage in probabilistic analyses of much higher complexity than what would be possible through traditional approaches that mostly rely on rather rigid, purely arithmetic developments. Moreover, the graphical nature of Bayesian networks facilitates the formal discussion and clarification of probabilistic arguments [173]. Along with accessory concepts, such as qualitative probabilistic networks and sensitivity analyses, the range of applications can be extended to problems involving an acute lack of numerical data [46].

7.2   Building Bayesian networks for inference

7.2.1   General considerations

Although there is now a remarkable diversity of Bayesian network applications that have been proposed for forensic purposes (Section 7.1.2), the respective reports usually restrict themselves to present models in their final version. Generally, only few explanations, if any, are given on how practitioners can use the method to build their own models. This is so even though it is acknowledged that the derivation of appropriate representations is crucial for reasons such as viability or computational routines [117].

This is a frequently encountered complication which is not specifically related to Bayesian networks, but appears to be a characteristic of modeling in general. To some extent, (statistical) modeling is considered an art-form, but one that can be guided by logical and scientific considerations [117]. Usually, it is not, however, that one can state explicit and universally applicable guidelines. Rather, there are some general considerations that may assist scientist in eliciting sensible network stuctures. Below, these are discussed and illustrated in some detail.

7.2.2   Structural aspects

In forensic contexts, part of a reasonable strategy for eliciting network structures should consist of a careful inspection of the inference problem at hand and the kind of scientific evidence involved.

For the purpose of illustration, consider classic types of evidence, such as fibers or toolmarks. Here, the number of relevant propositions that one can formulate may be limited because of the complexity of the processus underlying the generation of such evidence. The assessment of fiber evidence, for example, may require consideration of factors such as transfer, persistence and recovery – in particular where one's interest is to evaluate the evidence with respect to particular actions of interest (e.g., a contact between a victim and an assailant). According to the current state of knowledge, however, the nature of such parameters is not yet very well understood. A further example of this is mark evidence, where comparative examinations between a crime mark and marks obtained under controlled conditions with a suspect's tool may be fraught with complications because of the changes that the suspect's tool may have undergone (e.g., through ongoing use or unfavorable storage conditions). Within such contexts, the Bayesian networks models may thus involve rather local structures that capture essentially global (and the sufficiently well understood) aspects thought to be relevant to the inference problem of interest.

The scientist's starting point may differ, however, according to the domain of application. Consider, for instance, the problem of drawing inferences from results of DNA profiling analyses. Within the branch of DNA evidence, an extensive body of knowledge (accepted biological theory) is available and upon which one can rely during network construction. For example, the consideration of Mendelian laws of inheritance allows one to obtain clear indications on how nodes in a network ought

(i)   (ii)

Figure 7.1   Basic submodels representing (i) a child's genotype, *cgt*, with *cpg* and *cmg* denoting, respectively, the child's paternally and maternally inherited genes, and (ii), a child's maternal gene, *cmg*, reconstructed as a function of the mother's paternal and maternal genes, *mpg* and *mmg*, respectively *mgt* denotes the mother's genotype. The states of gene nodes represent the different forms (that is, alleles) that a genetic marker can assume whereas the states of the genotype nodes regroup pairs of alleles.

to be combined. In this way, basic submodels can be proposed and repeatedly reused for logically structuring larger networks. Sample network fragments of this kind, adapted from [117], are shown in Figure 7.1.

In the particular area of model specification for DNA evidence, it appears worth mentioning that it has also been found useful to follow a hierarchical approach, notably where analyses lead to large network topologies (e.g., when information pertaining to different genetic markers needs to be combined). For this purpose, object-oriented Bayesian networks, supported by certain probabilistic expert system software packages, such as Hugin, are reported to be particularly well suited [114].

Yet another modeling approach proposed for forensic inference from DNA evidence is based on a graphical specification language. Cowell [105] developed a software tool, called FINEX (Forensic Identification by Network Expert systems), which has built-in algorithms that use input in the specification language for the automatic construction of appropriate Bayesian networks, useable in probabilistic expert systems. Such an approach allows one to save time in setting up networks and to reduce the potential of error while completing large probability tables. FINEX appears to be among the sole reported forensic applications of Bayesian networks where the construction process is, to some degree, automated. The approach rests, however, restricted to the analysis of identification problems based on the results of DNA profiling analyses.

At this point, it should be noted that the comments so far made on the advantages that DNA evidence may present over more traditional kinds of evidence hold essentially when the target problem is of a particular form. Examples include inferences about genotypic configurations (e.g., during relatedness testing) or source-level propositions (e.g., 'the suspect is the source of the crime stain'). As pointed out in [148], for instance, as soon as the focus of attention shifts from questions of the kind 'whose DNA is this?' to questions of the kind 'how did this DNA get here?', a series of issues pertaining to the generation of DNA evidence may need to be addressed (e.g., modus operandi). This requires a careful examination of all relevant aspects of the framework of circumstances in the individual case under consideration [9]. As may be seen, model construction in such contexts tends to become less amenable to the invocation of repetitively useable building blocks. Here, structural issues such as the number of propositions, along with their definition (e.g., level of detail), will rather depend on the extent of case-specific knowledge that is available to the analyst. This then is an instance when model construction for DNA evidence may become comparable to the kinds of evidence mentioned at the beginning of this section.

For the most part, thus, the formal structuring of inference problems with a forensic connotation involves 'hand-constructed' models, derived by or in collaboration with domain experts. An imporant aspect of such personalized construction processes is the way in which the principal propositions of interest are formulated. In forensic science, there is established literature and practice that considers that relevant propositions can be framed at different hierarchical levels [99]. For example, in an inference from a source-level proposition (e.g., 'the crime stain comes from the suspect') to a crime-level proposition (e.g., 'the suspect is the offender'), consideration needs to be given to the relevance of the crime stain (that is, the probability of the crime stain being left by the offender). When addressing a proposition at the activity-level (e.g., 'the suspect physically attacked the victim'), factors such as transfer, persistence and recovery will need to be taken into account.

Arguably, if the aim is to derive models that are in agreement with established precepts of evidential assessment, then a consideration of hierarchical subtleties should be part of one's structural elicitation strategy (see also Section 7.2.4). Generally, scientists should also consult existing probabilistic inference procedures (notably, likelihood ratios) from scientific literature. These too may provide useful indications on the number of propositions, their definition as well as their relationships [174, 449].

7.2.3   Probabilistic model specification

According to a general principle in forensic science, analysts should approach both inferential and practical issues by following a procedure that goes from the general to the particular. With respect to an inference problem, this may be interpreted to mean that, prior to considering a full numerically specified network, one may start by assigning qualitative expressions of probability. A useful collection of concepts for this purpose is due to Wellman [490]. Qualitative probabilistic networks can rapidly provide valuable preliminary information on the direction of inference without the need to deploy possibly extensive elicitation efforts for specific numerical values. It may also be that the query of interest may be sufficiently well answered on a purely qualitative level [46].

Generally, the numerical specification of Bayesian networks for forensic inference problems can involve different kinds of probabilistic information. Among the more common of them are – in analogy to many other domains of application – logical assignments of certainty and impossibility, estimates derived from relevant databases, subjective expert opinions or combinations of these.

As an example for logical assignments of certainty and impossibility, consider the conditional probability $\mathbb{P}(S \mid G, H)$, associated with a node S of a network fragment $H \to S \leftarrow G$. Logically, $\mathbb{P}(S \mid G, H)$, the probability of the suspect being the source of the crime stain (proposition S), given that he is the offender (proposition H) and given that the crime stain is originating from the offender (proposition G), is one [174]. Given that S is defined as a binary proposition, $\bar{S}$, that is, the negation of S $\mathbb{P}(S \mid G, H)$, must therefore be zero.

The same network fragment $H \to S \leftarrow G$ is amenable to illustrate subjective assignments of probability. The (unconditional) probability required for the nodes H may be formed in the light of previously heard evidence concerning the suspect's guilt. An assignment of a probability to the node G (i.e., evidential relevance) may be the result of an evaluator's consideration of the position in which a crime stain was found, its freshness and abundance. Here, the proposition G relates to a non-replicable real-world event that happened in the past, that is, the crime. The respective probability is epistemic and can enjoy a wide inter-subjective agreement [174].

Relevant databases may be consulted when the probability of interest relates, in one way or another, to a countable phenomenon. Genetic data are typical examples of this. For the purpose of illustration, consider again the network fragments displayed in Figure 7.1. Let us suppose that the node mpg (short for 'mother paternal gene') covers the states 14, 16 and x, representing the number of short tandem repeats (STR) at the locus D18 (whereas x is an aggregation of all alleles, that is, repeat numbers, other than 14 and 16). The unconditional probabilities required for the various states of the node mpg can be interpreted as (estimates of) the frequencies of the respective alleles among members of a relevant population, that is, information that may be obtained from databases (or, literature).

There may also be occasions where the scientist seeks to employ variables whose probability distributions are thought to adhere to certain distributional and functional forms. Some Bayesian network software offer means that facilitate the specification of such forms through expressions, either manually or assisted by various interface boxes. In recent versions of Hugin, for instance, expressions can be constructed using certain statistical distributions, mathematical functions, relations as well as arithmetic and logical operators. An example of the use of an expression is given at the end of Section 7.3.1.

7.2.4   Validation

What is an adequate Bayesian network? How is a Bayesian network to be validated? Such questions are recognized and recurrent complications in many applications of Bayesian networks, not just within forensic science. Notwithstanding, a close inspection of the domain of interest may often allow one to set forth some viable procedural directions.

Throughout the previous sections, the application of Bayesian networks to forensic inference problems has been described as a reasoner's reflection of the essential aspects of a real-world problem or situation. There is no claim, however, that any proposed model will amount to a perfect problem description, or, in some sense, a 'true' model. As noted earlier in Section 7.2.2, inferential analyses in forensic science are, actually, strongly dependent on the extent of available information, the scientist's aims as well as his position within the evaluative process as a whole. These are factors that tend to preclude the possibility of a final or ultimate way in which an inference is decomposed. Scientists thus need to argue for the reasonabless of their models through the construction of arguments. These assume a crucial role in the course of deciding whether a scientist's analysis, based on a given model, is one that can be trusted.

One particular means through which a model's adequacy can be assessed are established likelihood ratio formulae, currently used for evaluating scientific evidence, and, accessible in specialized literature on the topic [9]. Extensive literature is typically available on the application of algebraic calculations for solving various scenarios of relatedness testing. One can regard such existing approaches as reference procedures and compare their output to that obtained for target nodes of a Bayesian network after evidence has been propagated. It should solely be noted that there is not necessarily a univocal correspondence between an existing probabilistic inference procedure and a particular Bayesian network. That is to say, one can obtain identical numerical results for specific queries of interest even though the respective Bayesian networks are structurally different (e.g., in terms of the number and/or definition of the propositions) [448].

Often, however, forensic scientists may not be in the comfortable situation of having a well established reference procedure at their disposal. Sometimes, there may be one, but the prototype Bayesian network for the problem of interest yields results that diverge from that norm. To some degree, this is expectable but does not necessarily need to be a cause of concern. The reason for this is that Bayesian networks are a method that allows one to construct models whose underlying probabilistic tenets may exceed the degree of complexity at which reference procedures operate. What is important is that the inference analyst is prepared and able to forward reasons to explain differences – if they occur.

This may be more difficult in contexts where no reference procedure exists for the problem of interest, that is, the respective domain has previously not been investigated from a probabilistic point of view, or, not in such a comprehensive manner as it is feasible through a Bayesian network. In such cases, it may however be possible to identify local network structures that can be separately validated as selected issues of a larger problem. In a subsequent step, the logic of the way in which the component network fragments are combined may be used to argue in support of the reasonabless of the respective Bayesian network as a whole.

7.3   Applications of Bayesian networks in forensic science

This section presents two applications of Bayesian networks with the intention to illustrate the principal ideas outlined so far. The first example (Section 7.3.1) discusses a Bayesian network for evaluating mark evidence according to an established probabilistic reference procedure. It will provide an instance of the use of an expression for defining particular statistical distributions, applicable as flexible means to account for results of comparative forensic examinations. The second example (Section 7.3.2) focuses on a scenario involving DNA evidence and discusses the construction of Bayesian networks through distinct building blocks.

7.3.1   Application 1: Issues in the evaluation of mark evidence

Forensic scientists routinely examine footwear marks collected on crime scenes and compare these with impressions obtained under controlled laboratory conditions from soles of suspects' shoes. For the ease of argument, and, in order to conform with habitual notation used in probabilistic evaluation of scientific evidence in forensic science, the variables y and x are retained here to denote observations relating to crime and comparison material respectively. In addition, each of these descriptors will be divided into a component relating to traits that originate from, respectively, the features of manufacture (subscript 'm') and acquired characteristics (subscript 'a') of the shoe that left the mark or impression at hand. Observations are thus described as $y = (y_m, y_a)$ and $x = (x_m, x_a)$. Let us note that shoes from the same line of production may share the 'same' features of manufacture (e.g., the general pattern of a shoe sole) whereas acquired characteristics (e.g., accidental damage through wear) are specific to each shoe.

When a suspect's shoe is available for comparative examinations, then a subsequent inference problem may consist in constructing a probabilistic argument to the proposition according to which that shoe (or some unknown shoe) is at the origin of the crime mark. The latter is a so-called 'source-level' proposition, commonly denoted F. On the basis of information pertaining to such a proposition, it may be of interest to reason about H, that is, the proposition 'the suspect (some unknown person) is the offender'. This inferential step will need to give consideration to evidential relevance, denoted G (a binary proposition of the kind 'the crime mark has been left by the offender'). Details are presented in [174]. A summary of the defninition of the principal nodes, that is F, G and H, is given in Table 7.1.

As noted in Section 7.1.1, a scientist's role in evaluating evidence usually consists in eliciting a likelihood ratio. For the application considered here, there is literature that proposed a probabilistic inference procedure involving variables with definitions as given above. For an inference to a crime level proposition (H), that procedure involves a likelihood ratio, V, of the following form [149]:

$$V = \frac{\mathbb{P}\left(y_m, y_a, x_m, x_a \mid H\right)}{\mathbb{P}\left(y_m, y_a, x_m, x_a \mid \bar{H}\right)} = rw\ V_m\ V_a + (1 - rw), \qquad (7.2)$$

Table 7.1   Summary of the definitions of the nodes F, G and H.

$F(\bar{F})$:	the crime mark was made by the suspect's shoe (some other shoe)
$G(\bar{G})$:	the crime mark has (has not) been left by the offender
$H(\bar{H})$:	the suspect (some other person) is the offender

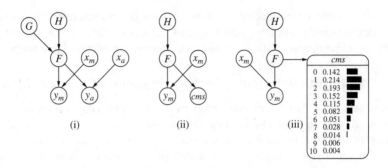

Figure 7.2   Bayesian networks for evaluating mark evidence (partially adapted from [47]). The definition of the nodes are given in the text. Reproduced with permission of Elsevier Ireland Ltd.

with r denoting $\mathbb{P}(G)$, w denoting $\mathbb{P}(F \mid G, H)$ and V_m and V_a representing, respectively, the following ratios:

$$V_m = \frac{\mathbb{P}(y_m \mid x_m, F)}{\mathbb{P}(y_m \mid \bar{F})}, \qquad V_a = \frac{\mathbb{P}(y_a \mid x_a, F)}{\mathbb{P}(y_a \mid \bar{F})}.$$

Figure 7.2(i) shows a Bayesian network that appropriately encodes the various conditional independence assumptions encapsulated in Equation (7.2).

A more formal demonstration of this can be found in [448].

A numerical specification of the Bayesian network is not pursued in further detail here because the assignment of specific numerical values makes only sense with respect to a particular scenario. The extent of considerations that this may amount to is outlined in [149].

Although, at first glance, the proposed model is of rather unspectacular topology, it captures a series of subtle considerations which may otherwise not be easy to convey. Examples include the following:

- The component observations pertaining to features originating from a shoe-sole's features of manufacture and acquired characteristics contribute to the overall likelihood ratio through multiplication.

- The absence of arcs between nodes with subscripts 'a' and 'm' reflects an assumption of independence between features of manufacture and wear. The validity of such an assumption depends on the type of wear and the way in which it has been described [149].

- If some other shoe is at the origin of the crime mark (that is, $\bar{F}$ being true), then the probability of the observations pertaining to the crime mark (nodes y_m and y_a) are not affected by knowledge of the patterns left by the suspect's shoe (nodes x_m and x_a).

The practical interest in the proposed network consists in its capacity to store a basic inferential structure and to offer a way to interface probabilistic judgements of different kind (including their updating as required). For example, the probability $\mathbb{P}\left(y_m \mid \bar{F}\right)$, that is, the probability of the observed manufacturing features given that the crime mark had been left by some other shoe, can be estimated from a relevant database and/or information on sales and distribution obtained from a manufacturer or supplier (provided that model and manufacturer has been identified). Notice that the probability $\mathbb{P}\left(y_m \mid \bar{F}\right)$ may also be investigated through sensitivity analyses because of the effect it has on the magnitude of the likelihood ratio [47].

At other instances, such as the nodes G (evidential relevance) or H ('prior' probability of guilt), the Bayesian network involves probabilistic assessments that typically require information in the hands of a recipient of expert information (e.g., a judge). Here, the Bayesian network can take the role of clarifying and delimiting the respective areas of competence of forensic scientists and recipients of expert information. It allows one to show how these actors may draw their attention on how to collaborate towards a meaningful and complementary probabilistic assessment without the need to care about various underlying lines of algebra.

Yet another feature of the model is its flexibility to accommodate – given only minor changes in network specification – inference problems of similar kind. One example is mark evidence on fired bullets [47]. Such marks are the result of a bullet's passage through a firearm's barrel, a processus which may provoke sets of distinct traits. These, too, can be described as originating from the respective barrel's manufacturing features and acquired characteristics. Figure 7.2(i) shows a Bayesian network that is amenable for evaluating such evidence. One would solely need to eliminate the node G (i.e., the relevance factor). In the context, uncertainty about relevance is, usually, not an issue since bullets from crime scenes have been selected precisely because they induced a particular damage.

It may be, however, that scientists wish to be more specific in the way in which marks originating from a barrel's acquired features are described. Specifically, a concept frequently invoked in settings involving firearms is known as the number of 'consecutive matching striations' (CMS), observable between marks present on, respectively, a bullet in question and a bullet fired under controlled conditions through the barrel of a suspect's weapon. Within a Bayesian approach, such a descriptor would require scientists to assess probabilities of observing a given number of matching striations conditional on knowing F and $\bar{F}$, respectively.

The Bayesian network shown in Figure 7.2(i) can be modified to accommodate CMS data by adopting a node with states numbered, for instance, $0, 1, 2, \ldots, 10$ [47]. Such a network is shown in Figure 7.2(ii). Following literature on the topic, the probability of observing a given number of CMS can be expressed in terms of a Poisson distribution. For the purpose of illustration, Figure 7.2(iii) shows a Bayesian network with an expanded node *cms* (initialized state). Using sample data from [69], the probability table of this node can be completed automatically by means of the following expression (using Hugin language):

```
if(F==true,Poisson(3.91),Poisson(1.325)).
```

This particular feature appears interesting insofar as it allows one to show that one can readily refine selected parts of an existing model without the need to review the derivation of the entire probabilistic inference procedure.

7.3.2   Application 2: Inference based on results of DNA profiling analyses

DNA profiling analyses performed on genetic markers, most often short tandem repeat (STR) markers (i.e., regions on DNA with polymorphisms that can be used to discriminate between individuals; Section 7.2.3), currently represent the standard means to obtain relevant information for investigating various questions of relatedness. For each marker included in an analysis, the genotype is noted. The latter consists of two genes whereas one is inherited from the mother and the other from the father (although one cannot observe which is which). Basic Bayesian network fragments (due to [117]) that capture an individual's genotype for a given marker or the transmission of alleles to descendants (children) have earlier been mentioned in Section 7.2.2 (Figure 7.1). The aim of this section is to illustrate how such submodels can be combined in order to approach more complex scenarios, typically encountered in day-to-day laboratory practice (i.e., casework) or proficiency testing (as part of laboratory quality assurance).

Let us imagine the following scenario. There are two individuals (offspring, denoted here child c1 and child c2, respectively) who share the same two parents (mother m1 and father f). A third individual, c3, known to have a mother m2 different from m1, is interested in examining the degree of relatedness with respect to c1 and c2 (e.g., half-sibship versus unrelated). Notice that f is considered as a putative father of c3. A particular complication of the scenario consists in the fact that f is deceased and unavailable for DNA testing.

The currently considered scenario can be studied through a Bayesian network as shown in Figure 7.3.

This network can accommodate DNA profiling results for a single marker. The structure of the model can be explained as a logical combination of submodels that themselves may be a composition of model fragments. Examplar submodels denoted (a), (b) and (c) are highlighted through rounded boxes with dotted lines (other submodels may be chosen).

The submodel (a) represents the genotypes of the individuals c1 and c2 as a function of the genotypes of the undisputed parents m1 and f. As may be seen, the submodel (a) is itself a composition of a repeatedly used network fragment earlier mentioned in Section 7.2.2 (Figure 7.1(i)). The same network fragment is invoked to implement the genotype of the individuals c3 (submodel (b)) and m2 (submodel (c)).

A noteworthy constructional detail connects the two submodels (a) and (b). As there is uncertainty about whether f is the true father of c3, the paternal gene of c3, $c3pg$, is not directly conditioned on f's parental genes (that is, fmg and fpg). Such uncertainty is accounted for through a distinct node $tf = f$? that regulates the degree to which f's allelic configuration is allowed to determine c3's true father's

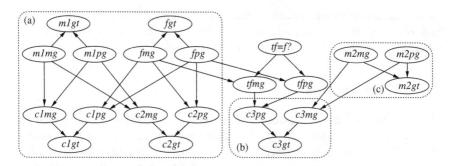

Figure 7.3 Bayesian network for evaluating a relatedness testing scenario: c, f, m (in the first place) and tf denote, respectively, child, father, mother and true father; nodes with names '...mg' and '...pg' denote, respectively, an individual's maternally and paternally inherited genes; nodes with names '...gt' represent an individual's genotype; the node $tf = f$? is binary with values 'yes' and 'no' in answer to the question whether the undisputed father f of the children c1 and c2 is the true father of the child c3.

parental genes, represented here as $tfmg$ and $tfpg$. Table 7.2 illustrates the way in which the conditional probability tables of the nodes $tfmg$ and $tfpg$ need to be completed. Notice further that the values γ_{14}, γ_{16} and γ_x are relevant allelic frequencies that are also used for specifying the unconditional probability tables of the nodes $m1mg$, $m1pg$, fmg, fpg, $m2mg$ and $m2pg$.

The use of Bayesian networks for the study of inference problems involving results of DNA profiling analyses currently is a lively area of research with important practical implications. In view of the fact that in forensic applications of DNA typing technology one may need to handle degraded or incomplete sets of evidence,

Table 7.2 Conditional probability table for the node $tfpg$. The nodes fpg and $tfpg$ denote father and true father paternal gene, respectively. The states 14, 16 and x are, as outlined in Section 7.2.3, sample numbers of STR repeats for a given marker. γ_{14}, γ_{16} and γ_x are relevant population frequencies of the alleles 14, 16 and x. The node $tf = f$? is a binary node with states 'yes' and 'no' as possible answers to the question whether the individual f is the true father of c3. An analogical table applies to the node $tfmg$.

$tf = f$? :		*yes*			*no*	
fpg :	14	16	x	14	16	x
$tfpg$: 14	1	0	0	γ_{14}	γ_{14}	γ_{14}
16	0	1	0	γ_{16}	γ_{16}	γ_{16}
x	0	0	1	γ_x	γ_x	γ_x

along with additional complications such as genetic mutation, inference modeling may often require case specific analyses. The versatility of Bayesian networks makes them particularly well suited for this purpose.

7.4 Conclusions

The evaluation of scientific evidence in forensic science requires the construction of arguments in a balanced, logical and transparent way. Forensic scientists need to clarify the foundations of such arguments and to handle sources of uncertainty in a rigorous and coherent way. These requirements can be appropriately conceptualised as reasoning in conformity with the laws of probability theory, even though the practical implementation of such a view may not always be straightforward – in particular where real-world applications need to be approached.

It is at this juncture that Bayesian networks can be invoked as a means to handle the increased complexity to which probabilistic analyses may amount in forensic applications. As a main argument, this chapter has outlined that the construction of viable Bayesian network structures is feasible and, by no means arbitrary, because of the ways through which the reasonabless of inference models may be argumentatively underlined. Bayesian networks should not be considered, however, as a self-applicable formalism and, generally, there is seldomly a uniquely 'right' model for an issue of interest. Often, a careful inspection of the inference problem at hand is required, along with consideration to be given to possibly preexisting evaluative procedures. These can provide useful indications on how to derive appropriate network structures.

A sketch of the wide range of considerations that may be considered in the construction of Bayesian networks for inference problems in forensic science is given in this chapter in both, a rather general formulation (Section 7.2) as well as in terms of two sample applications (Section 7.3). The proposed Bayesian networks illustrate – with the advent of powerful Bayesian network software – the capacity of the method as an operational tool and a way to provide insight in the static or structural aspects of a model as well as in the 'dynamics' of probabilistic reasoning.

8

Conservation of marbled murrelets in British Columbia

J. Doug Steventon

Ministry of Forests and Range, British Columbia, Canada

8.1 Context/history

In this chapter I describe the application of BNs to conservation planning for a threatened sea-bird, the Marbled Murrelet (*Brachyramphus marmoratus*), on the coast of British Columbia, Canada. While spending most of its life on the ocean this unusual species (family Alcidae) nests in forests, usually on large branches of old-growth trees, widely dispersed at low densities as far as 50 km inland [71]. During the breeding season they commute twice daily from the ocean to incubate or feed their young. This nesting strategy has resulted in a conflict with commercial logging, historically considered the primary threat to the species and the focus of the analyses presented here. Additional threats to the species while at sea include changes in forage abundance and distribution with climatic change and ocean exploitation, fishing by-catch, and oil spills. As a consequence of these threats, the murrelet is listed as threatened under the Canadian Species at Risk Act, and is the subject of conservation efforts and land-use planning in British Columbia.

Setting quantitative conservation objectives for murrelets and conservation in general has been an elusive goal [450]. To support conservation planning efforts and land-use decisions involving this species, I led a team that developed

---

Bayesian Networks: A Practical Guide to Applications Edited by O. Pourret, P. Naïm, B. Marcot
© 2008 John Wiley & Sons, Ltd

decision-support models using BNs as our analytical framework. The models were used to examine the implications to murrelet abundance and viability (chances of the population persisting through time) of alternative nesting habitat goals (amount and characteristics), or alternatively the amount and characteristics of habitat consistent with desired population size and viability targets. We applied the analyses at the coast-wide scale to assist the setting of goals among six large conservation regions [444, 445], and at a watershed scale (topographic units typically 10 000–50 000 hectares in size) to assess alternative land-use scenarios for a 1.7 million hectare area of the northern mainland coast region [443].

Our models are a form of habitat-linked population viability analysis (PVA). PVA is a widely used methodology for assessing conservation priorities and risks to species persistence [41]. PVA commonly uses population projection models to estimate population longevity (mean length of time the population is projected to persist). Our application is habitat-linked in that nesting habitat quantity and quality is directly linked to the population model. Usually, as I present here, mean persistence time is converted to the probability that the population will persist (persistence probability) over some particular time frame of interest [191].

We confronted many of the challenges found in other applications of PVA, such as defining the appropriate structure of the models, and a lack of data for parameterization [289]. PVA is therefore often best used to rank hazards and assess the relative effectiveness of policy choices [62] and that is the main way we used it. We treat persistence probability as a measure of population resilience: the ability of the population to recover from environmental variability and/or infrequent catastrophes.

There were multiple users of the modeling results, and application is continuing. Users included the Canadian Marbled Murrelet Recovery Team (CMMRT, a multi-agency technical committee tasked with planning conservation strategies), provincial resource management agencies, and the technical advisory team and planning table for the multi-stakeholder North Coast Land and Resource Management Plan (LRMP) land-use planning process.

We selected the BN approach as meeting several requirements. First, using a BN approach enforces explicit consideration of the logic and assumptions in modeling ecological systems and in making management recommendations. Second, it allows an intuitive, visual approach to model building through influence diagrams. Third, it provides a means for combining direct use of field data, results of statistical analyses, simulation output, analytical equations, and expert opinion in one summary model. Fourth, uncertainty is explicitly incorporated and displayed through the potential node state values (the values a node can assume), belief probability weightings for each state, and by the propagation of uncertainty through the network. Fifth, the BN structure and software include useful functionalities for sensitivity analyses, model updating/validation from case data, inducing reverse findings, processing of management scenarios as case data generated by other types of models, and rational model simplification through node absorption. I provide examples of each of these features.

8.2   Model construction

For the coast-wide policy assessment the modeling team first developed conceptual models as influence diagrams (Figure 8.1). We divided the analyses into three components:

1. a Habitat Quality Index BN applied at the forest site scale (1–1000 ha of relatively homogenous forest) to predict relative nesting probability or abundance;

2. a Nesting Capacity BN to estimate maximum nesting abundance at watershed and regional scales;

3. a Population Model BN (two versions) to estimate future population sizes and viability at the region and coast-wide scales. This phase went through multiple iterations until models acceptable to the team were in hand.

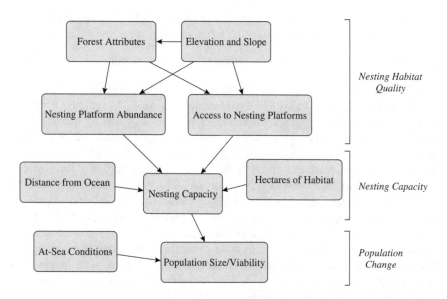

Figure 8.1 Preliminary influence diagram of the murrelet analysis. Starting from the bottom, risk to murrelet populations (*Population Size/Viability*) is a function of *At-Sea Conditions* (survival and breeding condition) interacting with on-shore *Nesting Capacity*, the number of nesting pairs the forest habitat can support. *Nesting Capacity* is in turn conditional on *Distance from Ocean*, *Hectares of Habitat*, *Nesting Platform Abundance* (occurrence of suitable nesting sites), and physical *Access to Nesting Platforms*, the latter two in-turn predicted by biophysical inputs such as *Forest Age* and *Forest Height*. The analyses was then broken into three separate components for developing detailed BNs; Nesting Habitat Quality, Nesting Capacity, and Population Change.

The team then held workshops with the CMMRT and other species experts to further revise the model structure and parameters. In the final stage, we conducted policy scenario experiments and formally reported the results [444].

For the northern mainland coast land-use planning application, I modified the coast-wide models which were then further reviewed by a technical committee, revised, and applied to land-use scenarios [443]. These results were incorporated into an overall environmental risk assessment presented to the land-use planning team.

We applied a variety of methods to build the node contingency probability tables (CPTs) in the models, including:

- expert opinion entered directly into CPTs or described by equations that were then converted into CPTs by Netica;

- probabilistic equations developed through external statistical analyses (e.g., regression equations and their standard errors);

- the use of case files (tab delimited text files) of field data that populate the CPTs based on observed state-value frequencies;

- the use of case files of results generated by external simulation modeling.

I illustrate the use of each of these approaches in the examples that follow.

The Habitat Quality Index (HQI) submodel (Figure 8.2) predicted a relative nesting habitat suitability index (representing likelihood of use by murrelets) for forest sites. The three key influences on nesting included abundance of potential nesting platforms (horizontal, large diameter, moss covered tree limbs), physical access to those platforms through the forest canopy (the upper layer of tree branches), elevation, and distance from the sea. The input nodes (those with no arrows leading to them or only from the *Biogeoclimatic Variant*[1] node) represent the biophysical inputs considered predictive of platform abundance and physical access, or are modifiers (i.e., *Distance from Sea* and *Elevation*) of over-all habitat quality. An arrow from one input node to another input node (e.g., forest *Age Class* → *Height Class*) indicates a correlation between them, with height class (the top height of the main forest canopy) in part conditional on forest age (the age of the trees dominating the main forest canopy).

If there were missing values for one variable, rather then assuming total ignorance the missing state value was inferred probabilistically based on those correlations. All the forest attribute input nodes were conditional on the *Biogeoclimatic Variant* which was also used as a predictor for elevation and distance from the sea if those parameters were not directly available.

Platform Abundance/ha is an example of an intermediate node which had the CPT populated directly from field data. Data sets from two areas on the coast relating observed platform abundance to forest characteristics were applied through

---

[1]Ecological land classification applied in British Columbia [314]. Variants are geographical units of similar climate.

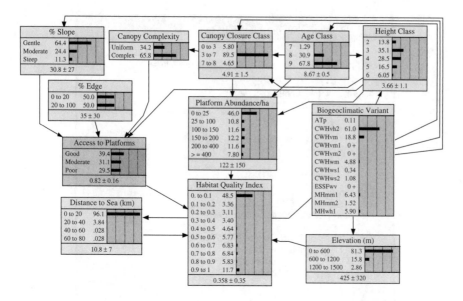

Figure 8.2 Habitat Quality Index BN. The probability distributions shown here are the starting conditions (circa 2001) for the North Coast planning land-use planning area [443].

a case file. An important issue was the weighting to apply to the case file vs. the prior probability of the node states, starting with equal probabilities representing complete ignorance. Netica prompts for a weighting to apply to the case file vs. prior probabilities. In this instance we chose to apply a very high weighting to the case file data (10^9) and a very low weighting to the uniform prior probabilities (10^{-9}), in effect causing the CPT to reflect only the case data. In contrast, the *Access to Platforms* node was parameterized using expert opinion entered directly in the CPT.

The final output node *Habitat Quality Index* was adapted from [32] as presented in [71, page 81]. We expressed the relationship as a probabilistic equation (Figure 8.3), modified based on expert opinion by physical access to platforms, distance to sea, and elevation.

For the spatially explicit application on the northern coast, each watershed was processed through the HQI model. Case files were first output from a spatially explicit landscape simulation model.[2] These files provided the number of hectares (the case weighting factor or NumCases field in Netica case file structure) by watershed, biogeoclimatic variant, forest age class, height class, and canopy closure class (the proportion of the land surface obscured by a vertical projection of the forest canopy). These files were then processed through the BN (process cases feature of Netica) creating an output case file of the expected value and standard deviation of HQI by watershed.

---

[2]Simulates landscape change through time, recording forest condition for each hectare of the landscape.

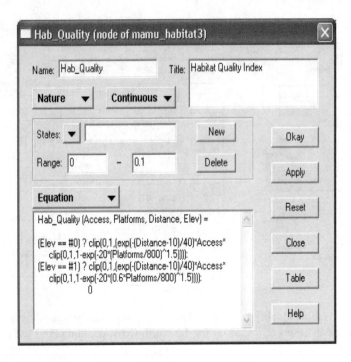

Figure 8.3   Node equations for Habitat Quality Index. Equations are similar to Java or C++ programming language format. HQI is the product of a negative exponential function of node *Distance to Sea* (equation variable Distance), linear function of node *Access to Platforms* (equation variable Access), and a Weibull function of node *Platform Abundance/ha* (equation variable Platforms). Platform density in the Weibull function declines with node *Elevation* (equation variable Elev), expressed as a 0.6 multiplier of platform abundance for *Elevation* state 1 (600–1200 m), and 0.8 for Elevation state 2 (1200 m +).

We then processed that case file through the Nesting Capacity BN (Figure 8.4). This model applied regression equations (including standard error of the estimates) to estimate the density of murrelets using each watershed (node *Commuting Birds/1000 ha (radar)*)[3] as a function of the expected value of HQI (*Mean Habitat Quality*) [443]. This was then multiplied by the proportion of birds that breed (*Proportion Nesting Females*) to estimate *Nesting Density*. Finally, nesting density was multiplied by both a weighting factor for watershed accessibility (existence of potential murrelet flight paths to sea, node *LU Accessible?*) and by the hectares of potential nesting habitat (*Total Hectares of Habitat*) to calculate the number of nesting pairs the watershed will support (*Nesting Capacity*).

---

[3]Murrelets can be counted using marine radar as they enter or exit valleys to attend the nest [72]. If the area accessed can be defined, an estimate of murrelet density can be inferred. Both sexes commute, and not all commuting birds are actually breeding.

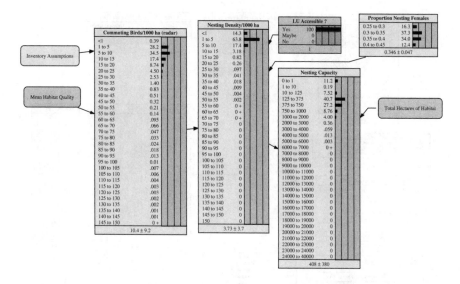

Figure 8.4   Nesting Capacity BN.

Alternative or competing opinions can be applied in the BNs to determine consistency of outcomes, or to allow explicit weighting of those competing views as a source of uncertainty. This was done in the northern coast analysis, where there was debate over the definition of potential habitat. One opinion was that only *Height Class* 4 or greater (≥ 28.5 m forest height) should be considered possible habitat. Evidence from elsewhere showed murrelets sometimes nested in lower stature forests, but there was no evidence directly from the study area to support or refute the notion. Rather then have a win–lose debate, the nesting carrying capacity BN included both suppositions (node *Inventory Assumptions*), equally weighted. As it turned out, this uncertainty made little difference to the relative ranking of the land-use scenarios.

We then used the nesting capacity estimates, along with murrelet survival and reproduction estimates, in the Population Model BNs to predict potential future population sizes and persistence probabilities. In [444] we populated the node CPTs in the population model BN (Figure 8.5) using results from an external population model (written in an Excel spreadsheet). In [445] we used a different approach to the population model (Figure 8.6). Rather than using an external Monte Carlo population simulation model and importing the results as a case file, we used analytical diffusion approximation equations to calculate persistence probability directly in the BN.

With the external population model approach we conducted 200 Monte Carlo population simulations for each combination of input node parameter values. The *Background Vital Rates* node represents combinations of at-sea demographic parameter probability distributions for juvenile, subadult, and adult survival, and

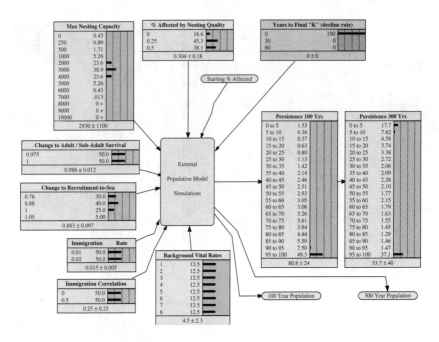

Figure 8.5   Population model BN of [444] parameterized using output from an external Monte Carlo model as case data.

proportion of females breeding (described in detail in [444]. Results were output as a case file with a record for:

1. each combination of input parameter values;

2. the resulting length of time 90% of the simulated populations persisted (the 90% persistence time truncated at 500 years, node *External Population Simulations*);

3. the associated population size and proportion (the NumCases field in Netica case file format) of simulations that persisted at 100 year intervals (in the BN we presented the 100 and 300 year results).

The analytical diffusion equations approach (Figure 8.6) conceptualized population trajectories as a stochastic process described by a mean (node *Lambda* converted in node *Mean r* to $\log_e(Lambda)$) and variance (node *Annual Variation (Vr)*) of annual population changes; the difference between annual birth and death rates [153, 215]. The variance represents environmental variability among years, modified (in node *Corrected Vr (Vrc)*) by temporal autocorrelation (node *Temporal Autocorrelation (p)*). It is this variability that makes the future outcome probabilistic (uncertain) rather than deterministic for any combination of input parameter values [58, 153]. The population trajectory is also constrained by a ceiling on the

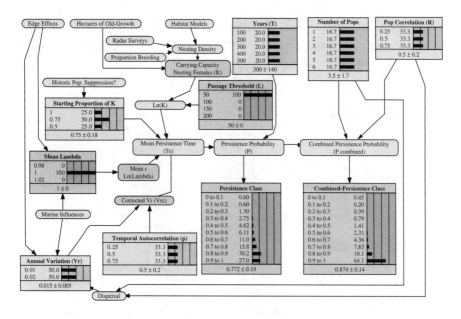

Figure 8.6 Diffusion analysis population model BN of [445].

maximum number of nesting pairs (nesting habitat capacity affected by land management policies), and a lower threshold (the extinction threshold) that acts as an absorbing boundary (i.e., if it hits the lower boundary it is deemed effectively extinct and cannot recover).

We also applied the bet hedging approach of [58] to combine estimates from multiple subpopulations into a coast-wide joint persistence probability (node *Combined Persistence Probability*, the probability that at least one subpopulation will persist). This approach views individual subpopulations as linked through correlated environmental variation (node *Pop Correlation (R)*) and dispersal. Greater correlation of environmental variation among regions decreases the combined persistence probability; i.e., the chance of all populations suffering the same fate simultaneously increases through coincident periods of poor survival or productivity at-sea, and through reduced compensating dispersal among populations.

An advantage of this analytical approximation approach was the ability to readily change parameter values directly in the BN, thus achieving immediate results. Especially when working in an interactive workshop setting, I found it highly advantageous to be able to modify the model and see the results almost immediately. With the external population simulation model approach, if a new parameter value was desired we had to rerun the external model for that value in combination with all the potential values of the other parameters. This resulted in delays of several days or longer to run the population simulations and the import the new results into the BN as a case file. A discussion of the merits of alternative population modeling approaches can be found in [445].

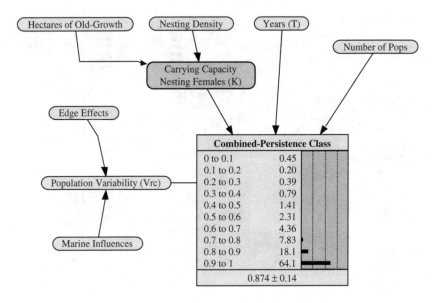

Figure 8.7 Simplified (using node absorption) version of the diffusion analysis BN of Figure 8.6.

Figure 8.7 shows a simplified summary version of the diffusion analysis PVA BN, created through node absorption. Node absorption allows the removal of nodes while maintaining the probabilistic relationship among the remaining nodes; the same results are obtained with findings applied in the absorbed model as in the full model. In this instance we removed most nodes that had static parameterization, represented intermediate calculations, or had little influence on the outcome. Users should be aware that equations are lost in this process, and if any subsequent changes are made to the node state values or CPTs, the resulting model will no longer represent the original model. The main use of node absorption was to reduce a visually complex model to a more comprehensible presentation, and in some cases to run faster when processing case files.

8.3 Model calibration, validation and use

8.3.1 Calibration and validation

Some relationships among nodes in our models were directly calibrated using field data. For example, between the expected value of Habitat Quality Index and the abundance of murrelets using a watershed was statistically calibrated in the Nesting Capacity BN using regression analyses external to the BN [443]. This relationship, including the estimated error around those predictions, was expressed as equations converted to CPTs by Netica.

Model validation requires testing the calibrated model against independent data, often followed by model updating [300, 301]. Burger et al. [73] conducted an independent aerial assessment by helicopter of 100 forest inventory polygons in the northern mainland coast study area. Experienced observers visually rated the quality of the habitat using a standardized protocol based on structural attributes such as average tree height and vertical complexity (variability of tree heights), and the number of trees with features suitable as potential nesting platforms. They gave the sites a ranking of from 1 to 6, with 1 being the poorest habitat and 6 being the best. I processed the forest attributes of 94 sites[4] through the HQI and Nesting Capacity BNs, comparing the predicted nesting density (expected value) of murrelets (node *Nesting Density/1000 ha*) with their helicopter-based habitat rankings (Figure 8.8).

There was a general relationship of increasing mean predicted abundance with helicopter-survey ranking, although ranks 3 and 4 were similar in predicted density, as were ranks 5 and 6.

I then added the helicopter ranking as an output node (*Helicopter Ranking*, Figure 8.13 below) in the Habitat Quality BN, conditional on *Habitat Quality Index*. To do this I processed each helicopter site through the Habitat Quality BN, outputting the probability of each *Habitat Quality Index* state (the belief vector) for each site. I then used that output to construct the CPT for *Helicopter Ranking* using the normalized mean probability of helicopter ranking class by HQI class across the 94 sites.

Direct validation could not be done for predictions of future population size or persistence, for management policies not yet implemented. In such situations, careful crafting of the models, peer review, and ensuring that a suitable range of

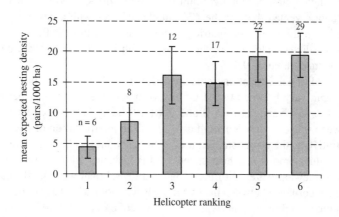

Figure 8.8   Comparison of the BN predicted murrelet nesting abundance (mean expected value and standard error) with independent helicopter-based habitat rankings of 94 forest sites in the North Coast study area.

---

[4]The remaining six sites did not have forest attribute data.

plausible parameter values and model structure are applied to capture uncertainty, was particularly important. I could, however, examine the degree of consistency between the two PVA analyses (the external Monte Carlo model approach and the diffusion equations approach). I generated 1000 random cases (using the Simulate Cases feature in Netica) from the Monte Carlo BN for 100-year persistence probability. That case file was then applied as test data to both that originating BN and the diffusion approximation BN (for 100-year persistence probability, to be consistent). The results in the form of confusion matrices are shown in Figure 8.9.

When the simulated files were tested through the BN that originated them there was a 57% error rate (the proportion of cases whose predicted value differed from the actual value). At first consideration this seemed odd; how could there be a classification error with the model that generated the test cases? This reflects the high uncertainty of persistence probability among individual population projections. The simulated cases represent individual population projections drawn from the probability distribution of all outcomes of the originating BN, each with a single actual persistence probability state value. The predicted values, in contrast, represent the most probable value of the persistence probability node given the combination of input node values associated with each simulated case.

When the diffusion BN was tested with the same simulated case file, the error rate was nearly the same (slightly higher). The error rate, however, doesn't distinguish between different degrees of error; either the actual and predicted values agree or they don't. Although both models had essentially the same error rate, there was a substantive qualitative difference between the two confusion matrices. With the diffusion BN, many cases that were in the lowest predicted persistence class (first column of confusion matrix) for the Monte Carlo BN, moved to the second highest predicted class (second column from the right). Closer examination of the case file data revealed that for the 476 simulated cases with a nesting capacity of 2000 or higher, the confusion matrices were identical with both BNs (error rate of 37%); the predicted value for all those cases, with both models, was a persistence probability state value of 0.9–1.

Three additional validation scores are shown in Figure 8.9. The Logarithmic loss score can vary between 0 and infinity with zero the best model performance; quadratic loss can vary between 0 and 2 with 0 being best; spherical payoff can vary between 0 and 1 with 1 being best [325]. The interpretation of these scores is not as intuitive as the error rate, but they better account for the degree of spread of predicted values among alternate states. In this instance, all three scores were (as expected) somewhat better with the Monte Carlo BN than with the diffusion BN.

Close concordance between the two models was not particularly expected, as the demographic assumptions (survival and reproductive rates) differed between them. In the first analysis, several scenarios assumed that the mean rate of population growth over 100 years could be either negative or positive, whereas in the second analysis we set long-term mean growth rate to zero, and varied the annual variation and temporal autocorrelation of the growth rate, to induce varying amplitude and duration of population decline. The result was that, in contrast to the diffusion

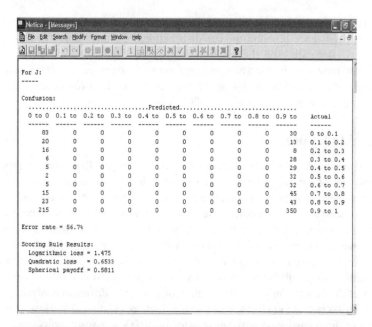

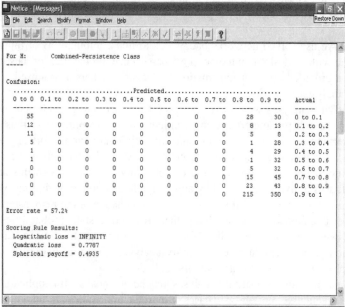

Figure 8.9   Confusion matrices and scoring rules results (as shown in Netica) for simulated cases tested through the Monte Carlo population model BN (upper) and diffusion model BN (lower). The Actual column is the simulated cases; Predicted is the most probable state value prediction from processing the cases through the model.

analysis assessment, at lower carrying capacities the Monte Carlo analysis created bi-modal persistence probability state distributions of either very low or very high persistence, depending on the at-sea demographics scenario. Both analyses were consistent, however, in the relative improvement in persistence with increasing carrying capacity and as noted above the results of the two models converged above a nesting carrying capacity of 2000 pairs.

8.3.2   Model use

We used the models in both reactive and proactive ways. Reactive use processed externally proposed policy or land-use scenarios through the BNs and presented the outcomes (nesting density and carrying capacity, median population size through time, and persistence probability). Because there was voluminous output (a probability distribution for each node of interest for each combination of input node values), we usually presented the expected value (the mean, or probability-weighted state value) of output nodes in map or graphical form (Figures 8.10 and 8.11). New scenarios were then proposed and similarly processed in an iterative manner. Results of the Monte Carlo based modeling of [444], the diffusion analysis of [445] and a multi-vertebrate analysis of minimum viable populations [390] were in broad agreement that coast-wide population sizes of 12 000+ nesting pairs are resilient (show high intrinsic tendency to persist) regardless of parameterization or model structure.

Proactive use solved networks backwards to find combinations of input values consistent with a desired outcome (i.e., persistence probability or nesting density). This provided up-front information to decision makers as opposed to strictly reactive analysis.

Sensitivity analyses were also very useful for examining model behavior and identifying which parameters are most influential on outcomes, and thus perhaps most deserving of management attention, calibration/validation efforts, or research. We applied the variance reduction measure [303, Appendix B] to examine the influence of individual nodes on persistence probability estimates. With this approach each node was individually varied by applying a large number of findings weighted according to the state probabilities and measuring change in the response node. All other nodes remained static, applying their respective state probabilities. Alternatively, we could have set beliefs for some or all nodes to some other desired distribution prior to conducting sensitivity analyses. Uniform distributions would have implied all state values are equally realistic.

In some situations, sensitivity analyses may be the most useful application, especially for speculative models or in cases of extreme uncertainty. In our example, we could seek on-shore nesting habitat policies that reduce *sensitivity* of persistence to conditions at sea (Figure 8.12) rather than focusing on the persistence probability estimates themselves that had high uncertainty (represented by the probability distribution among state values).

Proactive use proposes management solutions consistent with desired outcomes rather than assessing externally proposed management policies. When a finding

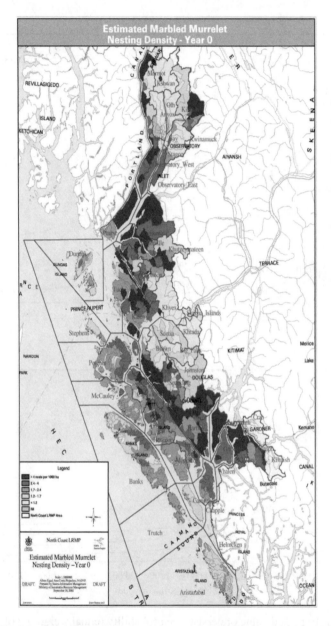

Figure 8.10 Map of the starting condition (circa 2000) for murrelet nesting density in the North Coast Land and Resource Management Plan area. The darkest shaded watersheds account for the highest density 20% of the landscape as predicted by the nesting density BN, the lightest shading the lowest density 20%. This map helped stratify the plan area in terms of importance as nesting habitat. Similar maps were produced for alternate future landscape scenarios.

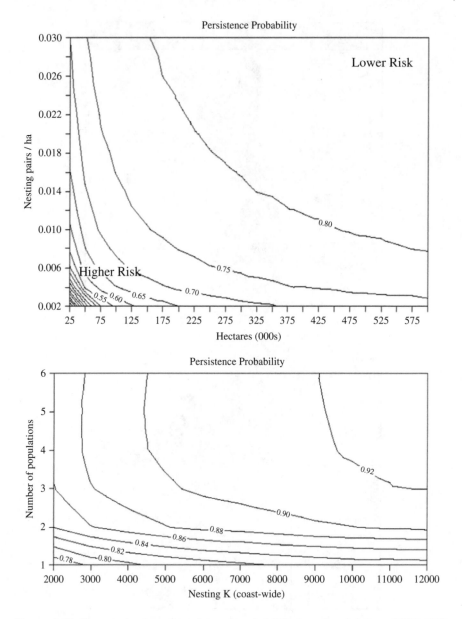

Figure 8.11 Expected value of persistence probability (equal weighting of 100–500 year time-frames) as a function of amount of habitat and nesting density for a single population (upper) and multiple populations (lower) for the diffusion analysis BN. Risk decreases from lower left to upper right with each contour representing an improvement in expected value of persistence of 0.05. Intervals between contours represent combinations of hectares and nesting density with similar persistence probability. Figures are adapted from [445].

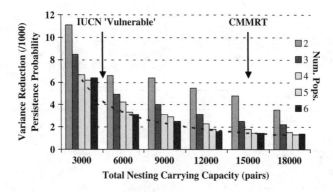

Figure 8.12 Sensitivity of persistence probability to the weighting of at-sea demographic parameters (for the diffusion PVA BN) as a function of total nesting capacity and number of subpopulations. Dividing the population among six subpopulations was clearly more efficient at reducing sensitivity to at-sea demographic uncertainty. With 4+ populations there appears to be minimal incremental benefit above about 12 000 pairs. The dashed line is a fitted power function for six subpopulations. The two vertical arrows represent the IUCN vulnerable threshold for population size (10 000 mature individuals or 5000 pairs), and interim advice of the Canadian Marbled Murrelet Recovery Team (CMMRT, 70% of 2001 estimated population). This was conducted using the Sensitivity to Findings feature in Netica.

is entered in an output node the beliefs (the marginal posterior probabilities) of intermediate and input nodes are automatically updated to be consistent with the finding. I illustrate this in Figure 8.13, where my purpose was to pose a feasible combination of input conditions consistent with a nesting capacity of 3500 pairs for total habitat amounts of either 400 000 or 100 000 hectares. These solutions could then be posed as potential management scenarios/targets in the northern coast spatial-temporal landscape model.

It is important to examine the proposed solutions closely. With the 100 000 ha scenario (Figure 8.13, bottom) we can see that the objective was only met if higher quality forest is retained (i.e., *Platform Density* and *Helicopter Ranking* nodes are weighted to higher value states). Height Class 3 forest was abundant in the circa 2000 landscape (Figure 8.2), but only 53% was predicted to be in the top four platform density classes and thus potentially suitable for inclusion in the 100 000 ha scenario. Confirmation of the suitability of retained areas would be important to achieving the objective, either using ground-based sampling of platform abundance and/or the helicopter-based habitat ranking assessment method. For the 400 000 ha scenario, the range of platform densities and helicopter-based rankings consistent with the outcome were wider, suggesting less importance (and cost) for field confirmation.

Netica also has a Most Probable Explanation (MPE) function that determines the relative likelihood for node states consistent with the chosen outcome. This

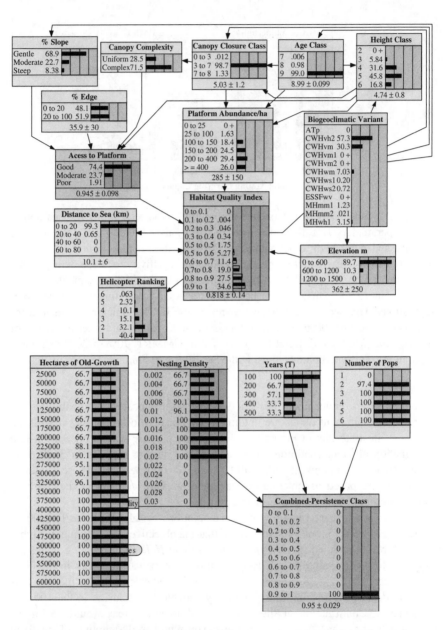

Figure 8.13 Example scenarios for 400 000 ha (top) and 100 000 ha (bottom) consistent with a nesting capacity of 3500 nesting pairs. Findings consistent with the objective, from the Carrying Capacity model, were entered in the Habitat Quality Index node using the node Calibration feature of Netica. Not surprisingly, with fewer hectares of habitat there was a necessary compensatory shift to higher quality habitat.

approach indicates the case state (or states) for each node that has the greatest likelihood (shown in Netica with a probability bar of 100%) given findings entered elsewhere in the network. Other state values have a smaller relative likelihood. There is often more than one state value in a given node with a 100% probability, indicating an ambiguous answer dependent on findings at other nodes, i.e., there is more then one combination consistent with the chosen outcome. By sequentially selecting 100% state-value findings in nodes of interest, one can explore possible policy combinations.

I illustrate MPE for the diffusion PVA BN in Figure 8.14 where I sought a policy (hectares of habitat, nesting density representing habitat quality, and number

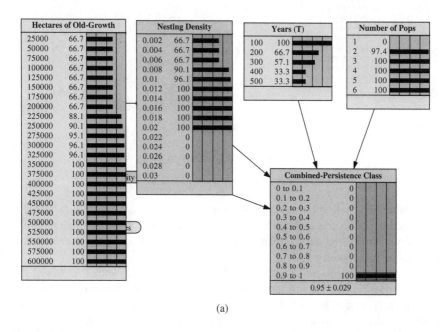

(a)

Figure 8.14 Two Most Probable Explanation scenarios consistent with a persistence probability ≥0.9. In both scenarios the at-sea demographics were assumed worst-case, namely high inter-year variability and temporal autocorrelation of the population change rate. Nesting densities >0.02/ha were given a likelihood of zero. In (a) to meet the IUCN criteria for avoiding Vulnerable designation (≥90% probability of persistence over 100 years) the results suggest that ≥3 subpopulations of ≥350 000 ha supporting an average nesting density of ≥0.012/ha would be required to maximise the likelihood of achieving the desired outcome. Also, we can see that high persistence is less certain with increasing time scale (Years). In (b) the options are narrowed by conservatively setting the time scale to 500 years. A minimum of five subpopulations of 200 000 ha in each would be required to have any chance of meeting the objective, with >300 000 ha (depending on habitat quality) required to maximise the likelihood of success.

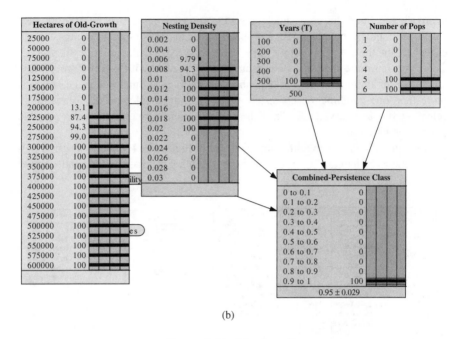

(b)

Figure 8.14   (*continued*)

of populations) that would most likely achieve a persistence probability ≥90%.[5] I conservatively set the at-sea conditions to the worst case scenario (high environmental variability and autocorrelation) to bound the worst case as a starting point towards setting a policy.

The MPE scenario was highly ambiguous (Figure 8.14a) until I narrowed the range of possibilities for several influential nodes. In Figure 8.14b I assumed that average nesting density will be less than or equal to half of recent coast-wide estimates (due to lower economic cost of preserving lower quality habitat) and we set the time frame to 500 years. The results narrowed such that at least 200 000 ha in each of five subpopulations was required, most likely ≥300 000 ha in each. Note that the MPE solution does not necessarily mean that high persistence probability is assured, but rather indicates the state values most consistent with the desired outcome.

Selecting a policy requires subjective choices of state values among the interacting nodes. There were multiple combinations of hectares of habitat, quality of habitat, and number of subpopulations consistent with the same outcome.

---

[5]The International Union for the Conservation of Nature (IUCN) threshold for Vulnerable status is <90% persistence probability over 100 years. However, they do not provide a definition for Least Concern. Persistence probability declines exponentially with the length of the assessment period, thus 90% persistence probability over 100 years implies only 59% persistence probability over 500 years $(0.9^{500/100})$.

8.4   Conclusions/perspectives

I have found BNs to be a useful tool in exploring and presenting ecological issues. They are generally intuitive to build, apply and explain; the ability to visually display and rapidly manipulate models is a very useful property of BN shells such as Netica. As with any modeling technique and associated software, however, there is a substantive learning curve to becoming proficient.

Many of the issues we encountered are common to modeling in general, centering on the credibility of the models rather than technical issues of BN construction. While the graphical representation of BNs is more intuitive than many other multivariate approaches, this does not necessarily mean they are credible with stakeholders. A quote from Varis and Kuikka, authors who have applied BNs in many fisheries related and other conservation problems, resonates with me: 'It is not enough that one has learned and applied a methodology; it has also to be comprehended and accepted by many others who often are not all that devoted to methodological challenges; and launched to responsible institutions' [473]. The model development guidelines of [303] are crucial in this regard: careful framing of objectives including upfront involvement of the targeted users of the results, involving influential peers in model development so they have ownership of the results, and formal peer review.

We also began to develop a formal decision model incorporating utility nodes [444], but found there was little appetite for this approach among stakeholders at the time. I think this stemmed, in part, from a reluctance to express utilities in advance of seeing outcomes, and then having to live with the results. This was perhaps perceived as ceding too much authority to the model and, by extension, to the modelers. It may also reflect the fact that most stakeholders had little or no training and experience with formal decision science. Despite those impediments, I think formal decision models could play a useful role in many applications [343] and is an area I would like to pursue further.

Some limitations we encountered specific to building BNs included the limitation to acyclic models (i.e., no feed-back loops), and inability to cross spatial/temporal scales within the same model. For simple models the former limitation can be overcome in Netica using the time-delay links feature, but in our experience this became unwieldy for practical application. Another solution is exporting the probability outcomes of a network as a case file and using those as input to the same BN in an iterative manner.

The spatial-scale limitation stems from the basic underlying concept of BNs representing cases or entities with probability distributions of state values for each node. Crossing scales (e.g., forest sites to watersheds) within a BN would make the case identity ambiguous. In multi-scale analyses, separate BNs are usually constructed for each spatial scale, with the results at one scale used as input to another scale [302]. For our northern mainland coast application, the HQI case scale was 1-ha raster cells output from a spatial simulation model. Each watershed was processed through the HQI BN with the expected value for HQI and watershed ID exported as an output file. These were then input as cases to the Nesting Capacity BN to estimate the nesting capacity for the plan area (expected value and standard deviation of the Nesting Capacity node). One potential limitation this may impose is applying decision models where the result at one scale may affect decision making at another scale, which cannot be done simultaneously in a single BN.

There is a significant learning curve for those unfamiliar with probabilistic reasoning. This may be worsened by cognitive factors, such as the difficulty many people have understanding probability [15]. These problems can be minimized through appropriate presentation (such as percentages rather than probabilities) and elicitation of expert belief weightings [15, 301]. We did not fully recognize these factors in our model building process, which may have hindered acceptance of results by some stakeholders.

While ours is the first BN application I am aware of that combined spatial modeling output, habitat assessment models, and quantitative population models, BN's have seen increasing application in environmental assessment and resource management, predicting habitat supply, species distributions, population modeling, and PVA. [472] provides a summary of the basic methodology and early history of application. [302] presented practical model building advice that greatly influenced our adopting the approach.

Raphael et al. [383] and Marcot et al. [301] are examples of expert-opinion driven PVA and species conservation status assessments, while McNay et al. [311] is an example of combining empirical data and expert models for assessing caribou habitat and populations, and possible forest management and predation impacts. Marcot et al. [301] is a good example of testing and updating expert models (predicting species occurrence) with field data. Lee and Rieman [276] used BNs to examine viability of salmon populations in the northwest USA, and Schnute et al. [417] used a BN-type approach to assess management options for Fraser River (British Columbia) salmon stocks, directly incorporating stock recruitment data. Reckhow [387] applied BNs to prediction of water quality, and gives a good discussion of their construction. Newton et al. [338] is an interesting example of ecological models and socio-economics combined in a BN approach, examining the impacts of the commercialisation of non-timber forest products on local livelihoods in tropical forests of Mexico and Bolivia.

Acknowledgment

I would like to acknowledge the contributions of Glenn Sutherland and Peter Arcese in development of the population models. Glenn Sutherland and Gina LaHaye provided editorial reviews of earlier drafts. The marbled murrelet research community and members of the Canadian Marbled Murrelet Recovery Team (CMMRT) provided much of the data, literature, and concepts drawn on in building the models. Bruce Marcot provided comment and inspiration in the use of BNs, and editorial review. The results and opinions expressed here, however, are solely my responsibility and do not necessary reflect the policy or opinions of others, including the members of the CMMRT or my employer (the government of the Province of British Columbia).

9

Classifiers for modeling of mineral potential

Alok Porwal

Center for Exploration Targeting, University of Western Australia, Crawley, WA and Department of Mines and Geology, Rajasthan, Udaipur, India

and

E.J.M. Carranza

International Institute for Geo-information Science and Earth Observation (ITC), Enschede, The Netherlands

9.1 Mineral potential mapping

9.1.1 Context

Classification and allocation of land-use is a major policy objective in most countries. Such an undertaking, however, in the face of competing demands from different stakeholders, requires reliable information on resources potential. This type of information enables policy decision-makers to estimate socio-economic benefits from different possible land-use types and then to allocate most suitable land-use. The potential for several types of resources occurring on the earth's surface (e.g., forest, soil, etc.) is generally easier to determine than those occurring in the subsurface (e.g., mineral deposits, etc.). In many situations, therefore, information on potential for subsurface occurring resources is not among the inputs to land-use decision-making

Bayesian Networks: A Practical Guide to Applications   Edited by O. Pourret, P. Naïm, B. Marcot
© 2008 John Wiley & Sons, Ltd

[85]. Consequently, many potentially mineralized lands are alienated usually to, say, further exploration and exploitation of mineral deposits.

Areas with mineral potential are characterized by geological features associated genetically and spatially with the type of mineral deposits sought. The term 'mineral deposits' means accumulations or concentrations of one or more useful naturally occurring substances, which are otherwise usually distributed sparsely in the earth's crust. The term 'mineralization' refers to collective geological processes that result in formation of mineral deposits. The term 'mineral potential' describes the probability or favorability for occurrence of mineral deposits or mineralization. The geological features characteristic of mineralized land, which are called *recognition criteria*, are spatial objects indicative of or produced by individual geological processes that acted together to form mineral deposits. Recognition criteria are sometimes directly observable; more often, their presence is inferred from one or more geographically referenced (or spatial) datasets, which are processed and analyzed appropriately to enhance, extract, and represent the recognition criteria as spatial evidence or predictor maps. Mineral potential mapping then involves integration of predictor maps in order to classify areas of unique combinations of spatial predictor patterns, called *unique conditions* [51] as either barren or mineralized with respect to the mineral deposit-type sought.

9.1.2   Historical perspective

Methods for mineral potential mapping, based on the Bayesian probability concept, exist. The PROSPECTOR expert system, which was developed by the Stanford Research Institute, uses a series of Bayesian inference networks for evaluating mineral deposits [140]. In PROSPECTOR, pieces of evidential information are propagated in a network by application of fuzzy Boolean operators for Bayesian updating of prior to posterior probability. Originally, spatial data were not supported by the PROSPECTOR, but it was later modified to do so. It has been demonstrated by Duda et al. [140] to predict occurrence of the Island copper deposit (British Columbia, Canada) and by Campbell et al. [81] to map potential for molybdenum deposits in the Mt. Tolman area (Washington State, USA). Implementations of the PROSPECTOR system using a geographic information system (GIS) were demonstrated by Katz [243] and Reddy et al. [388]. A more popular GIS-based technique for mineral potential mapping, based on Bayes' rule, is weights-of-evidence or WofE method [3, 52]. It was developed to make use of spatial exploration datasets to derive posterior probability of mineral occurrence in every unit cell of a study area. The WofE method has been applied by many workers, e.g., by Carranza and Hale [84] to map potential for gold deposits in Baguio district (Philippines), and by Porwal et al. [370] to map potential for base-metal deposits in Aravalli province (India). In WofE, prior probability of mineral occurrence is updated to posterior probability using Bayes' rule in a log-linear form under assumption of conditional independence (CI) among predictor patterns with respect to known occurrences of a target mineral deposit-type. It should be noted that although the PROSPECTOR and WofE are based on the Bayesian probability concept, they are not examples of Bayesian networks.

9.1.3   Bayesian network classifiers

The objective of this chapter is to explain and demonstrate applications of Bayesian networks as another tool for mapping and classifying potential for mineral deposit occurrence. In application of Bayesian network to mineral potential mapping, each unit cell of land is evaluated as barren or mineralized, in a continuous scale of [0,1] interval (i.e., completely barren = 0; completely mineralized = 1), with respect to occurrence of target mineral deposit-type based on a number of predictor patterns. These predictor patterns are spatial attributes representing the recognition criteria, and each unique condition is considered a feature vector containing unique instances of such spatial attributes. A Bayesian network classifier is trained with a set of a priori or pre-classified feature vectors (i.e., unique conditions that are associated with known mineralized unit cells and known barren cells), and the trained classifier is then used to process all feature vectors. The classification output is an estimate of mineral occurrence for each unit cell in the [0,1] interval, which then can be mapped either as continuous or binary variable, using a geographic information system (GIS) (see [51]), to portray spatial distribution of potentially mineralized land.

A simple Bayesian network classifier that can be used in mineral potential mapping is naïve classifier [139, 264]. This classifier assumes complete CI among predictor patterns with respect to target pattern, which is unrealistic for geological features associated with mineralization. A naïve classifier performs well, however, in several other application domains [132, 160]. It bas been shown, nonetheless, that the CI assumption can be relaxed in using a naïve classifier [161]. Several other Bayesian network classifiers unrestricted by the CI assumption are described in the literature: semi-naïve classifier [254]; multinet classifier [178, 201]; tree-augmented naïve classifier [161]; and augmented naïve classifier [161]. If the CI assumption must be obeyed strictly, then a selective naïve classifier could be used [265]. For mineral potential mapping, we describe below algorithms for three Bayesian network classifiers: (1) naïve classifier; (2) augmented naïve classifier; and (3) selective naïve classifier. These Bayesian network classifiers are demonstrated to map regional-scale base metal potential in part of the Aravalli province in western India.

9.2   Classifiers for mineral potential mapping

Consider a Bayesian network $B = \langle G, \Theta \rangle$ in which D is at the root, i.e., $\Pi_D = \emptyset$ (the notation Π denotes here the set of parents of a node) and every predictor has D as its one, and only one, parent, i.e., $\Pi_i = D$ and $\Pi_{i\bar{D}} = \emptyset$. The joint probability distribution of B is given by

$$\{\alpha * \mathbb{P}(D) * \prod_{i=1}^{I} \mathbb{P}(P_I|D)\}, \tag{9.1}$$

where α is a normalizing constant. This is a naïve classifier as defined for example in [264].

Let $\mathbf{f_m} = [p_{1j}, p_{2j}, \ldots, p_{Ij}]$ be an I-dimensional feature vector to be classified as either d_0 (barren) or d_1 (mineralised). A naïve classifier estimates the posterior probabilities of d_0 and d_1 for $\mathbf{f_m}$ by a sequential updating for every predictor:

$$\mathbb{P}(d_1|p_{1j}, p_{2j}..p_{Ij}) = \frac{\mathbb{P}(d_1|p_{1j}, p_{2j}\cdots p_{(I-1)j})\mathbb{P}(p_{Ij}|d_1)}{\mathbb{P}(d_1|p_{1j}\cdots p_{(I-1)j})\mathbb{P}(p_{Ij}|d_1)+\mathbb{P}(d_0|p_{1j}\cdots p_{(I-1)j})\mathbb{P}(p_{Ij}|d_0)}. \tag{9.2}$$

If $\mathbb{P}(d_1|p_{1j}, p_{2j}..p_{Ij})$ is greater than 0.5, $\mathbf{f_m}$ is classified as d_1, otherwise as d_0.

Augmented naïve classifiers are obtained from naïve classifiers by relaxing the restriction that every predictor can have the target mineral deposit-type as the one, and only one, parent, i.e., $\Pi_{i\bar{D}}$ need not necessarily be a null set. An augmented naïve classifier estimates the posterior probabilities of d_0 and d_1 for $\mathbf{f_m}$ using an updating procedure similar to the one used by a naïve classifier. However, while updating the probability for every P_i, an augmented naïve classifier also takes $\Pi_{i\bar{D}}$ into account:

$$\mathbb{P}(d_1|p_{1j}, p_{2j}..p_{Ij}) = \frac{\mathbb{P}(d_1|p_{1j}, p_{2j}\cdots p_{(I-1)j})\mathbb{P}(p_{Ij}|\Pi_{i\bar{D}}, d_1)}{\mathbb{P}(d_1|p_{1j}\cdots p_{(I-1)j})\mathbb{P}(p_{Ij}|\Pi_{i\bar{D}}, d_1)+\mathbb{P}(d_0|p_{1j}\cdots p_{(I-1)j})\mathbb{P}(p_{Ij}|\Pi_{i\bar{D}}, d_0)}. \tag{9.3}$$

Selective naïve classifiers are obtained by removing conditionally dependent predictors from a naïve classifier. They have an identical functional form as naïve classifiers. The above relations can be easily expanded for multi-state class variables. Bayesian network classifiers can therefore be applied to any generalized classification problem in earth sciences.

9.2.1   Training of Bayesian network classifiers

The training of B involves estimation of the parameters Θ and the DAG G that together provide the best approximation of conditional dependencies in $\mathbf{U}^*$. Obviously, naïve and selective naïve classifiers are special cases of augmented naïve classifiers when G is predefined and only Θ is required to be estimated.

9.2.1.1   Estimation of parameters

Consider the augmented naïve classifier described above. Assuming that G is given, Θ can be decomposed into $\{\Theta_i\}$, where $\Theta_i = \{\Theta_{i1},.., \Theta_{iK}\}$ is the set of parameters containing the conditional probability distribution of $P_i|\Pi_i$, and estimated using conjugate analysis [380]. Because $\Pi_i = \{D\}$ in the case of naïve and selective classifiers and $\Pi_i = \{D, \Pi_{i\bar{D}}\}$ in the case of augmented naïve classifiers, these probabilities can be directly used in Equations (9.2) and (9.3), respectively, to estimate the posterior probabilities of d_0 and d_1.

Let $\mathbf{T} = |\mathbf{t_1}, \ldots, \mathbf{t_M}|$ be a set of M $(I + 1)$-dimensional training vectors. Let $\Theta_{ik} = [\theta_{i1k},.., \theta_{iJk}]$ be the parameter vector containing conditional probability distribution of $P_i|\pi_{ik}$ and $\theta_{ijk} = \mathbb{P}(p_{ij}|\pi_{ik})$ be the conditional probability of $p_{ij}|\pi_{ik}$. Let $n(p_{ij}|\pi_{ik})$ be the frequency of pairs $p_{ij}|\pi_{ik}$ and $n(\pi_{ik}) = \sum_{j=1}^{J} n(p_{ij}|\pi_{ik})$ be the frequency of π_{ik} in $\mathbf{T}$. Assuming that Θ_{ik} and $\Theta_{i'k}$ are independent $\forall i \neq i'$ (*global independence*) and Θ_{ik} and $\Theta_{ik'}$ are independent $\forall k \neq k'$ (*local independence*), the prior probability of $p_{ij}|\pi_{ik}$ can be estimated as prior expectation of $\theta_{ijk}|\mathbf{T_0}$:

$$E(\theta_{ijk}|\mathbf{T_0}) = \mathbb{P}\left(p_{ij}|\pi_{ik}\right) = \frac{\alpha_{ijk}}{\alpha_{ik}},$$

where $\mathbf{T_0}$ symbolizes 'prior to seeing the training set $\mathbf{T}$', and the posterior probability of $(p_{ij}|\pi_{ik})$ can be estimated as posterior expectation of $\theta_{ijk}|\mathbf{T}$:

$$E(\theta_{ijk}|\mathbf{T}) = \mathbb{P}\left(p_{ij}|\pi_{ik}\right) = \frac{\alpha_{ijk} + n(p_{ij}|\pi_{ik})}{\alpha_{ik} + n(\pi_{ik})}, \qquad (9.4)$$

where α_{ijk} and α_{ik} are the prior hyper-parameters and local prior precisions, respectively, that encode the modeler's prior belief and are interpreted as frequencies of real or imaginary instances of $p_{ij}|\pi_{ik}$ the modeler has seen prior to the training set $\mathbf{T}$ [161]. These are estimated using the following equations:

$$\alpha_{ijk} = \frac{\alpha}{J * K}, \qquad (9.5)$$

$$\alpha_{ik} = \sum_{j=1}^{J} \alpha_{ijk}, \qquad (9.6)$$

where J is the total number of states of the predictor P_i, K is the total number of parents in Π_i and α is the global prior precision encoding the certainty in the prior belief. Friedman et al. [161] describe criteria for selecting the value of global prior precision. In order to avoid bias due to the prior precision, a value much smaller than the number of training occurrences should be used (a global prior precision of 1 is a reasonable starting point).

Thus the information conveyed by $\mathbf{T}$ is captured by a simple update of the prior hyper-parameters α_{ijk} by adding the frequency of the pairs (p_{ijk}, π_{ik}) in $\mathbf{T}$. Consequently, $\mathbb{P}\left(p_{ij}|\pi_{ik}\right)$ can be estimated directly from a contingency table of frequencies of child-parent dependencies (for example, Table 9.1) using the following algorithm.

Algorithm 1

1. Based on the confidence in the prior belief, select a value of the global prior precision (α).

2. Given G and $\mathbf{T}$, construct a contingency table for P_1 by collecting the frequency distribution of the child−parent dependencies.

3. Calculate prior hyper-parameters (α_{1jk}) using Equation (9.5).

4. Substitute every $[n(p_{1j}|\pi_{1k})]$ by $[\alpha_{1jk} + n(p_{1j}|\pi_{1k})]$ and recalculate marginal row totals.

5. Divide every $[\alpha_{1jk} + n(p_{1j}|\pi_{1k})]$ by the corresponding marginal row total. Substitute every $[\alpha_{1jk} + n(p_{1j}|\pi_{1k})]$ by the result to obtain Θ_i.

6. Repeat Steps 2−5 for every predictor P_i ($i = 2$ to I) to obtain Θ.

Table 9.1   Contingency table.

Π_i	P_i					Marginal row total
	p_{i1}	$\cdots$	p_{ij}	$\cdots$	p_{iJ}	
π_{i1}	$n(p_{i1}|\pi_{i1})$	$\cdots$	$n(p_{ij}|\pi_{i1})$	$\cdots$	$n(p_{iJ}|\pi_{i1})$	$n(\pi_{i1})$
$\vdots$			$\vdots$		$\vdots$	$\vdots$
π_{ik}	$n(p_{i1}|\pi_{ik})$	$\cdots$	$n(p_{ij}|\pi_{ik})$	$\cdots$	$n(p_{iJ}|\pi_{ik})$	$n(\pi_{ik})$
$\vdots$			$\vdots$		$\vdots$	$\vdots$
π_{iK}	$n(p_{i1}|\pi_{iK})$	$\cdots$	$n(p_{ij}|\pi_{iK})$	$\cdots$	$n(p_{iJ}|\pi_{iK})$	$n(\pi_{iK})$

9.2.1.2   Estimation of DAG

In the case of naïve and selective naïve classifiers, the DAG G is completely predefined, i.e., it is known that (a) $\Pi_D = \emptyset$ and (b) $\Pi_i = \{D\}$. In the case of augmented naïve classifiers, G is only partially predefined, i.e., it is known $\Pi_D = \emptyset$ and $\{D\} \in \Pi_i$, the members of $\Pi_{i\bar{D}}$ are not known.

In order to select the most probable DAG, it is sufficient to estimate and compare the marginal likelihood of all DAGs that model all possible dependencies in an augmented naïve classifier [380]. The marginal likelihood of the DAG G_g ($g = 0$ to G, where G is the total number of possible DAGs) is estimated (see also [103]) as:

$$\mathbb{P}\left(\mathbf{T}|G_g\right) = \prod_{i=1}^{I}\prod_{k=1}^{K}\frac{\Gamma(\alpha_{ik})}{\Gamma(\alpha_{ik} + n(\pi_{ik}))}\prod_{j=1}^{J}\frac{\Gamma(\alpha_{ijk} + n(p_{ij}|\pi_{ik}))}{\Gamma(\alpha_{ijk})},$$

where $\Gamma(\cdot)$ is the Gamma function [498]. The marginal likelihood of G_g can be decomposed into local marginal likelihood $\mathbf{g}(P_i, \Pi_i)$ of the predictor P_i given Π_i in G_g:

$$\mathbf{g}(P_i, \Pi_i) = \prod_{k=1}^{K}\frac{\Gamma(\alpha_{ik})}{\Gamma(\alpha_{ik} + n(\pi_{ik}))}\prod_{j=1}^{J}\frac{\Gamma(\alpha_{ijk} + n(p_{ij}|\pi_{ik}))}{\Gamma(\alpha_{ijk})}. \tag{9.7}$$

Because the value of $\mathbf{g}(P_i, \Pi_i)$ is very small, its natural logarithm can be used:

$$\ln[\mathbf{g}(P_i, \Pi_i)] = \left\{\sum_{k=1}^{K}\ln[\Gamma(\alpha_{ik})] + \sum_{j=1}^{J}\ln[\Gamma(\alpha_{ijk} + n(p_{ij}|\pi_{ik}))]\right\}$$
$$\left\{\sum_{k=1}^{K}\ln[\Gamma(\alpha_{ik} + n(\pi_{ik}))] + \sum_{j=1}^{J}\ln[\Gamma(\alpha_{ijk})]\right\}. \tag{9.8}$$

Local log-likelihood of each predictor given a set of parents can be calculated by substituting values for various frequencies in Equation (9.8). These values can be read directly from a contingency table of frequencies of various parent–child dependency (for example, Table 9.1).

Equation (9.7) decomposes the global search for the best DAG into the computationally more tractable local searches for the best sets of parents for individual predictors. However, for a large number of predictors, even the local search for parents can become intractable [94] and therefore conceptual genetic models are used to limit the search space (a) by specifying a search order on the predictors, so that the search space for the parents of a predictor is limited to its predecessors in the search order and (b) by forbidding certain dependencies. Additionally, an upper limit to the number of parents can be defined.

Let $P = \{P_1, P_2, \ldots, P_I\}$ be the set of I predictors and let $SO_P = \{P_1 \succ P_2 \succ P_3 \succ \ldots \succ P_I\}$, where $\Pi_i \subseteq \{P_1, P_2, \ldots, P_{i-1}\}$, be the search order on P. Let $F_{P_i} (\subset P)$ be a set of predictors that are forbidden to be parents of P_i. The best set of parents for P_i is estimated by adapting the K2 algorithm [103] as follows:

Algorithm 2 (Pseudocode)

Input: global prior precision (α); set of training occurrences of target deposit-type (**T**), target deposit-type (D), number of predictors (I); search order (SO_P); forbidden parents $(F_{P_i}, i = 1$ to $I)$ and maximum number of parents (MAX).
Output: $\Pi_i (i = 1$ to $I)$

1: START
2: set $\Pi_D = \emptyset$ {set D at the root of DAG}
3: **for** $i = 1; i = I; i + +$ **do** {starting with P_1, iterate for every predictor in SO_P}
4:     $\Pi_i = D$ {add directed arc from D to P_i}
5:     calculate α_{ijk} {use Equation (9.5) to calculate prior hyper-parameters}
6:     calculate α_{ik} {use Equation (9.6) to calculate local prior precision}
7:     calculate $\ln[\mathbf{g}(P_i, \Pi_i)]$ {use Eq. (9.8) to calculate likelihood of $\Pi_i = \{D\}$}
8:     $\max\{\ln[\mathbf{g}(P_i, \Pi_i)]\} = \ln[\mathbf{g}(P_i, \Pi_i)]$ {set current likelihood as maximum likelihood}
9:     **for** $i' = 1; i' < i; i' + +$ **do** {starting with P_1, iterate for every predecessor of P_i}
10:       **while** $n(\Pi_i) \leq MAX$ **do** {verify that current number of parents is less than maximum allowed}
11:         **if** $P_{i'} \notin F_{P_i}$ **then** {if $P_{i'}$ is not forbidden parent of P_i}
12:           $\Pi_i = \Pi_i + P_{i'}$ {add directed arc from $P_{i'}$ to P_i}
13:           calculate $\ln[\mathbf{g}(P_i, \Pi_i)]$ {use Equation (9.8) to calculate likelihood of current Π_i}
14:           **if** $\ln[\mathbf{g}(P_i, \Pi_i)] > \max\{\ln[\mathbf{g}(P_i, \Pi_i)]\})$ **then** {if current likelihood is more than current maximum likelihood}
15:             $\max\{\ln[\mathbf{g}(P_i, \Pi_i)]\} = \ln[\mathbf{g}(P_i, \Pi_i)]$ {set current likelihood as maximum allowed likelihood and save directed arc from $P_{i'}$ to P_i}
16:           **else**
17:             $\Pi_i = \Pi_i - P_{i'}$ {else remove directed arc from $P_{i'}$ to P_i}
18:           **end if**
19:         **else**
20:           $\Pi_i = \Pi_i$ {if $P_{i'}$ is forbidden parent, do not add directed arc from $P_{i'}$ to P_i}
21:         **end if**
22:       **end while** {if current number of parents is already equal to maximum allowed, abort nested FOR loop}
23:     **end for** {end of nested FOR loop}
24: **end for** {end of main FOR loop}

9.2.1.3   Validation of classifiers: n-fold cross-validation

Most published studies on mineral potential mapping use hold-back validation, which involves using a part (at least three quarters) of known occurrences of target deposit-type for model training and holding back the rest for model validation. The method, although computationally efficient, has several limitations, for example, (a) it requires a large number of training occurrences for minimizing uncertainty and avoiding over-fitting, (b) the validation is biased by the selection of training and validation occurrences and (c) it does not make an optimal use of available data. These limitations are addressed by leave-one-out validation, which involves leaving only one occurrence out, training a model with the rest of the occurrences and validating the model with the left-out occurrence. The process is implemented iteratively for all sets of training occurrences. This method is extremely accurate but computationally expensive and, in some situations, impracticable. It is proposed to use n-fold cross-validation, which retains the advantages of leave-one-out validation and, at the same time, is computationally more efficient. Given the set of training data ($\mathbf{T}$) containing M occurrences, the following algorithm can be used to implement n-fold cross-validation.

Algorithm 3

---

1. Partition $\mathbf{T}$ into n subsets $\mathbf{T}^i$ ($i = 1$ to n), each having (M/n) samples.

2. Leave $\mathbf{T}^1$ out and pool remaining $(n - 1)$ subsets to generate a new set $\mathbf{T}^{\bar{1}}$ for training a classifier.

3. Train the classifier on $\mathbf{T}^{\bar{1}}$.

4. Validate the classifier on $\mathbf{T}^1$ and record the number of correct classifications.

5. Repeat steps 2–4 for all $\mathbf{T}^i$ ($\forall i = 2$ to n).

6. Report percent correct classifications for all subsets.

Clearly, the higher the value of n, the higher the accuracy of validation (at $n = M$, the method becomes leave-one-out validation) and the higher the computational expense.

9.2.2   Software packages used

The above algorithms were implemented using the software package Bayesian Discoverer [382]. ArcView GIS package was used for spatial data compilation, processing and mapping.

9.3   Bayesian network mapping of base metal deposit

The methodology followed in this case study for base-metal potential mapping in the Aravalli province using Bayesian network classifiers involves several steps (Figure 9.1).

> **Step 1: Identification of base-metal deposit recognition criteria.** A conceptual approach [369] was used for identifying regional-scale (1:250 000) recognition criteria for base-metal deposits in the study area and representing them as predictor maps as inputs to the Bayesian network classifiers. A conceptual model of formation of base-metal deposits in the study area was defined based on published studies coupled with new interpretive syntheses of regional-scale exploration data [372]. Based on the conceptual model, controls on mineralization and recognition criteria for base metal deposits in the study area were identified and represented as predictor maps by processing, interpretation and reclassification of the exploration datasets in a GIS [369].

> **Step 2: Generation of unique conditions grid.** A 'unique conditions grid map' was generated by digital superposition of predictor maps using ArcSDM software [245]. An attribute table associated with a unique conditions grid map (unique conditions table) contains one record per unique condition class and one field for each predictor map. In the context of a Bayesian network classifier, each unique condition is considered a feature vector whose attributes are defined by the attributes of the unique condition. The predictor maps were input to the Bayesian network classifiers in the form of feature vectors.

> **Step 3: Modelling with Bayesian network classifiers.** The Bayesian network classifiers were implemented outside the GIS using the software package Bayesian Discoverer. Each Bayesian network classifier was first trained by estimating the model parameters from training data. The trained classifiers were then used to process all feature vectors. The outputs for each feature vector were imported back and joined to the respective unique condition in the GIS.

> **Step 4: Generation of continuous-scale favorability maps.** The outputs of the Bayesian network classifiers for the unique conditions were mapped in the GIS to generate continuous-scale [0,1] favorability maps. For each unique condition, the outputs were interpreted as relative favorability values.

> **Step 5: Generation of binary favorability maps.** Continuous-scale favorability maps are cumbersome to interpret for demarcating areas of

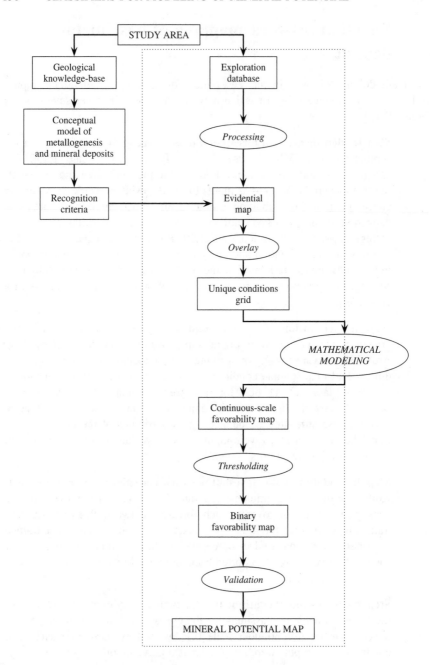

Figure 9.1 Flowchart of the steps used for base-metal potential mapping using Bayesian network classifiers. Small rectangular boxes contain knowledge or maps created at various stages and small elliptical boxes contain procedures/algorithms used for creating the objects. Area in the dotted box represents steps/outputs performed/created using a GIS environment.

base-metal potential because they represent favorability in a continuous scale from 0 (minimum) to 1 (maximum). A threshold of 0.5 was applied to reclassify the continuous scale favorability maps into binary favorability maps.

Step 6: Validation of binary favorability maps The binary favorability maps were validated by overlaying locations of known deposits on the binary favorability maps. This step determines the usefulness of a binary favorability map as a guide for further mineral exploration in a study area.

9.3.1 Study area

The study area forms a part of the Aravalli metallogenic province in the state of Rajasthan, western India (Figure 9.2). Its area is about 34 000 km² and it is located between latitudes 23°30′ N and 26° N and longitudes 73° E and 75° E.

The province is characterized by two fold belts, viz., the Palaeo-Mesoproterozoic Aravalli Fold Belt and the Meso-Neoproterozoic Delhi Fold Belt, which are

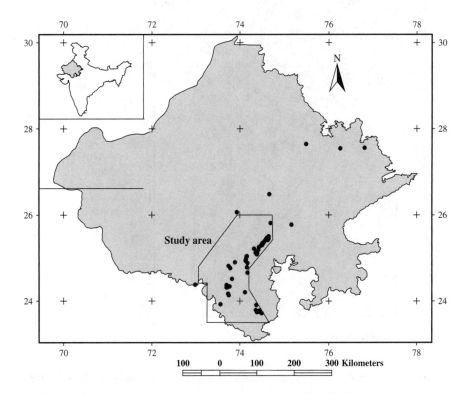

Figure 9.2   Location map of study area in state of Rajasthan, India. Small black circles are locations of occurrences of base metal deposits.

ingrained in a reworked basement complex that contains incontrovertible Archaean components [196, 211, 403]. Our work on geophysical data [372] indicates that the province comprises a number of subparallel and linearly disposed belts (Figure 9.3), which broadly coincide with the tectonic domains proposed in [446].

The Aravalli province holds substantial reserves of base metal deposits, particularly lead and zinc. The lead–zinc reserves in the province stand at 130 million tones with 2.2% lead and 9.2% zinc with an additional 30 million tonnes of resources in producing mines and deposits under detailed exploration [197].

A majority of the lead–zinc deposits of the province are contained in the study area (Figure 9.3). Rampura-Agucha is a world-class zinc–lead–(silver) deposit containing the highest amount of lead and zinc metals amongst all deposits of India. The Bhilwara belt contains a large majority of the base metal deposits of the study area, mainly in the Bethumni-Dariba-Bhinder and Pur-Banera areas. In the Aravalli belt, low-grade copper, gold and uranium mineralizations occur in the oldest rocks,

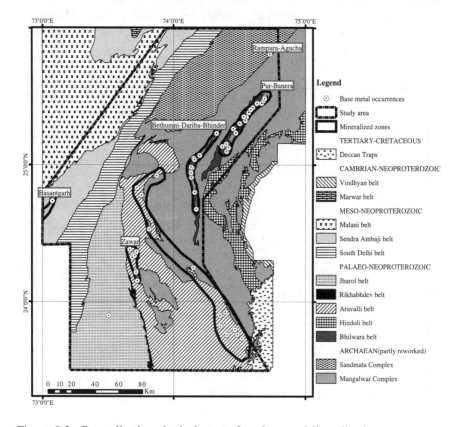

Figure 9.3 Generalized geological map of study area. Mineralized zones are outlined in white. White circles are locations of occurrence of base-metal deposits.

whilst the younger rocks host the zinc–lead deposits of the Zawar area. The Jharol belt hosts a copper–(zinc–lead) deposit at Padar Ki Pal. The South Delhi belt hosts copper–(zinc) deposits in the Basantgarh area. The deposits of the Bhilwara and Aravalli belts are so-called 'sedimentary-exhalative' (SEDEX)-type [187], while those of the South Delhi belt show a closer affinity to the so-called 'volcanic-hosted massive sulfide' (VMS)-type deposits [122]. Porwal et al. [372] identified (1) host rock lithology, (2) stratigraphic position, (3) association of synsedimentary mafic volcanic rocks,[1] (4) proximity to regional tectonic boundaries and (5) proximity to favorable structures as the most significant regional-scale recognition criteria for base-metal deposits in the province.

9.3.2   Data preprocessing

In the case of mineral potential mapping, the training target variable is generally binary with the labels 'mineralized' and 'barren'. The feature vectors associated with known mineralized or with known barren locations constitute training occurrences, which are referred to as deposit or nondeposit training occurrences, respectively.

Appropriate preprocessing of the exploration database and selection of training occurrences is important for a successful implementation of Bayesian classifiers. The following two factors required especial consideration in this case study.

Firstly, conditional dependencies are generally 'state specific' and seldom 'map specific'. Consider, for example, multi-state maps of lithologies and stratigraphic groups. In the absence of ubiquitous chronological data, stratigraphic classifications are generally made on the basis of lithological associations, which results in significant correlations between stratigraphic groups and lithologies. These correlations, however, are state specific, i.e., a particular stratigraphic group is correlated with specific lithologies. If each map is used as a single multi-state predictor in an augmented naïve classifier and Algorithm 2 estimates a significant likelihood of the map of lithologies being a parent of the map of stratigraphic groups, then every state (lithology[2]) of the map of lithologies is indiscriminately included in the set of parents of every state (stratigraphic group[3]) of the map of stratigraphic groups. This may result in a large number of erroneous dependencies. More importantly, it results in manifold increase in the number of parameters, which may lead to over-fitting. It is therefore preferable to use 1-of-n encoding [307] for transforming an n-state map into n binary maps before inputting into an augmented Bayesian classifier. This forces the algorithm to search for dependencies amongst individual states and hence only true dependencies are identified.

---

[1] Dark-colored rocks rich in magnesium and iron.
[2] A rock unit defined on the basis of diagnostic physical, chemical and mineralogical characteristics.
[3] A group of rock units closely associated in time and space.

Secondly, exploration data sets are highly imbalanced and biased towards the barren class. If deposit and nondeposit occurrences are represented in the training set in the same proportion as they are expected to occur in the general population, then the performance of a Bayesian classifier is optimized for recognizing nondeposit occurrences rather than deposit occurrences. This may give rise to a large number of type II errors, which have severe consequences in mineral potential mapping. The problem can be addressed by using 'one-sided selection' [257] to balance the number of deposit and nondeposit occurrences in the training set. Both data-driven and knowledge-driven approaches can be used to select nondeposit occurrences. In a data-driven approach, nondeposit occurrences can be randomly selected from feature vectors that have been modeled previously (e.g., via Weights of Evidence method) as having very low probability of occurrence for mineral deposit-type of interest. In a knowledge-driven approach, feature vectors that are least likely to be associated with the target mineral deposit-type are selected based on expert knowledge of genetic mineral deposit models.

9.3.3   Evidential maps

A regional-scale GIS was compiled by digitizing the lithostratigraphic map [194], the structural map [195] and the map of total magnetic field intensity [181]. Locations of 54 known base metal deposit/occurrences were compiled from various sources. The digitized maps were converted into grid format, processed, interpreted and reclassified to create evidential maps for the recognition criteria. In all, five evidential maps were generated: two multi-state evidential maps of (1) lithology and (2) stratigraphy; and three binary evidential maps of (1) mafic igneous rocks (2) buffered regional lineaments (3) buffered fold axes were generated.

The multi-state evidential maps of lithologies and stratigraphic groups were transformed into 13 binary evidential maps through 1-of-n encoding. Of these, 13 binary evidential maps, 11 were used in subsequent processing (two binary maps comprising lithologies and stratigraphic groups that have no known relationship with the base metal mineralization in the province were not used). The resulting 14 binary maps were superposed and unique combinations of the maps in unit areas of 1 km^2 were mapped to generate a map constituting 519 feature vectors. As the operation was carried out in a GIS environment, an associated database table was automatically generated, which held the components of the feature vectors. In the table, each feature vector is described by a unique identification number and 14 components representing each evidential map encoded as either present or absent.

9.3.4   Training of Bayesian network classifiers

The feature vectors associated with known occurrences of base metal deposits were extracted to create a subset of 54 deposit occurrences. An equal number of feature vectors, which were considered, on the basis of expert knowledge, least likely to be associated with base metal deposits were extracted to create a subset of 54 nondeposit occurrences. The two subsets were merged to generate a training set containing 108 deposit/nondeposit occurrences.

Algorithm 2 was implemented using Bayesian Discoverer software to train an augmented naïve classifier using the training set to determine the DAG that best simulates the dependencies in the data. To limit the search space, (a) dependencies amongst the binary maps of lithologies and amongst the binary maps of stratigraphic groups were forbidden, (b) the maximum number of parents for each predictor was set to three and (c) the following search order was specified:

Buffered regional lineaments ≻ Lithologies ≻ Buffered fold axes

≻ Mafic igneous rocks

≻ Stratigraphic groups.

The above search order is based on the following considerations:

- The regional lineaments represent fundamental tectonic features (extensional faults) in the Aravalli province [372]. Therefore there is little possibility of the map of buffered regional lineaments being dependent on any other predictor.

- A (metamorphosed)-sedimentary rock is a product of its basin environment, which, in turn, is controlled by basin tectonics [365]. Therefore there exists a possibility of dependence of the maps of lithologies on the map of buffered regional lineaments. However, there is little possibility of the maps of lithologies being dependent on any of the other predictors.

- Folding obviously postdates rifting and sedimentation and therefore there can be no possibility of the map of buffered fold axes being a parent of either the map of buffered regional lineaments or the maps of lithologies.

- The regional lineaments mark the extensional faults that could be possible conduits for the mafic rocks in the province [372]. Therefore there exists a possibility of the map of buffered regional lineaments being a parent of the map of mafic igneous rocks.

- Stratigraphic classification of the province in various groups is largely based on regional tectonic boundaries, lithological associations and deformation patterns [196]. Therefore there exists a strong possibility of the binary maps of stratigraphic groups being dependent on several of the other predictors.

After determining the DAG of the augmented naïve classifier, a selective naïve classifier was constructed by removing the conditionally dependent predictors. The DAGs of the trained naïve, augmented naïve and selective naïve classifiers are shown in Figures 9.4–9.6. In the figures, nodes and directed arcs represent binary predictors and conditional dependencies, respectively. Parameters associated with Rajpura-Dariba group are shown for illustration. The parameters (conditional

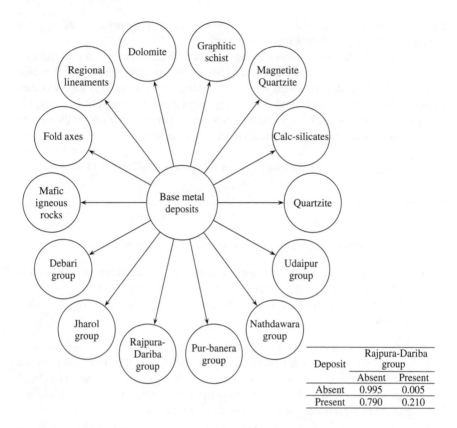

Deposit	Rajpura-Dariba group	
	Absent	Present
Absent	0.995	0.005
Present	0.790	0.210

Figure 9.4 Trained Bayesian network classifiers for base-metal potential mapping in study area: naïve classifier.

probabilities) associated with every node in each classifier were estimated using Algorithm 1. A 25-fold cross-validation was implemented to validate each classifier. The augmented naïve classifier performs the best, followed by the naïve classifier and then the selective naïve classifier (Table 9.2).

Table 9.2   Results of 25-fold cross validation.

Classifier	Correctly-classified deposits
Naïve	86.8%
Augmented naïve	88.7%
Selective naïve	83.0%

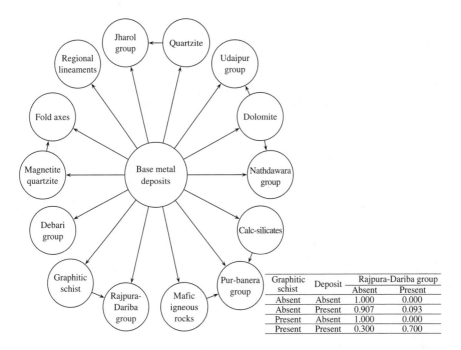

Graphitic schist	Deposit	Rajpura-Dariba group	
		Absent	Present
Absent	Absent	1.000	0.000
Absent	Present	0.907	0.093
Present	Absent	1.000	0.000
Present	Present	0.300	0.700

Figure 9.5 Trained Bayesian network classifiers for base-metal potential mapping in study area: Augmented naïve classifier.

9.3.5 Mineral potential maps

The trained classifiers were used to process all feature vectors. The output posterior probability of d_1 (mineralized class) for each feature vector is interpreted as a measure of favorability or potential for occurrence of base-metal deposit in every feature vector. The outputs of the three classifiers were mapped to generate continuous-scale favorability maps, which depict the posterior probabilities of the occurrence of a base-metal deposit in various parts of the study area in a scale of 0 to 1. However, to facilitate interpretation and validation of these continuous-scale maps, they were reclassified into binary favorability maps by using a threshold probability of 0.5 (Figures 9.7 to 9.9). Table 9.3 shows that (a) the naïve classifier demarcates favorable zones occupying 7% of the study area and containing 89% of the known deposit occurrences (b) the augmented naïve classifier demarcates favorable zones occupying 11% of the study area and containing 93% of the known deposit occurrences, and (c) the selective naïve classifier demarcates favorable zones occupying 11% of the study area and containing 83% of the known deposit occurrences.

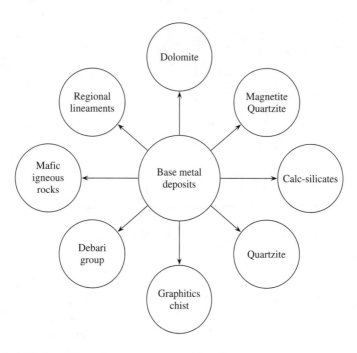

Figure 9.6 Trained Bayesian network classifiers for base-metal potential mapping in study area: Selective naïve classifier.

Table 9.3   Validation of favorability maps.

Classifier	Zone	Percent of study area (Total area: 34,000 km^2)	Percent of Deposits (Total deposits: 54)
Naïve	High favorability	7.1	88.9
	Low favorability	92.9	11.1
Augmented naïve	High favorability	11.3	92.6
	Low favorability	88.7	7.4
Selective naïve	High favorability	11.2	83.3
	Low favorability	88.8	16.7

9.4 Discussion

The formation and localization of mineral deposits are the end-results of a complex interplay of several metallogenetic processes that exhibit signatures in the form of geologic features associated with the mineral deposits. These geological features, called recognition criteria, are characterized by their responses in one or more data

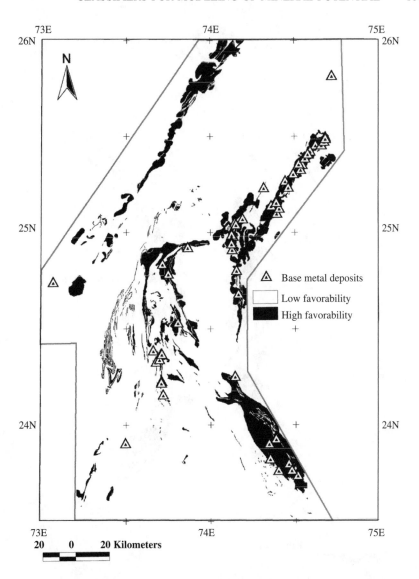

Figure 9.7 Binary base-metal potential map derived with a naïve classifier. Tri-angles are known occurrences of base-metal deposits. Reprinted with permission from Elsevier.

sets that are used as predictors in mineral potential mapping. It is unrealistic to assume conditional independence (CI) amongst the predictors with respect to target mineral deposit occurrences because (a) a particular geologic feature can partially respond in two or more geodata sets, (b) a particular metallogenetic process can be partially responsible for two or more geologic features or (c) two or more

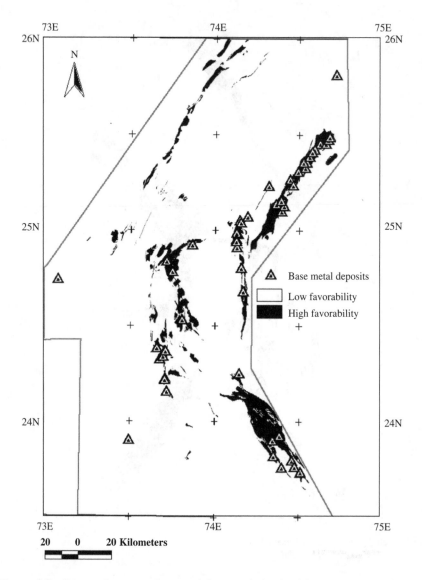

Figure 9.8   Binary base-metal potential map derived with an augmented naïve classifier. Triangles are known occurrences of base-metal deposits. Reprinted with permission from Elsevier.

metallogenetic processes can be related. In addition, the response of a geologic feature in one geodata set may be conditioned by the response of another geologic feature in a different geodata set. Considering that violations of the CI assumption is unavoidable in mineral potential mapping, the following paragraphs discuss the results of the Bayesian network applications described above in terms of CI violation in Bayesian approaches to mineral potential mapping.

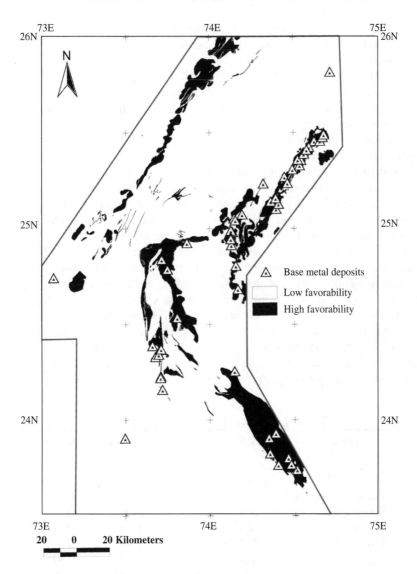

Figure 9.9   Binary base-metal potential map derived with a selective classifier. Triangles are known occurrences of base-metal deposits. Reprinted with permission from Elsevier.

In general, the naïve classifier performs well in the predictive mapping of base metal potential in the study area (Tables 9.2 and 9.3), which suggests that a naïve classifier can tolerate significant violations of the conditional independence assumption (see also [132, 133]). This also implies that a weights-of-evidence model, which can be compared to a naïve classifier (although with some significant differences and simplifications[4]), may not be seriously hampered by the violation of the conditional independence assumption provided that its output is interpreted as a measure of relative favorability rather than absolute posterior probability (see below). This also explains the widespread and generally successful application of weights-of-evidence models to mineral potential mapping.

The results (Tables 9.2 and 9.3) also indicate that the performance of the naïve classifier is improved if the CI assumption is relaxed by recognizing and accounting for some of the dependencies in the training data. The naïve classifier returns higher values, however, for the deposits of the Bhilwara belt (where there are strong dependencies amongst favorable predictors), which suggests a significant influence of dependencies amongst predictors on the output of a naïve classifier. The augmented naïve classifier, in contrast, identifies several zones of high favorability in the Jharol belt that are missed by the naïve classifier. Moreover, the high favorability zones tend to cluster around known deposits with the naïve and selective classifiers, which suggests that the augmented naïve classifier has better generalization capability compared to the naïve classifier. This is further evidenced by comparing the outputs of the three classifiers for the misclassified deposits (Table 9.4), which shows that the augmented naïve classifier returns a higher value for all deposits misclassified by the naïve classifier. Because the geological settings of the misclassified deposits are different in many respects from the geological settings of majority of deposits in the study area, it indicates that the generalization capability of a naïve classifier is improved by recognizing and accounting for dependencies amongst predictors.

Table 9.4   Outputs for misclassified* deposits.

Deposit	Belt	Naïve	Augmented naïve	Selective naïve
Padar-Ki-Pal	Jharol	0.002	0.534	0.009
Rampura-Agucha	–	0.056	0.091	0.102
Baroi	Aravalli	0.067	0.910	0.374
Anjani	Aravalli	0.076	0.206	0.196
Basantgarh	South Delhi	0.406	0.482	0.352
Bara	Aravalli	0.429	0.637	0.615
Wari Lunera	Bhilwara	0.946	0.931	0.200
Dariba	Bhilwara	0.985	0.973	0.470
Dariba Extn.	Bhilwara	0.985	0.973	0.470
Rajpura A	Bhilwara	0.985	0.973	0.470

* Threshold favorability score of 0.500 is used for classification.

---

[4]In a weights-of-evidence model, all feature vectors that are associated with the unit areas that do not contain a known mineral deposit are indiscriminately used as nondeposit samples. In a naïve classifier, on the other hand, only the feature vectors that are associated with the unit areas that are reasonably well known to be barren are used as nondeposit samples.

The selective naïve classifier misclassifies all but one deposit misclassified by the naïve classifier (Table 9.4, Figure 9.4). In addition, it also misclassifies several deposits of the Bhilwara belt, which is clearly a result of the rejection of conditionally dependent maps of stratigraphic groups. Evidently, conditionally dependent predictors could make a significant independent contribution to the information content of a naïve classifier (except when there is a perfect correlation) and therefore the rejection of conditionally dependent predictors affects its performance adversely. It is also possible that a dependent predictor makes only a minor independent contribution to the information content, yet that contribution is crucial for making a correct classification. Therefore, in order to minimize the bias due to dependencies amongst predictors, it is preferable to augment a naïve classifier by relaxing the CI assumption and accounting for the dependencies instead of abridging it by rejecting conditionally dependent predictors.

9.5   Conclusions

Bayesian network classifiers are efficient tools for mineral potential mapping. They are easy to construct, train and implement. In the study area, the Bayesian network classifiers successfully demarcated favorable zones that occupying 7–11% of the area and containing 83–93% of the known base metal deposits. This is a significant result both in terms of reduction in search area and number of deposits predicted.

Although a naïve classifier is based on the assumption of conditional independence of input predictor patterns with respect to known occurrences of mineral deposit-type of interest, it shows significant tolerance for the violations of the assumption and performs well in classification of areas as mineralized or barren.

The performance of a naïve classifier is significantly improved if the conditional independence assumption is relaxed by recognizing and accounting for dependencies amongst the predictor patterns in an augmented naïve classifier.

Rejection of conditionally dependent predictor patterns in a selective naïve classifier degrades the performance of a naïve classifier.

10

Student modeling

L. Enrique Sucar

Instituto Nacional de Astrofísica, Óptica y Electrónica, Tonantzintla, Puebla, 72840, México

and

Julieta Noguez

Tecnológico de Monterrey, Campus Ciudad de México, México D.F., 14380, México

10.1 Introduction

An intelligent tutoring system (ITS) tries to emulate a human tutor by adapting itself to the learner. A key element of an intelligent tutor is the student model, that provides to the ITS knowledge about each particular student, so its behavior can be adapted to the student needs. In the student model, the cognitive state (the student's knowledge of the subject matter) of a learner is inferred from: (i) previous data about the student, and (ii) the student's behavior during the interaction with the system. Student modeling, and in general user modeling, is a complex task which involves uncertainty. On one hand, there is still not a clear understanding on how people learn and how to represent their knowledge in a computer. On the other hand, usually the information available to build and update the student model is very limited. Although there have been several approaches for student modeling, in the last few years Bayesian networks have become one of the preferred methods.

Bayesian networks provide a natural framework for student modeling, which is basically a diagnosis problem. Under this framework, the different knowledge items

Bayesian Networks: A Practical Guide to Applications   Edited by O. Pourret, P. Naïm, B. Marcot
© 2008 John Wiley & Sons, Ltd

Table 10.1   Categories of student models based on Bayesian networks.

Type	Description
Expert-centric	An expert specifies either directly or indirectly the complete structure and conditional probabilities of the Bayesian student model.
Efficiency-centric	The model is partially specified or restricted in some way, and domain knowledge is 'fitted' to the model.
Data-centric	The structure and conditional probabilities of the network are learned from data.

related to the domain, and the information obtained from the student's interactions are represented as random variables; and their relations are represented as dependencies in a directed acyclic graph. Based on partial knowledge of the student given by a subset of variables, the cognitive state – represented by another subset – is inferred using probabilistic inference.

Several authors have developed students models based on Bayesian networks [70, 98, 305, 329]. Mayo [309] classifies these models into three categories, described in Table 10.1.

There are two main drawbacks of using standard Bayesian networks for student modeling:

1. Knowledge acquisition: building a Bayesian network model for a domain is a difficult and time consuming process.

2. Complexity: in some cases the model can become too complex, and consequently the inference process could be slow for some applications, in particular those that require a real time response, as virtual laboratories.

Probabilistic relational models (PRMs) [252] are an extension of Bayesian networks that help to overcome these problems. They provide a more expressive, object-oriented representation that facilitates knowledge acquisition and makes it easier to extend a model to other domains. In case of a very large model, only part of it is considered at any time, so the inference complexity is reduced.

A particularly challenging area for student modeling are virtual laboratories [478]. A virtual lab provides a simulated model of some equipment, so that students can interact with it and learn by doing. A tutor serves as virtual assistant in this lab, providing help and advice to the user, and setting the difficulty of the experiments, according to the student's level. In general, it is not desirable to trouble the student with questions and tests to update the student model. So the cognitive state should be obtained just based on the interactions with the virtual lab and the results of the experiments.

We have developed a student model based on PRMs for a tutor that helps students while they interact with a virtual laboratory. The model infers, from the student's interactions with the laboratory, the cognitive state; and based on this model, it gives personalized advice to the student. It has been applied in a virtual laboratory for mobile robotics, and has been evaluated in a robotics undergraduate course, showing that students that interact with the lab with help of the tutor have a better performance. The models for four different experiments in a virtual lab were easily adapted from a general template based on PRMs, showing the advantage of this type of model.

10.2   Probabilistic relational models

The basic entities in a PRM are objects or domain entities. Objects in the domain are partitioned into a set of disjoint classes $X_1, \ldots, X_n$. Each class is associated with a set of attributes $A(X_i)$. Each attribute $A_{ij} \in A(X_i)$ (that is, attribute j of class i) takes on values in some fixed domain of values $V(A_{ij})$ [252]. The dependency model is defined at the class level, allowing it to be used for any object in the class. PRMs explicitly use the relational structure of the model, so an attribute of an object will depend on some attributes of related objects. A PRM specifies the probability distribution using the same underlying principles used in Bayesian networks. Each of the random variables in a PRM, the attributes $x.a$ of the individual objects x, is directly influenced by other attributes, which are its parents. A PRM therefore defines for each attribute, a set of parents, which are the directed influences on it, and a local probabilistic model that specifies probabilistic parameters. An example of a PRM in the school domain, based on [252], is shown in Figure 10.1. There are four classes, with two attributes each in this example:

Professor: teaching-ability, popularity.

Student: intelligence, ranking.

Course: rating, difficulty.

Registration: satisfaction, grade.

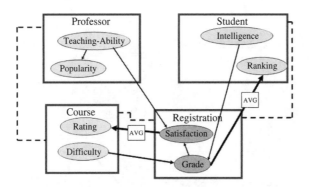

Figure 10.1 An example of a PRM structure for the school domain [252]. Dashed edges represent relations between classes, and arrows correspond to probabilistic dependency. The *AVG* in a link indicates that the conditional probabilities depend on this variable.

This representation allows for two types of attributes in each class: (i) information variables, (ii) random variables. The random variables are the ones that are linked in a kind of Bayesian network that is called a *skeleton*. From this skeleton, different Bayesian networks can be generated, according to other variables in the model. For example, in the student model we describe below, we define a general skeleton for an experiment, from which particular instances for each experiment are generated. This gives the model a greater flexibility and generality, facilitating knowledge acquisition. It also makes inference more efficient, because only part of the model is used in each particular case.

The probability distribution in the skeletons are specified as in Bayesian networks. A PRM therefore defines for each attribute $x.a$, a set of parents, which are the directed influences on it, and a local probabilistic model that specifies the conditional probability of the attribute given its parents. Once a specific network is generated from a skeleton, the inference mechanism is the same as for Bayesian networks.

10.3   Probabilistic relational student model

PRMs provide a compact and natural representation for student modeling. They allow each attending student to be represented in the same model. Each class represents the set of parameters of several students, like in databases, but the model also includes the probabilistic dependencies between classes for each student.

In order to apply PRMs to student modeling we have to define the main objects involved in the domain. A general student model oriented to virtual laboratories was designed, starting form a high-level structure at the class level, and ending with specific Bayesian networks for different experiments at the lower level. As shown in Figure 10.2, the main classes, related with students and experiments,

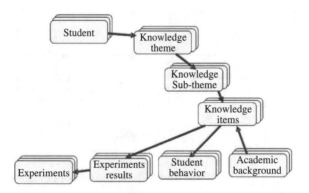

Figure 10.2 A high-level view of the PRM structure for the student model, showing the main classes and their relations.

were defined. In this case there are eight classes, with several attributes for each class, as listed below:

Student: student-id, student-name, major, quarter, category.

Knowledge Theme: student-id, knowledge-theme-id, knowledge-theme-known.

Knowledge Sub-theme: student-id, knowledge-sub-theme-id, knowledge-sub-theme-known.

Knowledge Items: student-id, knowledge-item-id, knowledge-item-known.

Academic background: previous-course, grade.

Student behavior: student-id, experiment-id, behavior-var1, behaviour-var2, . . .

Experiments results: student-id, experiment-id, experiment-repetition, result-var1, resultvar2, . . .

Experiments: experiment-id, experiment-description, exp-var1, exp-var2, . . .

The dependency model is defined at the class level, allowing it to be used for any object in the class.

Some attributes in this model represent probabilistic values. This means than an attribute represents a random variable that is related to other attributes in the same class or in other classes, as shown in Figure 10.3. The structure in Figure 10.3 shows a more detailed view of the general model in Figure 10.2, specifying some attributes for each class and their dependencies.

The main advantages of PRMs can be extended to relational student models. First, from a PRM model we can define a general Bayesian network, a skeleton, that can be instantiated for different scenarios, in this case experiments. Second, it is easy to organize the classes by levels to improve the understanding of the model. From the model in Figure 10.3, we obtain a hierarchical skeleton, as shown in Figure 10.4. We partitioned the experiment class, according to our object of interest, creating two subclasses: experiment performance and experiment behavior, which constitute the lowest level in the hierarchy. The intermediate level represents the different knowledge items (concepts) associated to each experiment. These items are linked to the highest level which groups the items in subthemes and themes, and finally into the students general category. We defined three categories of students: *novice, intermediate and expert*. Each category has the same Bayesian net structure, obtained from skeleton, but different CPTs are used for each one.

From the skeleton, it is possible to define different instances according to the values of specific variables in the model. For example, from the general skeleton for experiments of Figure 10.4, we can define particular instances for each experiment and student level, as it is shown in Figure 10.5. In this case, we illustrate the generation of nine different networks, for three experiments and three student levels.

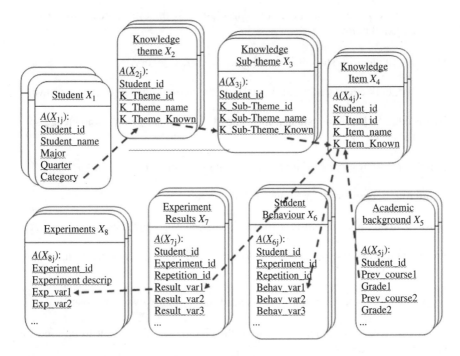

Figure 10.3   A more detailed view of the PRM structure for the student model, with some of the attributes and their dependencies.

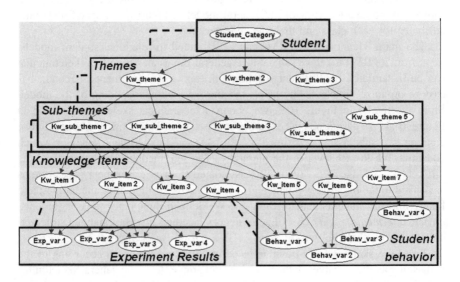

Figure 10.4   A general skeleton for an experiment derived form the PRM student model for virtual laboratories.

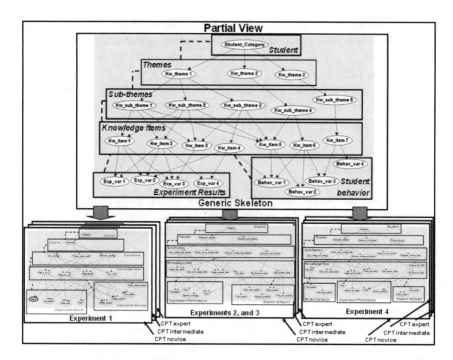

Figure 10.5 Different instances of Bayesian network models are generated from the skeleton. In this case we illustrate nine instances, for three different experiments and three student levels [341]. Reprinted with permission from IJEE.

The parameters of the model are defined with the help of an expert, that is an experienced professor in the area. There are basically two types of parameters that have to be defined:

- The influence of each aspect that is evaluated in each experiment (behavior and experiment result variables) on the basic concepts or items in the domain.

- The relations between themes, subthemes and items in the hierarchical knowledge structure.

Both are relatively easy to specify by an experienced teacher in the area covered by the experiment. We consider that the exact value of each parameter is not very important, although a sensitivity analysis experiment will be required to verify this hypothesis.

Once a specific Bayesian network is generated, it can be used to update the student model via standard probability propagation techniques. In this case, it is used to propagate evidence from the experiment evaluation to the knowledge items, and then to the knowledge subthemes and to the knowledge themes. Based on this evidence and the accumulated evidence from previous experiments, the system decides to re-categorize the student. After each experiment, the knowledge level at the different level of granularity – items, subthemes and themes – is used by the tutor to decide if it should provide help to the student, and at what level of detail. For example, if in general the experiment was successful, but some aspect was not very good, a lesson on a specific concept (item) is given to the student. While if the experiment was unsuccessful, a lesson on a complete theme or subtheme is recommended. Based on the student category, the tutor decides the difficulty of the next experiments to be presented to the student.

10.4   Case study

We have applied the student model based on PRMs to a virtual laboratory for mobile robotics. The objective of this virtual lab is that the students explore the basic aspects of mobile robots, such as mechanical configuration, kinematics, sensors and control. The students can easily explore different mechanical and sensors configurations before they start building their robot. The laboratory also includes facilities for practicing basic control programming. The virtual laboratory is designed as a semi-open learning environment and incorporates an intelligent tutor. Learners can experiment with different aspects and parameters of a given phenomenon inside an open learning environment, but in this kind of environment the results strongly depend on the learner ability to explore effectively. A semi-open learning environment provides the student with the opportunity to learn through free exploration, but with specific performance criteria that guide the learning process. That is, it provides the flexibility of open learning environments; but at the same time, it has specific goals that the student has to achieve, enabling a more effective and guided exploration. A view of the user interface for one experiment is depicted in Figure 10.6.

To define the experiments, we consider a line following competition, which requires some basic knowledge on mechanical design, sensors, control theory and programming from the students. We defined a sequence of specific experiments so that the students learn incrementally; and at the same time to enable the assessment by the tutor of the knowledge items, step by step. There are four experiments:

Experiment 1: the robot has to follow the line with manual control, exploring different types and configurations of robots. It involves mechanical design and kinematics.

Experiments 2 and 3: the robot has to follow the line based on infrared sensors, the number and position of the sensors can be modified. Knowledge about sensors is required.

Experiment 4: the student has to write a control program so that the robot follows a line. Basic knowledge on robot programming and control theory is required, as well as the integration of the concepts practiced in the previous experiments.

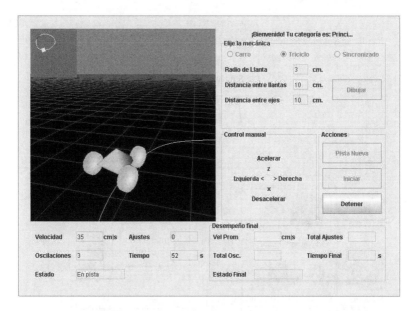

Figure 10.6   A screen view of the interface of the virtual lab for mobile robotics. The bottom windows show the performance of the student, during (left) and at the end (right) of the experiment. The right windows provide the exploration (top) and control (buttom) commands.

For all the experiments, a set of behavior (exploration) and result (performance) variables are defined, which are evaluated and sent to the student model.

This environment is designed as a general architecture for virtual laboratories that incorporate intelligent tutors, so it can be easily extended to other domains. The generic architecture is depicted in Figure 10.7. There are three basic elements in this architecture:

• User interface: it provides a 3-D visualization of the experiment, a set of commands to explore the lab, and feedback from the tutor.

• Simulator: it includes a mathematical model of the equipment (in this case a robot and its environment).

• Intelligent tutoring system: an intelligent assistant that monitors the performance of the student in each experiment and gives personalized help.

The core of the intelligent tutor is the student model based on PRMs, as described in Section 10.3. The model includes two parts: (i) initial categorization, and (ii) experiments.

To provide a personalized environment from the beginning, the model does an initial categorization. Following the philosophy of virtual laboratories of being

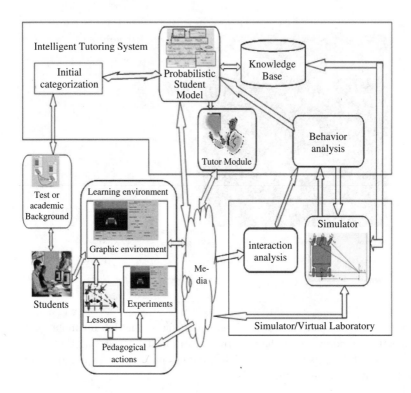

Figure 10.7   A generic architecture for virtual laboratories. Three main components are shown: learning environment, simulator/virtual laboratory and intelligent
tutoring system.

noninvasive (avoiding direct questions or tests), we used the academic background
of the student for an initial categorization as: novice, intermediate or expert. This
categorization is based on a probabilistic model that links previous courses with
the different knowledge concepts relevant for the experiments in the laboratory.
An initial category is obtained for each student, and this is updated after each
experiment based on the second part of this model. This category is used to define
the exercise complexity for each experiment and to select the type of lesson given
after each experiment, if required.

The second part of the model updates the student model according to the results,
in terms of exploration and performance, in each experiment. This part is based on
the structures described in the previous section.

10.5   Experimental evaluation

We incorporated the robotics virtual laboratory to an introductory course in mobile
robotics [341]. The course is based on project-oriented learning, so the students

have to build and program a small mobile robot for a competition. They form multidisciplinary teams and work during the term in the project. The virtual lab is used in the first part of the course to practice the basic concepts and explore different robot configurations, before they start building their robot.

The virtual lab was evaluated in an undergraduate course with 20 students, with majors in Computer Science or Electrical Engineering. The class was divided into two groups, 10 students each: *control group* and *test group*. The control group experimented in the virtual laboratory without a tutor. The test group used the virtual laboratory with the help of the intelligent tutor (based on the PRM student model).

In the experiment, all subjects used the virtual laboratory. We introduced the academic background of each student to the system. The system, using the probabilistic model, applied the initial categorization process for each student. Both, control and experimental group students were divided in two categories: novice and intermediate. We then applied a pretest after a 60 minutes lecture on basic robotics concepts. The pretest is a paper and pencil test designed to evaluate the learners knowledge of the objects target by the virtual laboratory. It consisted of 25 questions organized in the same way as the knowledge objects of the student model. Both, control and experimental groups participated in several experimental sessions with the virtual laboratory. The post-test consisted of a test analogous to the pretest.

Two aspects were analyzed in this evaluation:

- Initial categorization. We compared the knowledge level predicted by the student model based on the student background, to the scores in the pretest.

- Intelligent tutor. We compared the performance in the post test of the students that experimented in the lab without tutor (control group) vs. the students that have the help of the tutor.

We present the results for the intermediate students, which are similar to the other categories. Figure 10.8 shows the initial categorization results (predicted) versus the pretest. The predictions of the model are very good for almost all the knowledge items. Figure 10.9 summarizes the results after experiments 1, 2, 3 and 4, for the control and experimental groups. The results show that the students who explore the virtual environment with the help of the intelligent tutor have a significant improvement in their knowledge of the relevant concepts, in comparison with the students that use the virtual lab without assistant.

We consider that there are two main factors that determine this difference. We recorded the number of experiments performed by each student, and the ones that have the tutor, performed two or three times more experiments. It seems that the students who experiment without any help or feedback, get bored and do not continue using the lab; while the students that have a virtual tutor are motivated to continue until they reach a higher category. The other factor could be that the lessons and questions given by the tutor help the students connect their experience

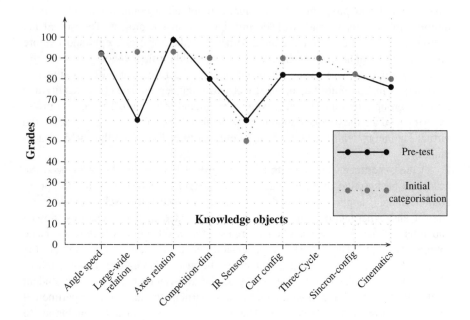

Figure 10.8   A comparison of the pretest scores and the predictions of the PRM model for each knowledge item. The graphs show the average grade (scale 0 to 100) per knowledge item.

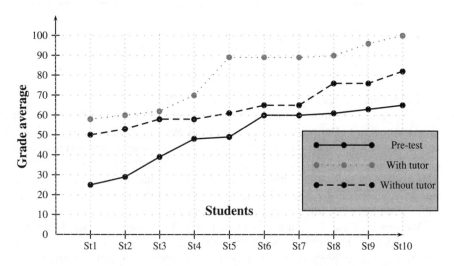

Figure 10.9   A comparison of the post-test results for students with tutor vs. students without tutor, the pretest scores are also shown for comparison. The graphs show the average grade (scale 0 to 100) per student, ordered by grade.

in the lab with the concepts in the domain. Although these lessons are available to all students, we think that the students without tutor did not access them frequently.

These results depend not only on the student model, but in general in the tutor and learning environment. However, the student model is a key element in this system, and the object oriented representation of Bayesian networks, provides a good framework to design and use these models.

10.6   Conclusions and future directions

Bayesian networks provide a general and effective framework for knowledge representation and reasoning under uncertainty. However, in some complex and large domains this representation is not sufficient. The object oriented paradigm facilitates building (the subject is singular) large complex systems via abstraction and reuse. PRMs combine the advantages of BNs and object oriented representations, by providing the effective management of uncertainty of BN and the organizational benefits of object oriented systems.

In this chapter we have shown the use of PRMs for student modeling, exploiting the advantages of this framework for defining a general student model for virtual laboratories. By defining a general framework based on the PRM student model, this can be easily adapted for different experiments, domains and student levels. This reduces the development effort for building and integrating ITS for virtual laboratories.

Based on the general model, we designed and implemented a tutor, together with a virtual lab, for mobile robotics. The students are categorized using PRMs to provide a personalized environment from the beginning. The model keeps track of the students's knowledge at different levels of granularity, combining the performance and exploration behaviour in several experiments, to decide the best way to guide and to recategorize the student. The virtual lab with the tutor were used and evaluated in a robotics course, with promising results.

We are currently implementing other virtual labs based on the same framework, for multiple robots and physics. In the future we plan to create an *authoring tool* (which helps to build intelligent tutors for different applications, in an analogous way as expert systems *shells* for expert systems) based on these models so that teachers can build their own tutors and virtual labs in different domains.

11

Sensor validation

Pablo H. Ibargüengoytia

Instituto de Investigacions Eléctricas, Av. Reforma 113, Cuernavaca, Morelos, 62490, México

L. Enrique Sucar

Instituto Nacional de Astrofísica, Óptica y Electrónica, Tonantzintla, Puebla, 72840, México

and

Sunil Vadera

University of Salford, Salford M5 4WT, United Kingdom

11.1   Introduction

Computers are playing an increasingly important role in many activities related to human daily life. Environments like finance in banks, medicine in hospitals, communication in telephones are obvious examples of this. Other domains using computers, less related with our daily life are industry, transportation and manufacturing. Also, other applications described in this book are good examples of computer-controlled processes. In general, all these computer controllers receive information, process it and provide an output or take a decision.

Imagine a decision taken or a command issued based on erroneous information. Consider for example an intensive care unit of a hospital. Complex equipment and instrumentation are used to constantly monitor the status of a patient. Suppose that the body temperature and the blood pressure must be kept under certain levels.

Bayesian Networks: A Practical Guide to Applications   Edited by O. Pourret, P. Naïm, B. Marcot
© 2008 John Wiley & Sons, Ltd

However, given that the sensors are working constantly under different patient conditions, there is some potential for them to produce erroneous readings. If this happens two situations may arise:

- a sensor indicates no changes in temperature even if it increments to dangerous levels;

- a sensor indicates a dangerous level even if it is normal.

The first situation may cause severe damage to the patient's health. This is called in the literature a missing alarm. The second situation may cause emergency treatment of the patient that can also worsen his/her condition. This is called in the literature a false alarm.

This chapter presents a sensor validation algorithm based on Bayesian networks. The algorithm starts by building a model of the dependencies between sensors represented as a Bayesian network. Then the validation is done in two phases. In the first phase, potential faults are detected by comparing the actual sensor value with the one predicted from the related sensors, via propagation in the Bayesian network. In the second phase, the real faults are isolated by constructing an additional Bayesian network based on the Markov blanket property [356]. This isolation is made incrementally or *any time*, so the quality of the estimation increases when more time is spent in the computation, making the algorithm suitable for use in real-time environments. The sensor validation algorithm is applied to validate the temperature sensors of a gas turbine in a power generation unit.

11.2   The problem of sensor validation

In the context of this work, a sensor is a device that receives an input V_v which is considered unknown and inaccessible, and provides an output which is a measurement V_m. A sensor is considered faulty when the measurement V_m gives an incorrect representation of V_v. In the case of power plants, V_v may be temperature, or pressure, or flow of gas of a turbine. In other domains, there can be different sources of information like human reports or complex instrumentation. Thus, the challenge is the determination of the validity of V_m in real time. This section presents an overview of the different forms proposed for the determination of the validity of V_m.

11.2.1   Traditional approaches

The traditional method to validate information is the use of *redundancy*. This means the use of different sources of the same information in order to make some kind of majority voting. The sources that provide similar values are considered correct. The question is, what kind of different sources can be used? There are four basic approaches:

1. Hardware redundancy: duplicate physical instrumentation.

2. Analytical redundancy: calculate the value using mathematical models.

3. Temporal redundancy: repeat measurements and apply statistics.

4. Knowledge based methods: use of human expert knowledge.

Hardware redundancy is not always possible for economical, spatial or security reasons. For example, adding new sensors might weaken the walls of pressure vessels.

Analytical redundancy refers to the creation of mathematical models based on the static and dynamic relations between measurements. This technique predicts a sensor's value by using measures from other sensors in known or empirical relations among all the variables. However, this approach becomes inefficient when there are many sensors, and when the complexity of the model increases. Additionally, the validation of sensors using analytical redundancy is adapted exclusively for a given process. A slight modification is extremely expensive and demands an enormous amount of expertise.

Temporal redundancy is difficult to apply in dynamic environments where parameters vary in time and there is no sense in repeating a measurement in different time periods.

11.2.2   Knowledge based and neural network approaches

An alternative approach is to use knowledge based techniques to detect inconsistencies in measured data. Knowledge based systems model the knowledge that a human expert employs to recognize inconsistencies in the reported measurements. Using this knowledge, the system can *reason* to infer the state of the set of sensors. In this classification, there are also approaches from computational intelligence like neural networks or fuzzy logic.

A neural network consists of a network whose nodes are arranged in layers. The arcs resemble the interconnection between neurons (the nodes) in the brain. Figure 11.1, left, exemplifies this method. The first layer is the input layer with the number of nodes as the number of input variables. The output layer produces

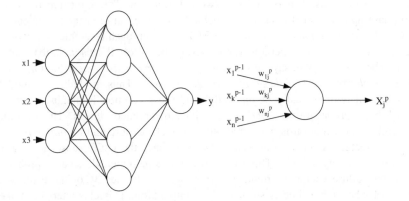

Figure 11.1   A basic neural network. Left: a typical network. Right: a *perceptron* or basic neuron.

the output of the network. In the figure there is only one output: the predicted signal. There can be none or several hidden layers. Figure 11.1, right, represents the jth neuron of the p layer. This neuron receives n inputs from layer $p - 1$ and produces one output that will be connected to neurons in layer $p + 1$. The output level will be the sum of all the inputs multiplied by a weight w_{ij}^p, and modified by a sigmoidal function.[1] These weights are learned during a training phase. To train the network, several examples of a learned concept are presented and several examples of what is not the concept are also presented. Consequently, the weights are defined so that future examples receive a similar conclusion. This is a typical method for parameter estimation. Several authors utilize this method in their sensor validation approach [308, 511].

Since knowledge based sensor validation involves imprecise and uncertain human knowledge, some authors utilize fuzzy logic. Here, measurements are assigned a membership function where certain sensor value belongs to a certain fuzzy set in some degree and maybe, belongs to another set with different degree. For example, a set of cool, and a set of hot temperatures can be defined. Using these sets, fuzzy rules can be used to infer the membership of a measurement to the fuzzy set of correct values [214].

Other approaches include hybrid methods that combine some of the approaches mentioned above. For example, Alan and Agogino developed a methodology for sensor validation that includes four steps [12]. The first step creates redundancy in the signals with some method, e.g., analytical redundancy or neural networks. The second step includes a state prediction with time series. The third step fuses this information and detects faulty behavior, and finally the fault is detected comparing the fused estimate and the sensor reading.

11.2.3   Bayesian network based approach

The approach presented in this chapter estimates the value of a variable given the related ones, using a Bayesian network formalism. The algorithm utilizes a two phase approach analogous to the methodology used in industry for diagnosis, namely *Fault Detection and Isolation* or FDI. This approach resembles the reasoning process of an expert operator when revising the set of all sensors in the control room. The operator may say: *this probably is a wrong measurement from this sensor given the readings of these other related sensors.* Thus, probabilistic models are captured (automatically or manually) and the prediction is obtained through probabilistic propagation of one sensor given the related signals. The resulting probabilistic distribution is compared with the real measure and an *apparent fault* can be detected. A second Bayesian network isolates the faulty sensor among all the apparent faulty sensors. This isolation is made incrementally, i.e., a probability of the failure vector is provided at *any time*, so the quality of the estimation increases when more time is spent in the computation. This characteristic makes the algorithm suitable for use in real time environments.

---

[1] A sigmoidal is an S-shaped function equivalent to $1/(1 + e^{-x})$.

The contributions of this approach are several as follows.

1. The model can be easily learned automatically with data obtained from the process. It is important to note that we need only data from the process working properly for the operating range. It is not necessary to simulate faults to recognize them later. On the contrary, neural networks require a training phase with data without faults and data with faults, in order to classify the output.

2. Most of the learning algorithms for Bayesian networks allow the participation of human experts in the formation of the models. This means that it is possible to build very efficient models relating all the variables, using human experience. Neural networks are closed in this sense.

3. All the sensors that are represented in a model can be validated. In the neural network case, every single sensor requires its own network. Also, our approach can detect multiple faulty sensors.

4. Bayesian network models work well even with incomplete information. The resulting probabilities may not be very accurate but they often provide useful information. On the contrary, neural networks require complete input information to provide good responses.

5. The isolation of faults is made with a second Bayesian network that is directly constructed from the first one using the Markov blanket property. This phase produces a vector with the probability of fault for all the sensors.

6. The algorithm proposed here performs in real time environments thanks to an *any-time* scheme. We can get an idea of the state of the sensors even with little time and incomplete information from other sensors.

11.3   Sensor validation algorithm

This section presents the proposed sensor validation algorithm based on Bayesian networks. The basic idea is the estimation of a certain sensor value in order to compare it with the real observed value. In this chapter, estimation means the calculation of a posterior probability distribution of a single variable, when related variables have been read and considered evidence in the Bayesian network.

It is assumed that it is possible to build a probabilistic model relating all the variables in the application domain. Consider for example the network shown in Figure 11.2. It can represent the most basic function of a gas turbine. The power generated in a gas turbine (node MW) depends on the temperature (node T) and pressure in the turbine (node Pt). Temperature depends on the flow of gas (node Fg) and this flow depends on the valve of gas position (node Pv) and the gas fuel pressure supply (node Ps). The pressure at the turbine depends on the pressure at the output of the compressor (node Pc). This model can be obtained from domain experts or by automatic learning algorithms based on historical data [336].

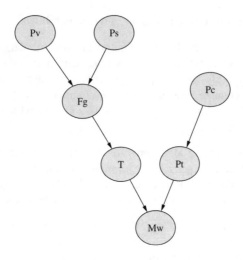

Figure 11.2 A basic probabilistic model of a gas turbine.

Suppose that it is required to validate the temperature measurements in the turbine. Reading the values of the rest of the sensors, and applying a propagation algorithm, it is possible to calculate a posterior probability distribution of the temperature given all the evidence, i.e., $P(T \mid MW, Pt, Fg, Pc, Pv, Ps)$. We obtain a certain probability distribution. We assume that all the variables are discrete or discretized if continuous. Thus, if the real observed value falls in this probability distribution, then the sensor is considered correct, and faulty otherwise.

The basic validation algorithm or detection function is the following:

Algorithm 4 Basic validation or detection algorithm

Require: A node n.
Ensure: Either correct or faulty.
 1: assign a value (instantiate) to all nodes except n
 2: propagate probabilities and obtain a posterior probability distribution of n
 3: read real value of sensor represented by n
 4: **if** $P(real\ value) \geq p\ value$ **then**
 5:    return(correct)
 6: **else**
 7:    return(faulty)
 8: **end if**

The parameter $p\ value$ is a threshold chosen appropriately for each application.

This procedure is repeated for all the sensors in the model. However, if a validation of a single sensor is made using a faulty sensor, then a faulty validation can be expected. In the example above, what happens if T is validated using a faulty MW sensor? How can we know which is the liar sensor? Thus, making this validation procedure, we may only detect a faulty condition but we are not able to identify the real faulty sensor. This is called an *apparent fault*. An isolation stage is needed.

The isolation phase utilizes a property of Bayesian networks called the *Markov Blanket* (MB). By definition [356], the Markov blanket of a node X, $(MB(X))$, is the set of nodes that make X independent of all other nodes in a network. In Bayesian networks, the next three sets of nodes conform the Markov blanket of a node: (i) the set of parents, (ii) the set of children, and (iii) the other parents of the children nodes (spouses). For example in the network of Figure 11.2, the Markov blanket of T, $MB(T)$, is $\{MW, Fg, Pt\}$, and the Markov blanket of Pv, $MB(Pv)$, is $\{Fg, Ps\}$. The set of nodes that constitute the MB of a sensor can be seen as a protection of the sensor against changes outside its MB. Additionally, we define the *Extended Markov Blanket* of a node X, $(EMB(X))$, as the set of sensors formed by the sensor itself plus its MB. For example, the extended Markov blanket of Fg, $EMB(Fg)$, is formed by $\{Fg, Pv, Ps, T\}$.

Utilizing this property, if a fault exists in one of the sensors, it will be revealed in all the sensors on its EMB. On the contrary, if a fault exists outside a sensors' EMB, it will not affect the estimation of that sensor. It can be said then, that the EMB of a sensor acts as its protection against others faults, and also protects others from its own failure. We utilize the EMB to create a *fault isolation* module that distinguishes the *real faults* from the apparent faults. A more formal description of the algorithm can be found in [223].

After a cycle of basic validation of all sensors is completed, a set S of apparent faulty sensors is obtained. Thus, based on the comparison between S and the EMB of all sensors, the theory establishes the following situations:

1. If $S = \phi$ there are no faults.

2. If S is equal to the EMB of a sensor X, and there is no other EMB which is a subset of S, then there is a *single real fault* in X.

3. If S is equal to the EMB of a sensor X, and there are one or more EMBs which are subsets of S, then there is a real fault in X, and possibly, real faults in the sensors whose EMBs are subsets of S. In this case, there are possibly *multiple indistinguishable* real faults.

4. If S is equal to the union of several EMBs and the combination is unique, then there are *multiple distinguishable* real faults in all the sensors whose EMB are in S.

5. If none of the above cases is satisfied, then there are multiple faults but they cannot be distinguished. All the variables whose EMBs are subsets of S could have a real fault.

These situations occur after the basic validation function described in algorithm 4. This algorithm utilizes the model like the one shown in Figure 11.2, and is called the detection network. For example, considering the Bayesian network model in Figure 11.2, some of the following situations may occur (among others):

- $S = \{T, Pt, MW\}$, which corresponds to case 2, and confirms a single real fault in MW;

- $S = \{T, Pc, Pt, MW\}$, which corresponds to case 3, so there is a real fault in Pt and possibly in Pv and MW;

- $S = \{Pv, Ps, Fg\}$, which corresponds to case 4, so there are real faults in Pv and Ps.

The isolation of a real fault is carried out in the following manner. Based on the EMB property described above, there will be a real fault in sensor X if an apparent fault is detected in all its EMB. Also we can say that an apparent fault will be revealed if there exists a real fault in any sensor of its EMB. With these facts, we define the isolation network formed by two levels. The root nodes represent the real faults, and there is one per sensor or variable. The lower level is formed by one node representing the apparent fault for each variable. Notice that the arcs are defined by the EMB of each variable. Figure 11.3 shows the isolation network for the detection network of Figure 11.2. For instance, the apparent fault node corresponding to variable MW (node A_{mw}) is connected with the nodes R_{mw}, R_T and R_{Pt}, which represent the real faults of the EMB nodes of MW. At the same time, node R_{mw} is connected with all the apparent faults that this real fault causes, i.e., to nodes A_{mw}, A_T, and A_{Pt}. This is carried out by the isolation function described in algorithm 5.

Algorithm 5 Isolation algorithm using the isolation network

Require: A sensor n and the state of sensor n.
 1: assign a value (instantiate) to the apparent fault node corresponding to n
 2: propagate probabilities and obtain a posterior probability of all nodes *Real fault*
 3: update vector $P_f(sensors)$

The vector $P_f(sensors)$ updated in this function corresponds to the set of real fault nodes in the isolation network. In the example developed in this section,

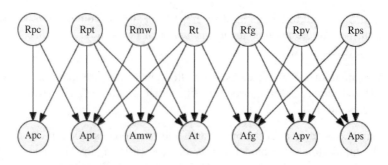

Figure 11.3 Isolation network of the example in Figure 11.2.

$P_f(Pc)$ corresponds to the posterior probability of node R_{pc} in Figure 11.3, $P_f(Pt)$ to R_{pt} and so on.

This algorithm works well in batch mode, i.e., a conclusion can be obtained after the whole validation has been executed. However, for real time environments, this sometimes is not possible, as a response could be required before the validation cycle is completed. In these situations, it is better to obtain an answer at any time, even if the *quality* of the results is not the best. *Any-time algorithms* provide a result at any time and the quality of the result increments as more time is spent in the computation. The sensor validation algorithm can also work in an any-time fashion.

Algorithm 6 Any time sensor validation algorithm

1: Initialize $P_f(s_i)$ for all variables s_i.
2: **while** there are unvalidated variables **do**
3:      select the next variable to validate
4:      validate the selected variable
5:      update the probability of failure vector P_f
6:      measure the quality of the partial response
7: **end while**

First, at the beginning of a validation cycle, the ignorance condition is assumed and the vector of $P_f(s_i)$ for all variables s_i are set at 0.5. Next, we need to put emphasis on the any-time behavior of the algorithm. This is achieved if we provide a partial response at each step. And this response must be the best possible. This can be done if we select the sensor that provides more information. For example, in Figure 11.2 it is easy to see that validating node Fg would provide more information than the validation of Pc. If no apparent fault is presented in Fg, then it is considered that there will be no fault in Pv, Ps, Fg, nor in T. On the contrary, if an apparent fault is found in Fg, then the suspicious real fault is between Pv, Ps, Fg, or in T.

One efficient way to measure the amount of information provided by an experiment is the *entropy* function I proposed by Shannon for communication theory [425]:

$$I(p_1, p_2, \ldots, p_n) = \sum_{i=1}^{n} P_i \log_2 P_i$$

where p_i represents the probability of the outcomes of an experiment. If a single validation of a sensor s_i is taken as an experiment, then the entropy function $I(s_i)$ is defined by:

$$I(s_i) = -p \log_2(p) - (1 - p) \log_2(1 - p) \tag{11.1}$$

where p refers to the failure probability $P_f(s_i)$ of sensor s_i (the value of node R_i of the isolation network). When $I(experiment) = 0$ the function can safely assumed to be 0. This function has its maximum when the ignorance is at a maximum, and it is zero when the information is maximum and ignorance is minimum.

In the sensor validation problem, the system's entropy function $IVAL(S)$ is defined as:

$$IVAL(S) = -\frac{1}{n} \sum_{i=1}^{n} P_f(s_i) \log_2 P_f(s_i) + (1 - P_f(s_i)) \log_2(1 - P_f(s_i))$$

Notice that $IVAL(S) = \sum I(s_i)$, so its maximum is when all functions I in the system are at their maximum, and the minimum $IVAL(S)$ is when all I are zero. Therefore, in case of single failures, the expected value of the function $IVAL(S)$ is zero, either because there were no faulty sensors ($P_f = 0$ for all sensors), or one of the sensors is faulty ($P_f = 1$ for some sensor). Consequently, $IVAL(S)$ can be utilized as a scheduling criterion for the selection of the sensor that, after validation, provides the more information, i.e., reduces the updated $IVAL(S)$. Also, $IVAL(S)$ can be used as a quality measure for the evaluation of partial results provided by the validation of single sensors.

For example, consider the network shown in Figure 11.2 and its corresponding isolation model shown in Figure 11.3. Suppose that there is a failure in the value of sensor Fg. Table 11.1 shows the probability vector, P_f, for all the stages in this example. The first row indicates the initial state in a cycle when assumed maximum ignorance. Assume that the selection of a sensor to validate follows this ordering: Pt, T, Fg, MW, Pv, Ps, and finally Pc. First, the basic validation of Pt will produce an apparent fault given that its estimation was made using an erroneous value. This implies the assignment of the value *true* to node Apt in Figure 11.3. By propagating in the isolation model, the probability of a real fault for all the sensors is updated, as shown in the second row in Table 11.1. Then, the second sensor is validated, T, and the real failure probabilities are updated. Following the procedure for all nodes will result in the detection of a definite real fault in node Fg with 99% of probability, and a suspicious fault in Pv and Ps with 54% of probability. This confirms the conditions mentioned above about indistinguishable double faults. However, it is clear in this case that the faulty sensor is simply Fg.

Table 11.1   Example of the values of the probability vector P_f (in %).

Step / P_f	(Pv)	(Ps)	(Fg)	(Pc)	(T)	(Pt)	(MW)
initial state	50	50	50	50	50	50	50
$Pt = OK$	50	50	50	9	9	9	9
$T = Fault$	50	50	80	9	15	15	15
$Fg = Fault$	52	52	83	9	15	13	13
$MW = OK$	51	51	97	9	2	2	2
$Pv = Fault$	53	53	98	9	2	2	2
$Ps = Fault$	54	54	98	9	2	2	2
$Pc = OK$	54	54	99	1	2	0	2

11.4   Gas turbines

Gas turbines have become one of the most popular methods throughout the world for generation of electricity due to their efficiency and low pollution. This section briefly describes what kind of power plants utilize gas turbines, how gas turbines operate and the importance of measuring temperatures around the turbine.

11.4.1   Power plants using gas turbines

Gas turbines generate electricity in two kinds of units. The first kind and more efficient is the combined cycle power plant [355]. This kind of plant uses the gas turbines coupled with electric generators to obtain part of the production. In addition, the hot gases expelled from the turbine are recovered in a special boiler called the *heat recovery steam generator* or HRSG. This equipment generates steam and then feeds a steam turbine also coupled with other generator. Usually, there are two gas turbines per steam turbine. The second kind consists of turbogas units, i.e., the gas turbine alone coupled to the generator. These units are mainly used as backup in peak hours given their fast response. In Mexico for example, there are 12 combined cycle plants, generating 42 000 gigawatthour (GWh) a year, corresponding to 45% of the national production. The turbogas units generate 350 GWh a year, corresponding to just 0.4% of the national production.[2]

11.4.2   Operation of a gas turbine

Figure 11.4 shows a schematic of a gas turbine. The turbines generate power as follows. Air at ambient conditions enters the compressor through inlet air valves. Air is compressed to reach some higher pressure. At this point, compression raises the air temperature so the air at the discharge of the compressor is at higher

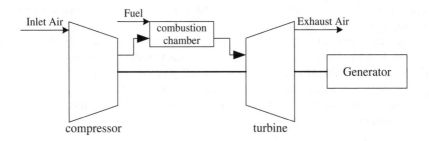

Figure 11.4 Schematic diagram of a gas turbine. Air enters and is compressed at the compressor and combined with fuel at the combustion chamber. The combustion produces hot gases flowing through the blades of the turbine producing work.

---

[2]Information obtained from the Federal Commission of Electricity (CFE) in Mexico, at www.cfe.gob.mx.

temperature and pressure. When leaving the compressor, air enters the combustion chamber, where fuel is injected and mixed with the air and combustion occurs. Here at the combustion chamber, high temperatures are reached. Thus, by the time the combustion mixture leaves the combustion chamber and enters the turbine, it is at its maximum temperature. In the turbine section, the energy of the expanded hot gases is converted into work. This conversion actually takes place in two steps. In the nozzle section of the turbine, the hot gases are expanded and a portion of the thermal energy is converted into kinetic energy. In the subsequent bucket section of the turbine, a portion of the kinetic energy is transferred to the rotating buckets and converted to work. Some of the work developed by the turbine is used to drive the compressor, and the remainder is available for useful work at the output flange of the gas turbine. Typically, more than 50% of the work developed by the turbine sections is used to power the axial flow compressor. The rest is used to drive the electric generator [63].

Summarizing, electric power is obtained through the conversion of mechanical work in the electric generator. Mechanical work is obtained with the expansion of hot gases moving through the blades of the turbine. Hot gases are obtained with the combustion of the mixture of gas and compressed air. In this cycle, thermal efficiency is the key for optimum turbine performance.

11.4.3   Temperature measures of a gas turbine

Increasing the temperature provides power increases and hence, a decrement in the fuel costs. However, there exist security levels for the maximum allowed temperatures. If these limits are exceeded, the operating life span of some components is reduced, and for extreme conditions the entire equipment is at risk. Additionally, there are many external factors that also contribute to the thermal efficiency of the turbine. Some of them are air ambient temperature and humidity, site elevation, input fuel temperature, and the heat energy associated to the kind of fuel used. Thus, temperatures in several parts of the turbine are the most important variables for the control of the plant.

Several temperature sensors participate in the decisions taken by the control system. If one or more sensors fail, this could make the control system take incorrect decisions that could affect the efficiency or safety of the power plant. So it is very important to detect and isolate errors in the temperature sensors.

The next section presents the empirical evaluation of the algorithm in the power plant domain.

11.5   Models learned and experimentation

This section describes the application of the sensor validation algorithm explained in Section 11.3, with the application domain revised in Section 11.4. First, experience with the learning process are presented. Later, the experiments and results are commented.

11.5.1   Learning models of temperature sensors

The sensor validation algorithm was applied to a set of 21 temperature sensors of a gas turbine in the CFE's Gómez Palacio combined cycle plant in Mexico.

Figure 11.5 shows the physical location of some of the temperature sensors used in the turbine. It shows six sensors across the beadings of the shaft (CH1, CH2, ..., CH6), i.e., the places where the shaft is supported. Three sensors are included in the turbine blades (EM1, EM2 and EM3), and two sensors of the temperature of the exciter air (AX1 and AX2). The experiments were carried out over a set of 21 sensors (though not all are shown in Figure 11.5). These sensors can be grouped into the following sets of measurements: six beadings (CH1–CH6), seven disk cavities (CA1–CA7), one cavity for air cooling (AEF), two exciter air (AX1–AX2), three blade paths (EM1–EM3), and two lube oil (AL1–AL2).

The instrumentation of the plant provides readings of all the sensors every second. The data set utilized in the experiments corresponds to the temperature readings taken during approximately the first 15 minutes after the start of the combustion. This corresponds to the start-up phase of the plant, where the thermodynamic conditions change considerably. Therefore, the data set consists of 21 variables and 870 instances of the readings.

The first step in the experiments was the acquisition of the model using the data set and the Hugin software package [14]. For learning, Hugin includes the NPC algorithm [441]. NPC stands for *Necessary Path Condition*, and it is a method that generates a proposed structure derived through statistical tests for conditional independence among all the nodes. In this process, some uncertain links may appear that are presented to the user for taking the final decision. With no human participation and the Hugin NPC algorithm, we obtain the network shown in Figure 11.6.

However, the structure is a multiply connected network, and the validation phase may take too long time to compute for real time applications. Notice, for example, node CH1 in Figure 11.6, that has six parents, two children and eight spouses, so its EMB has 17 nodes. To avoid the problem of computational complexity in the validation phase, we opted for simpler structures, in particular for tree-structured networks. In this kind of network, all the nodes can have at most one parent. This allows using propagation algorithms that are linear with respect to the number of nodes. Consequently, we wrote a computer program that uses the tree learning algorithm [96] for acquisition of the network shown in Figure 11.7.

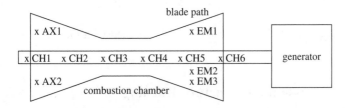

Figure 11.5   Location of the sensors in the turbine.

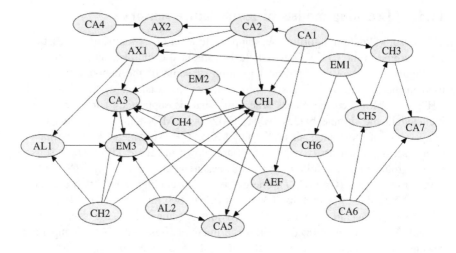

Figure 11.6 Bayesian network learned with the data set.

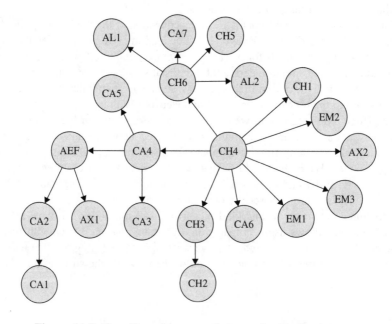

Figure 11.7 Tree Bayesian network learned with the data set.

11.5.2 Discussion on structure learning

Structure learning for on-line applications like the sensor validation, becomes a central issue. Expressive models represent closely the probabilistic relations between

the variables in the application. However, these models are usually very interconnected. This interconnection makes bigger *extended Markov blankets* resulting in
computationally expensive models. On the other hand, a tree algorithm is faster to
compute but some important probabilistic relations may be missing. In the middle,
most of the learning algorithms commercially available allow human participation.
For example, in the Hugin·s NPC algorithm mentioned above, the human expert
is presented with some uncertain links for the final decision: to delete a link, or to
include it with certain direction. Beside this, the expert may indicate some present
links (and some absent links) that the application demands even if these relations
are not represented in the data.

For the sensor validation application, we required a tree model, given the speed
requirements. The experiments presented next give reasons for this choice.

11.5.3   Experimental results

Two measures were considered in the empirical evaluation of the algorithm. First,
we evaluated the number of false alarms. This means the number of times that
the validation algorithm detected an alarm that did not exist. Second, we evaluated
the missing alarms. This means the number of times that the validation algorithm
did not detect an alarm that did exist. Table 11.2 shows the percentage of missing
alarms and false alarms for two kind of faults. A severe fault was simulated by
changing the real measure in 50% of its value, while a mild fault consisted of a
change of 25% of its original value.

False alarms imply that most of the sensors in a EMB present apparent false
alarms. This is more common as can be seen in Table 11.2. That is, there are
cases where the existence of an invalid apparent fault, together with the valid ones,
completes the EMB of a misdiagnosed sensor. Hence, a false alarm is produced.
On the contrary, missing alarms are detected at this stage when most of the sensors
of a EMB present misdiagnosed apparent faults. This is very improbable as the
results of Table 11.2 confirm. The percentages are obtained comparing the average
number of errors with the total number of experiments.

The parameter for changing these numbers is the *p value* defined in algorithm 4,
and may change according to the application. There may be applications where a
false alarm is not very important compared with a missing alarm of an important
abnormal event. On the contrary, some applications prefer to miss some alarms but
when an alarm is detected, drastic action can be taken.

Table 11.2   Percentage of errors for severe
and mild faults.

Fault	False alarms	Missing alarms
Severe	2.9 %	0.4 %
Mild	6.5 %	5.8 %

The any-time characteristic of the sensor validation algorithm was also considered in the experiments. We compared the performance of the sensor validation algorithm with the entropy criteria for selecting the next sensor to validate, against a random selection. This comparison indicates that, when the random criterion reaches 20% of accuracy, the entropy criteria had already reached 50% of accuracy in the results. When the random criterion reaches 60%, the entropy criterion reaches 80%.

11.6   Discussion and conclusion

This chapter has presented an application of Bayesian networks in the validation of sensors in an industrial environment. Specifically, the experiments considered the temperature sensor of a gas turbine, one of the most popular machines for electricity generation around the world.

We utilize two Bayesian networks for the any-time sensor validation system. The first network is a model representing the probabilistic relationship between all the signals in the application. This is called the detection network and can be learned from historical data, and expert participation if needed. The second Bayesian network is a causal model formed by real and apparent faulty nodes. If the detection function observes a fault, the corresponding apparent fault node is instantiated and propagation is called to update the real fault nodes. This set of nodes forms the output vector. At every step of validation, a more accurate response is produced, giving the system an *any-time* behavior, appropriate for real time applications.

One advantage of this approach is that it is only necessary to acquire information while the process (a turbine in this case) is working properly. It is not necessary to simulate faults in order to identify them later. Another advantage in this validation is in the complexity of the system when the number of nodes increases. As established in Section 11.3, the problem of complexity is located in the size of the largest Markov blanket among the models. This signifies that we can have as many nodes or variables that the process may require, if the interconnection between them is not too high. This may be a characteristic of the domain, or it can be forced by the learning algorithm, considering simple models such as trees.

The sensor validation presented in this chapter can be also applied in a general diagnosis of equipment where the requirement is the detection of abnormal behavior of the system. This abnormal behavior is immediately shown in the behavior of the signals. Other applications include the detection of false information even when this information is not provided by sensors. It can detect liar agents that manipulate information. Additionally, this technique is being used in the construction of *virtual sensors*, i.e., by estimating the value of an unobservable sensor based on the readings of related sensors.

12

An information retrieval system for parliamentary documents

Luis M. de Campos, Juan M. Fernández-Luna, Juan F. Huete, Carlos Martín, Alfonso E. Romero

Departamento de Ciencias de la Computación e Inteligencia Artificial, E.T.S.I. Informática y de Telecomunicaciones, Universidad de Granada, 18071 – Granada, Spain

12.1 Introduction

One of the main objectives of a democracy is that citizens know what their representatives in the parliament are dealing with at each moment. Therefore, national and regional parliaments, in particular those in Spain, are obliged to publish the work developed in these houses of representatives in order to make public all the matters being discussed. They have to offer the transcriptions of all the sessions, containing all the participations (reports of proceedings or session diaries), as well as all the text and documents generated by the parliamentary activity whose publication is mandatory (official bulletins). A few years ago, these texts were printed versions sent to all the official organizations or libraries, but nowadays the new information technologies have changed the method of publishing: given the development of new standards to represent and exchange documents electronically and the applications to work with them, as well as the spread of Internet use, parliaments

Bayesian Networks: A Practical Guide to Applications   Edited by O. Pourret, P. Naïm, B. Marcot
© 2008 John Wiley & Sons, Ltd

have combined these two complementary facets to create an alternative publishing method: electronic documents published on the web. This means three important advantages with respect to the old system: a faster spreading of the discussions carried out in the sessions, a cheaper way of publishing and an easier way of accessing the diaries and bulletins.

In the specific case of the regional Parliament of Andalucía, the southern Spanish region, when any of the different types of sessions (plenary, commissions, . . .) are held, a transcription of all the speeches given in the sessions is obtained. After an editing process, a document in Portable Document Format (PDF) is obtained and published on the Parliament website.[1] The official bulletins are also edited, converted to PDF and published in the website. Then, these documents can be accessed through the web by means of a database-like query, knowing the number of session, the date of holding or the legislature to which they belong to. This is an important drawback for a fast and easy access to the required documents. The user must know all that information to get a set of documents and after (s)he has to inspect them in order to find what (s)he is looking for.

Therefore, in order to fully exploit all the potential of this new system, we need not only an easy way of accessing the diaries and bulletins but also an easy way of *accessing the specific information contained in them*. So, the goal is to endow the website with a real search engine based on content, where the user expresses her/his information need using a natural language query (for instance, 'money given to agriculture in the budgets of 2005') and obtains immediately the set of relevant documents, sorted according to its degree of relevance. This problem falls within the scope of the so-called *Information Retrieval* field. Moreover, we also want to take advantage of the structure of the session diaries and official bulletins, in order to better determine which documents or which parts of these documents are truly relevant. So, the output of the system for a given query (e.g., 'information about bills relative to infrastructure and transport policies' or 'oral questions answered by Mr. Chaves – the prime minister of the autonomous government – in plenary sessions of the sixth legislature') will be a set of *document components* of varying granularity (from complete document to a single paragraph, for example), also sorted depending on its degree of relevance. In this way the user only sees the more relevant parts, avoiding the (manual) search of the requested information within the complete documents. The subfield of information retrieval dealing with documents having a well-defined structure is known as *structured information retrieval* or *structured document retrieval*.

This work describes both the theoretical foundations and practical aspects of a software system that has been designed and implemented for storing and accessing the document collections of the Parliament of Andalucía, taking into account their specific features, with the aim of bringing these texts nearer to the people, in an effective and efficient way. An outstanding feature of our system is that its search engine is based on probabilistic graphical models, namely Bayesian networks and influence diagrams.

---

[1] www.parlamentodeandalucia.es

12.2   Overview of information retrieval and structured information retrieval

Information retrieval (IR) is concerned with representation, storage, organization, and accessing of information items [406]. An information retrieval system (IRS) is the software that implements these tasks in a computer. A great number of retrieval models, i.e., specifications about how to represent documents and queries and how they are compared, have been developed [31]. Given a set of documents in their original format, the first step is to translate each document to a suitable representation for a computer. That translation is called *indexing*, and the output is usually a list of words, known as terms or keywords, extracted from the text and considered significant. Sometimes, and previously to the indexing, a preprocess is run, removing words that are not useful at all for retrieval purposes[2] (stopwords), and/or converting each word to its morphological representative[3] (stemming), in order to reduce the size of the vocabulary. The terms usually have associated weights expressing their importance (either within each document or within the collection, or both). The IRS stores these representations instead of the complete documents. The user expresses his/her information needs formulating a query, using a formal query language or natural language. This query is also indexed to get a query representation and the retrieval continues matching the query representation with the stored document representations, using a strategy that depends on the retrieval model being considered. Finally, a set of document identifiers is presented to the user sorted according to their relevance degree.

Standard IR treats documents as if they were atomic entities, so usually only entire documents constitute retrievable units. However, more elaborate document representation formalisms, like XML, allow us to represent so-called *structured documents*, whose content is organized around a well defined structure that enables us to describe the semantics of complex and long documents [93]. Examples of these documents are books and textbooks, scientific articles, technical manuals, etc, and also official bulletins and session diaries in a parliament. The structure of documents is therefore 'flattened' and not exploited by classical retrieval methods. Structured information retrieval views documents as aggregates of interrelated structural elements that need to be indexed, retrieved, and presented both as a whole and separately, in relation to the user's needs. In other words, given a query, a structured IR system must retrieve the set of document components that are most relevant to this query, not just entire documents. Structured IR models exploit the content and the structure of documents to estimate the relevance of document components to queries, usually based on the aggregation of the estimated relevance of their related components [262].

---

[2]Because they carry little meaning, as articles, prepositions, conjunctions, etc.

[3]For example the words conspirant, conspirator, conspirators, conspire, conspired, conspirers, conspires, conspiring are all converted to their common morphological root 'conspir.'

12.3   Bayesian networks and information retrieval

Probabilistic models constitute an important kind of IR model. They have been long and widely used [108], and offer a principled way of managing the uncertainty that naturally appears in many elements within this field. These models compute the probability of relevance given a document and a query (the probability that a document satisfies a query). Bayesian networks, which nowadays are one of the most important approaches for managing probability within the field of artificial intelligence, have also been used within IR as an alternative to classical probabilistic models.

Bayesian networks were first applied to IR at the beginning of the 1990s [456, 457], with the so-called *inference network model* developed by Croft and Turtle. From this original model, the InQuery retrieval system was developed [80], which at present is one of the main experimental and commercial software packages in this area. Since then, many models and applications have been developed. We shall mention only another two of the most representative approaches (for a more detailed account of the use of BNs within IR, see [119]): Ribeiro-Neto and co-researchers developed the *belief network model* [394] and, more recently, we designed the *Bayesian network retrieval model* [1, 118].

These three models have several features in common: each index term and each document in a given collection are represented as nodes in the network, and there are links connecting each document node with all the term nodes that represent the index terms associated to this document. Figure 12.1 displays an example of this basic topology. However, these models also differ in many aspects: the direction of the arcs (from term nodes to document nodes or vice versa), the existence of additional nodes and arcs (e.g., to represent term relationships as well as document relationships), the way of computing the probability distributions stored in the nodes of the graph, and the way in which they carry out the retrieval of documents (computing the probability of a query node being satisfied by a given document or computing the probability of a document node being relevant given either the query node or the index terms appearing in the query).

In any case, the development of solutions for IR problems has been very challenging and imaginative, because of two, a priori, drawbacks of Bayesian networks: firstly, the time and space required to assess the distributions and store them (the number of conditional probabilities per node is *e*xponential with the number of the

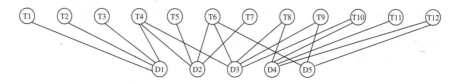

Figure 12.1 Basic network topology common to several BN-based retrieval models, connecting terms (T_i) and documents (D_i) where they appear.

node's parents); secondly, the efficiency of carrying out inference, because general inference in BNs is an NP-hard problem [100]. For example, if we want to represent a document collection using a BN, where the nodes represent documents, terms and queries, and there are, among others, arcs linking terms with the documents where they appear, this representation will be very expensive in terms of space and time: if a document is indexed by, say 500 terms, or a term is used to index 500 documents (depending on the direction of the arcs), the computation and storage of all the probability distributions are simply intractable tasks. But even in the case of coping with it, the retrieval of the documents relevant for a given query would take a long time, a fact that is not allowed in interactive environments. This means that a direct approach where we propagate the evidence contained in a query through the whole network representing a collection is clearly unfeasible even for small collections. Therefore, research effort has been necessary to overcome these problems, in order to obtain efficient and effective BN-based solutions for the IR context. The first problem can be fixed by defining and using new canonical models of multicausal interaction, whereas the second requires the development of inference algorithms able to take advantage of both these canonical models and the restricted network topologies being considered.

Given that BNs are powerful tools to represent and quantify the strength of relationships between objects, they have also been applied to structured document retrieval. The hierarchical structure of the documents may be qualitatively represented by means of a directed acyclic graph, and the strength of the relations between structural elements may be captured by a quantitative probabilistic model [107, 120, 189, 368]. The propagation through the network would give the relevance probability of all the structural elements comprising the document as a result. This problem is even more challenging, because the BN representing a collection of structured documents is much greater than that representing a 'flat' document collection.

In the next section we shall describe in more detail the model underlying the search engine implemented in our application, which is based on Bayesian networks and influence diagrams [120, 121].

12.4   Theoretical foundations

First, we describe a BN model for structured document retrieval and next extend it to an influence diagram.

12.4.1   The underlying Bayesian network model for structured documents

We start from a document collection containing M *documents*, $\mathcal{D} = \{D_1, \ldots, D_M\}$, and the set of the *terms* used to index these documents (the glossary or lexicon of the collection). Each document D_i is organized hierarchically, representing structural associations of elements in D_i, which will be called *structural units*. Each structural unit is composed of other smaller structural units, except some 'terminal' units which do not contain any other unit but are composed of terms (each term used to index the complete document D_i will be assigned to all the terminal units containing it). Conversely, each structural unit, except those corresponding to the complete documents, is included in only one structural unit. Therefore, the structural units associated to a document D_i form an (inverted) tree.

For instance, a scientific article may contain a title, authors, abstract, sections and bibliography; sections may contain a title, subsections and paragraphs; in turn subsections contain paragraphs and perhaps also a title; the bibliography contain references; titles, authors, paragraphs, abstract and references would be in this case the terminal structural units (see Figure 12.2).

The BN modeling the document collection will contain three kinds of nodes, representing the terms and two different types of structural units. The former will be represented by the set $\mathcal{T} = \{T_1, T_2, \ldots, T_l\}$. The two types of structural units are: *basic structural units*, those which only contain terms, and *complex structural units*, that are composed of other basic or complex units, $\mathcal{U}_b = \{B_1, B_2, \ldots, B_m\}$ and $\mathcal{U}_c = \{S_1, S_2, \ldots, S_n\}$, respectively. The set of all the structural units is $\mathcal{U} = \mathcal{U}_b \cup \mathcal{U}_c$. To each node T, B or S is associated a binary random variable, which can take its values from the sets $\{t^-, t^+\}$, $\{b^-, b^+\}$ or $\{s^-, s^+\}$, respectively, representing that the term/unit is not relevant (−) or is relevant (+). A unit is relevant for a given query if it satisfies the user's information need expressed by it. A term is relevant in the sense that the user believes that it will appear in relevant units/documents.

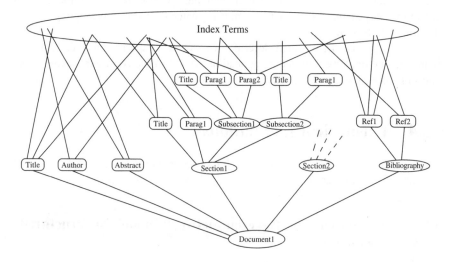

Figure 12.2 The structure of a scientific article.

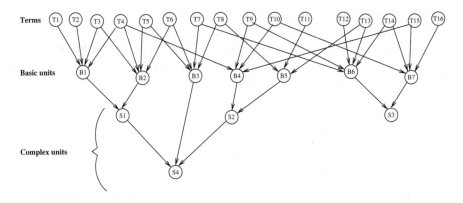

Figure 12.3 Bayesian network representing a structured document collection.

Regarding the arcs of the model, there is an arc from a given node (either term or structural unit) to the particular structural unit node it belongs to.[4] It should be noticed that terms nodes have no parents and that each structural unit node $U \in \mathcal{U}$ has only one complex structural unit node as its child, corresponding to the unique structural unit containing U, except for the leaf nodes (i.e., the complete documents), which have no child. We shall denote by $S_{ch(U)}$ the single child node associated with node U (with $S_{ch(U)} = null$ if U is a leaf node). Formally, the network is characterized by the following parent sets, $Pa(.)$, for the different types of nodes:

- $\forall T \in \mathcal{T}$, $Pa(T) = \emptyset$.

- $\forall B \in \mathcal{U}_b$, $\emptyset \neq Pa(B) \subseteq \mathcal{T}$.

- $\forall S \in \mathcal{U}_c$, $\emptyset \neq Pa(S) \subseteq \mathcal{U}_b \cup \mathcal{U}_c$, with $Pa(S_1) \cap Pa(S_2) = \emptyset$, $\forall S_1 \neq S_2 \in \mathcal{U}_c$.

Figure 12.3 displays an example of the proposed network topology.

The numerical values for the conditional probabilities have also to be assessed: $\mathbb{P}(t^+)$, $\mathbb{P}(b^+|pa(B))$, $\mathbb{P}(s^+|pa(S))$, for every node in $\mathcal{T}$, $\mathcal{U}_b$ and $\mathcal{U}_c$, respectively, and every configuration of the corresponding parent sets ($pa(X)$ denotes a configuration or instantiation of $Pa(X)$).

In our case, the number of terms and structural units considered may be quite large (thousands or even hundreds of thousands). Moreover, the topology of the BN contains multiple pathways connecting nodes (because the terms may be associated to different basic structural units) and possibly nodes with a great number of parents (so that it can be quite difficult to assess and store the required conditional probability tables). For these reasons, to represent the conditional probabilities, we

---

[4]Therefore the relevance of a given structural unit to the user will depend on the relevance values of the different elements (units or terms) that comprise it.

use the *canonical model* proposed in [118], which supports a very efficient inference procedure. We have to consider the conditional probabilities for the basic structural units, having a subset of terms as their parents, and for the complex structural units, having other structural units as their parents. These probabilities are defined as follows:

$$\forall B \in \mathcal{U}_b, \; \mathbb{P}\left(b^+|pa(B)\right) = \sum_{T \in R(pa(B))} w(T, B), \quad (12.1)$$

$$\forall S \in \mathcal{U}_c, \; \mathbb{P}\left(s^+|pa(S)\right) = \sum_{U \in R(pa(S))} w(U, S), \quad (12.2)$$

where $R(pa(U))$ is the subset of parents of U (terms for B, and either basic or complex units for S) relevant in the configuration $pa(U)$, i.e., $R(pa(B)) = \{T \in Pa(B) \,|\, t^+ \in pa(B)\}$ and $R(pa(S)) = \{U \in Pa(S) \,|\, u^+ \in pa(S)\}$. So, the more parents of U are relevant the greater the probability of relevance of U. $w(T, B)$ is a weight associated to each term T belonging to the basic unit B and $w(U, S)$ is a weight measuring the importance of the unit U within S. These weights can be defined in any way, the only restrictions are that $w(T, B) \geq 0$, $w(U, S) \geq 0$, $\sum_{T \in Pa(B)} w(T, B) \leq 1$, and $\sum_{U \in Pa(S)} w(U, S) \leq 1$.

With respect to the prior probabilities of relevance of the terms, $\mathbb{P}\left(t^+\right)$, they can also be defined in any reasonable way, for example an identical probability for all the terms, $\mathbb{P}\left(t^+\right) = p_0, \forall T \in \mathcal{T}$, as proposed in [120].

12.4.2   The influence diagram model

Once the BN has been constructed, it is enlarged by including decision and utility nodes, thus transforming it into an influence diagram. We use the following topology:

(a) *Chance nodes*: those of the previous BN.

(b) *Decision nodes*: one decision node, R_U, for each structural unit $U \in \mathcal{U}$. R_U represents the decision variable related to whether or not to return the structural unit U to the user. The two different values for R_U are r_U^+ and r_U^-, meaning 'retrieve U' and 'do not retrieve U', respectively.

(c) *Utility nodes*: one of these, V_U, for each structural unit $U \in \mathcal{U}$, will measure the value of utility of the corresponding decision.

In addition to the arcs between chance nodes (already present in the BN), a set of arcs pointing to utility nodes are also included, employed to indicate which variables have a direct influence on the desirability of a given decision, i.e., the profit obtained will depend on the values of these variables. As the utility function of V_U obviously depends on the decision made and the relevance value of the structural unit considered, we use arcs from each chance node U and decision node R_U to the utility node V_U. Another important set of arcs are those going from $S_{ch(U)}$ to V_U, which represent that the utility of the decision about retrieving the unit U also depends on the relevance of the unit which contains it.[5] This last kind of arc allows us to represent the context-based information and can avoid redundant information being shown to the user. For instance, we could express the fact that, on the one hand, if U is relevant and $S_{ch(U)}$ is not, then the utility of retrieving U should be large; on the other hand, if $S_{ch(U)}$ is relevant, even if U were also relevant the utility of retrieving U should be smaller because, in this case, it may be preferable to retrieve the largest unit as a whole, instead of each of its components separately.

We use another utility node, denoted by Σ, that represents the joint utility of the whole model, having all the utility nodes V_U as its parents. These arcs represent the fact that the joint utility of the model will depend (additively) on the values of the individual utilities of each structural unit. Figure 12.4 displays an example of the topology of the proposed influence diagram model, associated to the BN model shown in Figure 12.5.

Moreover, the influence diagram requires numerical values for the utilities. For each utility node V_U we need eight numbers, one for each combination of

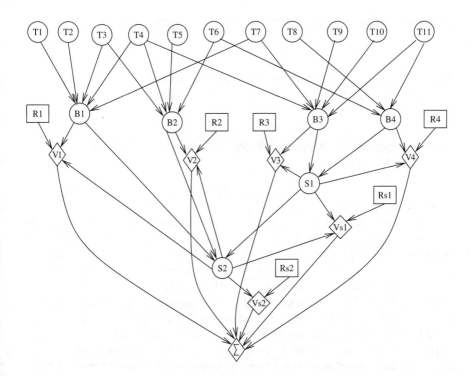

Figure 12.4 Topology of the influence diagram model associated with the Bayesian network in Figure 12.5.

[5]Obviously, for the units which are not contained in any other unit these arcs do not exist.

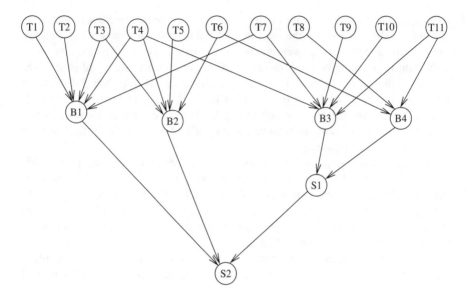

Figure 12.5 Basic Bayesian network topology underlying the influence diagram in Figure 12.4.

values of the decision node R_U and the chance nodes U and $S_{ch(U)}$. These values are represented by $v(r_U, u, s_{ch(U)})$, with $r_U \in \{r_U^-, r_U^+\}$, $u \in \{u^-, u^+\}$, and $s_{ch(U)} \in \{s_{ch(U)}^-, s_{ch(U)}^+\}$.

12.4.3   Retrieving structural units: Solving the influence diagram

To solve an influence diagram, the expected utility of each possible decision (for those situations of interest) has to be computed, thus making decisions which maximize the expected utility. In our case, the situation of interest corresponds to the information provided by the user when (s)he formulates a query. Let $\mathcal{Q} \subseteq \mathcal{T}$ be the set of terms used to express the query. Each term $T \in \mathcal{Q}$ will be instantiated to t^+; let q be the corresponding configuration of the term variables in $\mathcal{Q}$. We wish to compute the expected utility of each decision given q, i.e., $EU(r_U^+ \mid q)$ and $EU(r_U^- \mid q)$. In the context of a typical decision making problem, once the expected utilities are computed, the decision with greatest utility is chosen: this would mean retrieving the structural unit U if $EU(r_U^+|q) \geq EU(r_U^-|q)$, and not retrieving it otherwise. However, our purpose is not only to make decisions about what to retrieve but also to give a ranking of those units. The simplest way to do it is to show them in decreasing order of the utility of retrieving U, $EU(r_U^+|q)$, although other options would also be possible. In this case only four utility values have to be assessed, and only the computation of $EU(r_U^+|q)$ is required.

As we have assumed a global additive utility model, and the different decision variables R_U are not directly linked to each other, we can process each independently. The expected utility for each U can be computed by means of

$$EU(r_U^+ \mid q) = \sum_{\substack{u \in \{u^-, u^+\} \\ s_{ch(U)} \in \{s_{ch(U)}^-, s_{ch(U)}^+\}}} v(r_U^+, u, s_{ch(U)}) \, \mathbb{P}\left(u, s_{ch(U)} \mid q\right). \qquad (12.3)$$

According to Equation (12.3), in order to provide to the user with an ordered list of structural units, we have to be able to compute the posterior probabilities of relevance of all the structural units $U \in \mathcal{U}$, $\mathbb{P}\left(u^+ | q\right)$, and also the bi-dimensional posterior probabilities, $\mathbb{P}\left(u^+, s_{ch(U)}^+ | q\right)$. Notice that the other required bi-dimensional probabilities, $\mathbb{P}\left(u^+, s_{ch(U)}^- | q\right)$, $\mathbb{P}\left(u^-, s_{ch(U)}^+ | q\right)$ and $\mathbb{P}\left(u^-, s_{ch(U)}^- | q\right)$, can be easily computed from $\mathbb{P}\left(u^+, s_{ch(U)}^+ | q\right)$, $\mathbb{P}\left(u^+ | q\right)$ and $\mathbb{P}\left(s_{ch(U)}^+ | q\right)$. The specific characteristics of the canonical model used to define the conditional probabilities will allow us to efficiently compute the posterior probabilities.

The unidimensional posterior probabilities can be calculated as follows [118, 120]:

$$\forall B \in \mathcal{U}_b, \; \mathbb{P}\left(b^+ | q\right) = \mathbb{P}\left(b^+\right) + \sum_{T \in Pa(B) \cap Q} w(T, B)\left(1 - \mathbb{P}\left(t^+\right)\right). \qquad (12.4)$$

$$\forall S \in \mathcal{U}_c, \; \mathbb{P}\left(s^+ | q\right) = \mathbb{P}\left(s^+\right) + \sum_{U \in Pa(S)} w(U, S)\left(\mathbb{P}\left(u^+ | q\right) - \mathbb{P}\left(u^+\right)\right). \qquad (12.5)$$

So, the posterior probabilities of the basic units can be computed directly, whereas the posterior probabilities of the complex units can be calculated in a top-down manner, starting from those for the basic units. However, it is possible to design a more direct inference method for the complex units. Let us define the set $A_b(S) = \{B \in \mathcal{U}_b \mid B \text{ is an ancestor of } S\}$, $\forall S \in \mathcal{U}_c$. Notice that, for each basic unit B in $A_b(S)$, there is only one path in the graph going from B to S. Let us define the weight $w'(B, S)$ as the product of the weights of the arcs in the path from B to S. Then, the posterior probabilities of the complex units can be calculated as follows [121]:

$$\forall S \in \mathcal{U}_c, \; \mathbb{P}\left(s^+ | q\right) = \mathbb{P}\left(s^+\right) + \sum_{\substack{B \in A_b(S) \\ Pa(B) \cap Q \neq \emptyset}} w'(B, S)\left(\mathbb{P}\left(b^+ | q\right) - \mathbb{P}\left(b^+\right)\right). \qquad (12.6)$$

Equations (12.4) and (12.6) are the basis for developing an inference process able to compute all the posterior probabilities of the structural units in a single traversal of the graph, starting only from the instantiated terms in Q, as we shall see later.

The other required probabilities are the posterior bi-dimensional probabilities

$$\mathbb{P}\left(u^+, s_{ch(U)}^+ | q\right),$$

for any structural unit $U \in \mathcal{U}$ and its unique child $S_{ch(U)}$, provided that $S_{ch(U)} \neq$ *null*. Although they can be computed exactly and in a way much more efficient than using a classical BN inference algorithm [121], even this process could be expensive in terms of memory and time for very large document collections. For that reason we can use two approximations: the simpler is to assume the independence between each structural unit and the one which contains it [120], i.e.,

$$\mathbb{P}\left(u^+, s^+_{ch(U)}|q\right) = \mathbb{P}\left(u^+|q\right) \mathbb{P}\left(s^+_{ch(U)}|q\right). \tag{12.7}$$

The other, which is finer and can be computed as efficiently as the previous one, is based on the exact formulas developed in [121]:

$$\mathbb{P}\left(u^+, s^+_{ch(U)}|q\right) = \mathbb{P}\left(u^+|q\right) \mathbb{P}\left(s^+_{ch(U)}|q\right)$$
$$+ w(U, S_{ch(U)}) \mathbb{P}\left(u^+|q\right)\left(1 - \mathbb{P}\left(u^+|q\right)\right). \tag{12.8}$$

In any case the computation of these approximations of the bi-dimensional probabilities can be done starting from the corresponding unidimensional probabilities and the weight of each unit U within its child unit $S_{ch(U)}$.

Example

To illustrate the inference mechanism, let us consider a simple example, where there are three documents, corresponding with Sections 12.2, 12.3 and 12.4 of this chapter. Moreover, we use as indexing terms only the words appearing in the titles of these sections and the corresponding subsections (performing stemming[6] and excluding stopwords and words appearing only once). The Bayesian network representing this 'collection' is displayed in Figure 12.6. This collection contains eight terms, five basic and one complex structural units (excluding the leaf node representing the complete collection). We shall use the weighting scheme proposed in [120] (see Subsection 12.5.2 for details). The resulting weights of the arcs are also displayed in Figure 12.6, and the prior probability of relevance of all the terms has been set to 0.1. The utility values are $v(r^+_U, u^+, s^+_{ch(U)}) = 0.5$,

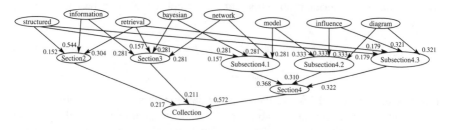

Figure 12.6 The Bayesian network representing part of this chapter (number 12 is omitted).

---

[6]So that the words 'retrieving' and 'retrieval', 'networks' and 'network', 'structural' and 'structured' become single terms.

Table 12.1   Posterior probabilities and expected utilities for queries Q_1 and Q_2.

	Section 12.2	Section 12.3	Subsec. 12.4.1	Subsec. 12.4.2	Subsec. 12.4.3	Section 12.4	Collection
$\mathbb{P}(u^+\mid q_1)$	0.863	0.494	0.100	0.700	0.839	0.524	0.591
$EU(r_U^+\mid q_1)$	**0.540**	0.076	-0.381	0.392	**0.557**	0.159	–
$\mathbb{P}(u^+\mid q_2)$	0.100	0.606	0.606	0.700	0.678	0.658	0.526
$EU(r_U^+\mid q_2)$	-0.390	0.265	0.191	0.305	0.278	**0.368**	–

$v(r_U^+, u^+, s_{ch(U)}^-) = 1$, $v(r_U^+, u^-, s_{ch(U)}^+) = -1$, $v(r_U^+, u^-, s_{ch(U)}^-) = 0$ for all the structural units.

Let us study the output provided by the model for two queries Q_1 and Q_2, where Q_1 is 'information retrieval, influence diagrams' and Q_2 is 'Bayesian networks, influence diagrams'. After instantiating to relevant these terms, we propagate this evidence through the network. The posterior probabilities of the structural units are displayed in Table 12.1. For Q_1, Section 12.2 and Subsection 12.4.3 are the most relevant structural units (as Section 12.2 speaks mainly about information retrieval[7] and Subsection 12.4.3 speaks about influence diagrams and retrieval). For Q_2, Subsection 12.4.2 is the most relevant unit (it is devoted exclusively to influence diagrams), although Section 12.3 and Subsection 12.4.1 (which speak about Bayesian networks) and Subsection 12.4.3 (dealing with influence diagrams) also gets a relatively high probability (lower than that of Subsection 12.4.2 because they also speak about other topics outside the query). However, for Q_2 it seems to us that retrieving Section 12.4 as a whole would be more useful for the user than retrieving its subsections. If we compute the expected utilities (Table 12.1 also displays all the utility values) using the approximation in Equation (12.8), we can see that Section 12.4 gets the highest value for Q_2, whereas Section 12.2 and Subsection 12.4.3 maintain the highest value for Q_1, as desired.

12.5   Building the information retrieval system

PAIRS[8] is a software package specifically developed to store and retrieve documents generated by the Parliament of Andalucía, based on the probabilistic graphical models described in the previous section. Written in C++, following the object-oriented paradigm, it offers a wide range of classes and a complete set of utility programs. Basically, this system allows us, on the one hand, to index the document collections and manage the created indexes and, on the other hand, retrieve relevant (parts of) documents given a query formulated by a user. Figure 12.7 shows a general scheme of PAIRS.

In the following sections we shall study some characteristics of the document collections as well as the architecture of PAIRS from the indexing and querying points of view.

---

[7]Notice that the words 'information' and 'retrieval' appear twice in Section 12.2.

[8]Acronym of Parliament of Andalucía Information Retrieval System.

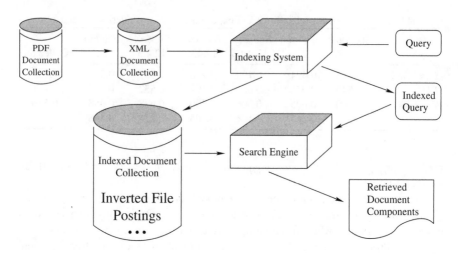

Figure 12.7 General scheme of PAIRS.

12.5.1   The document collections

As we mentioned earlier, the documents we are dealing with (session diaries and official bulletins) have a high degree of structuring, so that we decided to represent them using XML (Extensible Markup Language), which at present is the most standard format for structured documents and data interchange on the web.

The internal structure of all the session diaries is basically as follows:

(a) Information about the session: date, number, legislature, type, etc.

(b) Agenda of the session: headlines of all the points to be discussed, organized by types of initiatives.

(c) Summary: detailed description of the agenda, specifying those deputies who took part in the discussions and the results of the votes (added once the session is finished).

(d) Development of the session: exact transcription of all the speeches.

On the other hand, the official bulletins are composed of two sections. The first one is a summary of the content included in the second part. A brief reference to each parliamentary initiative developed in the main body of the document is included in the first part, as a kind of table of contents. There exists a well defined and static hierarchical taxonomy of initiatives designed by the Parliament (e.g., law projects, oral and written questions, motions, ...) so all the initiatives presented in each document are arranged in the corresponding place of the hierarchy. The second part, the body of the document, is organized around the same taxonomy presented in the table of contents, but developing the content of each initiative. Inside each text, basically the structure is the following: number of the initiative, title, who presents it, who answers (in case of a question to the regional government, for example), date, and the text itself explaining the request or the answer. Then, we designed the corresponding DTDs (document type definitions) to capture the structure of both types of documents in detail.

As all the documents are in PDF format but we need to manage plain text, we have used tools that extract the text of a PDF document. PDFBox and Pdftotext, are two free source libraries able to manipulate PDF documents. Once we have extracted the text of the documents, the next step is to identify the textual content associated with the structural units found in these documents, i.e., to transform the plain text into structured XML text conforming the specifications of the previously defined DTDs. The identification of this information is a quite difficult task which is carried out basically by means of the use of regular expressions trying to find the corresponding patterns in the text.

12.5.2   Indexing subsystem

The objective of the indexing process is to build the data structures that will allow fast access to those documents and structural units where query terms occur and their associated weights, in order to enable efficient retrieval.

Within the IR field, there are several approaches for tackling the problem of indexing structured documents,[9] mainly marked up in XML [290]. Basically, they either store the tree representing the document collection or use the classical approach of inverted file containing the occurrences of the terms in XML tags and any kind of data structure to represent the internal organization (hierarchy) of the documents, as in [455]. This is the philosophy that PAIRS implements, differing in the data structure used.

12.5.2.1   Application level

Obviously PAIRS is able to manage heterogeneous collections (having documents with different DTDs), and also different indexes over the same collection. Thus, it provides a software module, `makeIndex` to do this task. It can choose among different stopword lists and use (if desired) a stemming algorithm. This task is carried out by the SnowBall linguistic analyzer for the Spanish language, which applies a set of rules adapted to this language in order to remove prefixes and suffixes and extract the stems.

The performance on an IRS depends heavily of the weighting scheme being used, in our case the weights $w(T, B)$ of terms in the basic structural units and the weights $w(U, S)$ of units in the broader units containing them. In PAIRS several valid weighting schemes can coexist. As a consequence, indexing does not compute the weights (setting all of them to be zero). Instead of that, there is the possibility to calculate weights (following a certain weighting scheme) for previously built indexes without inserting into them, and store them in files – the so-called *weight*

---

[9]Other alternatives are based on storing documents in tables of relational databases and using the tools given by database management systems to retrieve documents, for instance, the SQL language.

files. So, records of that precomputed weight files are kept, and a fast way to insert one into the index itself is provided, in order to carry out the retrieval using it. To achieve these two tasks two separate program modules, `makeWeightFile` and `insertWeightFile`, are provided.

The specific weighting scheme that we are currently using is the following [120]: first, $\forall U \in \mathcal{U}$, let us define $A_t(U) = \{T \in \mathcal{T} \mid T$ is an ancestor of $U\}$, i.e., $A_t(U)$ is the set of terms that are included in any of the basic units which form part of U.[10] Given any set of terms C, let $tf(T, C)$ be the *frequency* of the term T (number of times that T occurs) in the set C and $idf(T)$ be the *inverse document frequency* of T in the whole collection, which is defined as $idf(T) = \ln(m/n(T))$, where $n(T)$ is the number of basic units that contain the term T and m is the total number of basic units. We define $\rho(T, C) = tf(T, C) \cdot idf(T)$. Then, the weighting scheme is:

$$\forall B \in \mathcal{U}_b, \ \forall T \in Pa(B), \quad w(T, B) = \frac{\rho(T, Pa(B))}{\sum_{T' \in Pa(B))} \rho(T', Pa(B))}; \quad (12.9)$$

$$\forall S \in \mathcal{U}_c, \ \forall U \in Pa(S), \quad w(U, S) = \frac{\sum_{T \in A_t(U)} \rho(T, A_t(U))}{\sum_{T \in A_t(S)} \rho(T, A_t(S))}. \quad (12.10)$$

It should be observed that the weights in Equation (12.10) are the classical tf-idf weights, normalized to sum one up. The weights $w(U, S)$ in Equation (12.10) measure, to a certain extent, the proportion of the content of the unit S which can be attributed to each one of its components U.

12.5.2.2   Physical level and data structures

In PAIRS we have emphasized querying time over indexing time, even storing some redundant information. Let us briefly describe the data structures associated to indexing.

To store textual information (terms and identifiers of the basic units where they appear), we use inverted indexes [502]. While the lexicon is kept entirely in memory (both while indexing and querying), the postings or lists of occurrences of terms are read from disk (for each term, a list of identifiers of basic units where this term appears, together with information about frequency and weight of the term within each unit). We use another file to write the list of relative positions of each term inside a unit, in order to answer queries containing either proximity operators or phrases.[11]

To maintain information about the structural units, we use several files, the two most important being: (1) A large direct access file that contains data of each unit itself (identifier, tag, position, ...), as well as the identifier of the container unit (i.e., its single child unit) and the corresponding weight; (2) another file containing the XPath route of each unit, i.e., the complete path from this unit to the root of the hierarchy (the unit representing the complete document). The first file is necessary

---

[10]Notice that $A_t(B) = Pa(B), \ \forall B \in \mathcal{U}_b$.

[11]This functionality is not yet implemented.

to efficiently traverse the network at propagation time, whereas the second is useful to provide to the user access to the units selected for retrieval after propagation.

To reduce the storage requirements of the files representing the indexed collection, some of these files are compressed (including both the postings and the XPath files, among others).

12.5.3   Querying subsystem

The query subsystem, once it receives a query, carries out the following steps:

1. The query is parsed, and the occurrences of the query terms are retrieved from disk.

2. For each occurrence, the implied basic units and their descendants are read to memory (if they were not already there).

3. Inference is carried out, computing first the posterior probabilities of the structural units and next the expected utilities; then units are sorted in descending ordering of their expected utility, and the result is returned.

It is worth noticing that the Bayesian network and the influence diagram underlying the search engine are never explicitly built; their functionality can be obtained by using the data structures that contain the structural and textual indexes.

The propagation algorithm used to efficiently compute the posterior probabilities, traversing only the nodes in the graph that require updating (those whose posterior probability is different from the prior), is shown here (Algorithm 7). It is assumed that the prior probabilities of all the nodes are stored in prior[X]; the algorithm uses variables prob[U] which, at the end of the process, will store the corresponding posterior probabilities. Essentially, the algorithm starts from the items in $\mathcal{Q}$ and carries out a width graph traversal until it reaches the basic units that require updating, thus computing $p(b^+|q)$ progressively, using Equation (12.4). Then, starting from these modified basic units, it carries out a depth graph traversal to compute $p(s^+|q)$, only for those complex units that require updating, using Equation (12.6).

The algorithm that initializes the process by computing the prior probabilities prior[U] (as the items $T \in \mathcal{T}$ are root nodes, the prior probabilities prior[T] do not need to be calculated, they are stored directly in the structure) is quite similar to the previous one, but it needs to traverse the graph starting from all the items in $\mathcal{T}$ instead from the items in $\mathcal{Q}$.

In order to compute the expected utilities, the parameters representing the utility values of the structural units have to be assessed. One easy way to simplify this task is to assume that these values do not depend on the specific structural unit being considered, i.e., $v(r_U^+, u, s_{ch(U)}) = v(r_{U'}^+, u', s_{ch(U')})$. In this way only four parameters are required.[12] However, PAIRS also uses another information source

---

[12]Another option so far not implemented in PAIRS would be to use different utility values for different types of units, reflecting user preferences about the desirability of more or less complex structural units.

Algorithm 7 Algorithm implemented in PAIRS to compute the posterior probabilities of the basic and complex structural units, $\mathbb{P}\left(b^+|q\right)$ and $\mathbb{P}\left(s^+|q\right)$

```
1:  for each item T in Q do
2:     for each unit B child of T do
3:        if (prob[B] exists) then
4:           prob[B] += w(T,B)*(1 − prior[T]);
5:        else
6:           create prob[B]; prob[B] = prior[B]+w(T,B)*(1 − prior[T]);
7:        end if
8:     end for
9:  end for
10: for each basic unit B s.t. prob[B] exists  do
11:    U = B; prod = prob[B] − prior[B];
12:    while (Ch(U) is not NULL) do {Ch(U) is the child of U}
13:       S = Ch(U);
14:       prod *= w(U,S);
15:       if (prob[S] exists) then
16:          prob[S] += prod;
17:       else
18:          create prob[S];prob[S] = prior[S]+prod;
19:       end if
20:       U = S;
21:    end while
22: end for
```

---

to dynamically define the utility values: the query itself. We believe that a given structural unit U will be more useful (with respect to a query Q) as more terms indexing U also belong to Q. We use the sum of the idfs of those terms indexing a unit U that also belong to the query Q, normalized by the sum of the idfs of all the terms contained in the query:

$$nidf_Q(U) = \frac{\sum_{T \in A_t(U) \cap Q} idf(T)}{\sum_{T \in Q} idf(T)}. \tag{12.11}$$

These values $nidf_Q(U)$ are used as a correction factor of the previously defined constant utility values, so that:

$$v^*(r_U^+, u, s_{ch(U)}) = v(r_U^+, u, s_{ch(U)}) \cdot nidf_Q(U). \tag{12.12}$$

Finally, in order to reduce at maximum the amount of disk accesses while processing a query, thus giving a shorter response time, PAIRS also stores in memory several unit objects (containing information about some units) in two caches, one for basic and other for complex units. Using a hash-function-like scheme,[13] we

---

[13]Identifier *mod* N, with N the size of the cache.

store in each cache slot either one basic or complex unit selected from all the candidates obtained doing the inverse of the hash function of its identifier. For complex units we choose the one closest to a leaf unit, whereas for basic units we select that one containing the greatest number of terms. In both cases the idea is to store in memory those units having more prior chance of being visited during the graph traversal carried out by the propagation algorithm.

12.5.4   Performance measures

In order to analyze the retrieval capabilities of an IR system, in terms of effectiveness, the standard approach is to generate a representative set of queries, determine the corresponding relevance judgements, i.e., the set of documents/units which would be truly relevant for each query, and compare the sets of retrieved documents/units against the relevant ones, thus obtaining several performance measures, such as *precision* (the proportion of retrieved documents that are relevant) and *recall* (the proportion of relevant documents retrieved).

From the point of view of efficiency for the tasks of indexing and retrieving, we have evaluated PAIRS in terms of disk space required by the indexes, and time to index the collection and to retrieve. The current number of documents in the collection of the Parliament of Andalucía (it is continuously growing) is: 3 097 official bulletins and 2 248 session diaries. The PDF versions of these documents take up around 6 gigabytes. The corresponding XML versions occupy around 1.2 gigabytes.

The last row in Table 12.2 displays several measures about the whole collection concerning sizes and times: the size of the collection and also the sizes of the weight file and the index file, all of them measured in Mbytes; the times required to index the collection, compute the weights and insert them into the index, measured in minutes, as well as their sum (the total indexing time); the average time (and the standard deviation) required to process a query and retrieve, measured in seconds. The first three rows in the table show the same measures for subcollections comprising around 25%, 50% and 75% of the whole collection, in order to get some insight about how these measures would increase when the size of the collection also increases. Figures 12.8 and 12.9 display plots of the collections sizes versus the total indexing times and average retrieval times, respectively.

Table 12.2   Index sizes, indexing and retrieval times for the whole collection (last row) and several sub-collections.

Size (Mbytes)			Time (min.)				Time (sec.)	
XML collection	Weight file	Index file	In-dexing	Computing weights	Inserting weights	Total indexing	Average retrieval	Std. dev. retrieval
301	34	262	4.67	0.95	0.67	6.28	0.12	0.32
611	96	826	15.10	3.27	2.08	20.45	0.44	0.65
913	126	1080	21.83	5.40	2.35	29.58	0.82	1.15
1240	174	1481	31.12	8.10	3.25	42.47	1.26	1.18

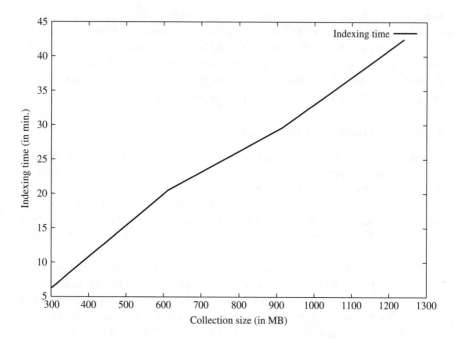

Figure 12.8 Collection size vs. total indexing time.

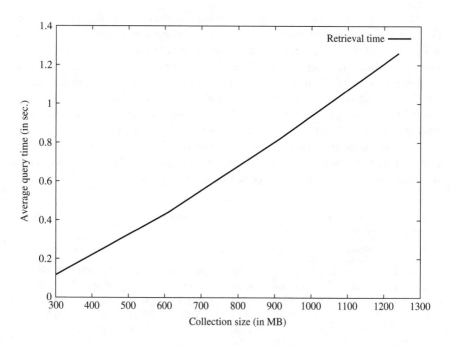

Figure 12.9 Collection size vs. retrieval time.

We can observe that the size of the indexes is only a bit larger than the size of the original XML collections. The indexing times are clearly affordable (remember that indexing is a task that is not very frequently carried out) and the retrieval times are quite low. It can also be observed that both the indexing and retrieval times are roughly linear with the size of the collections, so that the collection can still grow considerably without compromising the efficiency of the system.

12.6 Conclusion

In this chapter we have presented a system to retrieve legislative documents belonging to the collections of the Parliament of Andalucía, based on probabilistic graphical models. The main objective of this development is to bring nearer what politicians discuss in this chamber to general society, allowing normal users access to the documents by means of a query expressed in natural language. The system, which has been proved to be efficient in terms of indexing and retrieval times, also serves to illustrate how, through a combination of theoretical research effort and careful implementation, BN-based technologies can be employed in problem domains whose dimensionality would a priori prevent its use.

The system should not be regarded as a finished product; it is still open to several possible improvements that we plan to incorporate in the near future. Among them, we would stress the development of a more powerful query language, which allows users to formulate queries using exact phrases and also structural restrictions.

Acknowledgment

This work has been supported by the Spanish 'Consejería de Innovación, Ciencia y Empresa de la Junta de Andalucía', under Project TIC-276.

13

Reliability analysis of systems with dynamic dependencies

Andrea Bobbio, Daniele Codetta-Raiteri, Stefania Montani and Luigi Portinale

Dipartimento di Informatica, Università del Piemonte Orientale, Via Bellini 25 g, 15100 Alessandria, Italy

13.1 Introduction

Recent work performed by several researchers working in the dependability field have shown how the formalism of Bayesian networks can offer several advantages when analyzing safety-critical systems from the reliability point of view [49, 57, 266, 454, 483]. In particular, when the components of such systems exhibit dynamic dependencies, dynamic extensions of BN can provide a useful framework for the above kind of analysis [320, 321, 322, 483].

Traditionally, in order to deal with dynamic dependencies, the reliability engineer can resort to two different approaches:

- exploiting a state-space approach based on the generation of the underlying stochastic process;

- exploiting a dynamic extension of a combinatorial model.

Among the first category of approaches, we can find Markov chain models (both discrete and continuous), as well as the use of high level models like Petri nets or BNs where primitive entities are not the (global) states of the modeled system,

Bayesian Networks: A Practical Guide to Applications   Edited by O. Pourret, P. Naïm, B. Marcot
© 2008 John Wiley & Sons, Ltd

but partial state descriptions like places, transitions and tokens in Petri nets or random variables in BNs. In the case of Petri nets (and their stochastic extensions like *generalized stochastic Petri nets* (GSPN) [11] and *stochastic colored well-formed nets* (SWN) [95]) the global system states are obtained by generating all the reachable combinations of tokens in the places according to the enabling and firing rules of the Petri net semantics. In BN the global system states are obtained by considering the Cartesian product of the primitive random variables. For this reason, BN models are often referred to as 'factorized models' as well.

The most important approach in the second category is probably the formalism of a *dynamic fault tree* (DFT) [37] where, in addition to the usual combinatorial gates of a plain *fault tree* (FT) [416], several kinds of so-called 'dynamic gates' are introduced, in order to capture dynamic dependencies.

One of the advantages of the second category stands in the fact that the complexity underlying the generation (and the analysis) of a state-space model is restricted only to the portion of the model showing the dynamic dependencies. However, if dynamic dependencies involve a large portion of the model, the state space generation and analysis can still be problematic. This may be alleviated if the state-space model is generated resorting to a Petri net, either GSPN or SWN [50]. The standard measures that are computed from a fault tree or a dynamic fault tree DFT are the unavailability (or unreliability) of the root of the tree (the top event), or of any subtree, given the failure probability of each basic event, and criticality indices for the basic components given the top event has occurred [143].

In order to augment the modeling capabilities of the DFT approach and to provide a more extensive set of characterizing measure, we have developed an approach based on the generation of a *dynamic Bayesian network* (DBN) [327] model of the system to be analyzed. DBNs are a factorized representation of a stochastic process, where dependencies are modeled at the level of the net primitives, without the need of providing a global state description.

The approach has been implemented in a software tool named Radyban (Reliability Analysis with DYnamic BAyesian Networks) that is described in detail in [321]. The tool is able to:

- allow the user to implement the dependability model by resorting to an extension of the DFT formalism;

- translate the user model into an underlying DBN model;

- provide the analyst with a set of DBN inference algorithms implementing several reliability analysis.

A screenshot of the Radyban tool is shown in Figure 13.5 on page 237.

The present chapter aims at showing the advantages of such an approach, by illustrating several possible computations of reliability measures, as well as a comparison among some such measures as computed by our tool and by other reliability tools relying on standard DFT analysis (Galileo [38] or DRPFTproc [50, 379]).

13.2 Dynamic fault trees

13.2.1 Presentation

Dynamic fault trees (DFT) [37, 38], are a rather recent extension to FTs able to treat several types of dependencies. In particular, DFTs introduce four basic (dynamic) gates: the warm spare (WSP), the sequence enforcing (SEQ), the functional dependency (FDEP) and the priority AND (PAND) gate. A WSP dynamic gate models one primary component that can be replaced by one or more backups (spares), with the same functionality (see Figure 13.1(a), where spares are connected to the gate by means of 'circle-headed' arcs). The WSP gate fails if its primary component fails and all of its spares have failed or are unavailable (a spare is unavailable if it is shared and being used by another spare gate). Spares can fail even while they are dormant, but the failure rate of an unpowered (i.e., dormant) spare is lower than the failure rate of the corresponding powered one. More precisely, λ being the failure rate of a powered spare, the failure rate of the unpowered spare is $\alpha\lambda$, with $0 \le \alpha \le 1$ called the dormancy factor. Spares are more properly called 'hot' if $\alpha = 1$ and 'cold' if $\alpha = 0$.

A SEQ gate forces its inputs to fail in a particular order: when a SEQ is found in a DFT, it never happens that the failure sequence takes place in a different order. SEQ gates can be modeled as a special case of a cold spare [295].

In the FDEP gate (Figure 13.1b), one trigger event T (connected with a dashed arc in the figure) causes other dependent components to become unusable or inaccessible. In particular, when the trigger event occurs, the dependent components fail with probability $p_d = 1$; the separate failure of a dependent component, on the other hand, has no effect on the trigger event. FDEP has also a nondependent output, that simply reflects the status of the trigger event and is called a dummy output (i.e., not used in the analysis).

We have generalized the FDEP by defining a new gate, called the probabilistic dependency (PDEP) gate [320]. In the PDEP, the probability of failure of dependent components, given that the trigger has failed, is $p_d \le 1$.

Finally, the PAND gate reaches a failure state if and only if all of its input components have failed in a preassigned order (from left to right in the graphical

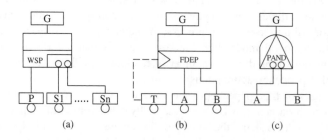

(a)   (b)   (c)

Figure 13.1 Dynamic gates in a DFT.

notation). While the SEQ gate allows the events to occur only in a preassigned order and states that a different failure sequence can never take place, the PAND does not force such a strong assumption: it simply detects the failure order and fails just in one case (in Figure 13.1c) a failure occurs iff A fails before B, but B may fail before A without producing a failure in G).

13.2.2   DFT analysis

The quantitative analysis of DFTs typically requires us to expand the model in its state space, and to solve the corresponding continuous time Markov chain (CTMC) [37]. Through a process known as modularization [142, 192], it is possible to identify the independent subtrees with dynamic gates, and to use a different Markov model (much smaller than the model corresponding to the entire DFT) for each one of them. In [50, 379], the modules are first translated into a GSPN or a SWN by a suitable graph transformation technique, and then the corresponding CTMC is automatically generated from the Petri net. Nevertheless, there still exists the problem of state explosion.

In order to alleviate this limitation, as stated above, we propose a conversion of the DFT into a *dynamic Bayesian network* (DBN) [327]. With respect to CTMCs, the use of a DBN allows one to take advantage of the factorization in the temporal probability model. As a matter of fact, the conditional independence assumptions implicit in a DBN enable us to confine the statistical dependence to a subset of the random variables representing component failures, providing a more compact representation of the probabilistic model. The system designer or analyst is faced with a more manageable and tractable representation where the complexity of specifying and using a global-state model (like a standard CTMC) is avoided; this is particularly important when the dynamic module of the considered DFT is significantly large.

13.3   Dynamic Bayesian networks

13.3.1   Presentation

DBN extends the standard Bayesian network formalism by providing an explicit discrete temporal dimension. They represent a probability distribution over the possible histories of a time-invariant process; the advantage with respect to a classical probabilistic temporal model like Markov chains is that a DBN is a stochastic transition model factored over a number of random variables, over which a set of conditional dependency assumptions is defined.

Given a set of time-dependent state variables $X_1, \ldots, X_n$ and given a BN N defined on such variables, a DBN is essentially a replication of N over two time slices t and $t + \Delta$ (Δ being the so-called discretization step), with the addition of a set of arcs representing the transition model. A DBN defined as above is usually called a 2-TBN (2-*time-slice temporal Bayesian network*). Letting X_i^t denote the copy of the variable X_i at time slice t, the transition model is defined through a distribution $P[X_i^{t+\Delta}|X_i^t, Y^t, Y^{t+\Delta}]$ where Y^t is any set of variables at slice t other than X_i (possibly the empty set), while $Y^{t+\Delta}$ is any set of variables at slice $t + \Delta$ other than X_i. Arcs interconnecting nodes at different slices are called interslice edges, while arcs interconnecting nodes at the same slice are called intraslice edges. For each internal node, the conditional probabilities are stored in the conditional probability table (CPT) in the form $P[X_i^{t+\Delta}|X_i^t, Y^t, Y^{t+\Delta}]$.

The conversion of the dynamic gates of Figure 13.1 into a DBN is considered at length in [321, 322], where it is shown that $Y^{t+\Delta}$ is nonempty only in the case of the PDEP gate conversion.

Of course a DBN defined as above (i.e., a 2-TBN) represents a discrete Markovian model. The two slices of a DBN are often called the *anterior* and the *ulterior* layer. Finally, it is useful to define the set of *canonical variables* as

$$\{Y : Y^t \in \bigcup_k \text{parents}[X_k^{t+\Delta}]\}; \tag{13.1}$$

they are the variables having a direct edge from the anterior layer to another variable in the ulterior layer. A DBN is in *canonical form* if only the canonical variables are represented at slice t (i.e., the anterior layer contains only variables having an influence on the same variable or on another variable at the ulterior layer).

Given a DBN in canonical form, interslice edges connecting a variable in the anterior layer to the same variable in the ulterior layer are called *temporal arcs*; in other words, a temporal arc connects the variable X_i^t to the variable $X_i^{t+\Delta}$. The role of temporal arcs is to connect the nodes representing the copies of the same variable at different slices. It follows that no variable in the ulterior layer may have more than one entering temporal arc.

In previous work [320, 321, 322], we have shown that a DFT characterized as above can be translated into a DBN in canonical form and the software tool Radyban [321] has been developed to automate this process, as well as to edit and work with the resulting DBN for possibly augmenting the modeling features.

13.3.2   Algorithms for DBN analysis

Concerning the analysis of a DBN, different kinds of inference algorithms are available. In particular, let X^t be a set of variables at time t, and $y_{a:b}$ any stream of observations from time point a to time point b (i.e., a set of instantiated variables Y_i^t with $a \leq t \leq b$). The following tasks can be performed over a DBN:

- **Filtering** or **monitoring**: computing $\mathbb{P}\left(X^t|y_{0:t}\right)$, i.e., tracking the probability of the system state taking into account the stream of received observations.

- **Prediction**: computing $\mathbb{P}\left(X^{t+h}|y_{0:t}\right)$ for some horizon $h > 0$, i.e., predicting a future state taking into consideration the observation up to now (filtering is a special case of prediction with $h = 0$).

- **Smoothing**: computing $\mathbb{P}\left(X^{t-l}|y_{0:t}\right)$ for some $l < t$, i.e., estimating what happened l steps in the past given all the evidence (observations) up to now.

In particular, the difference between a filtering and a smoothing inference relies on the fact that in the former case, while computing the probability at time t ($0 \leq t \leq T$), only the evidence gathered up to time t is considered; on the contrary, in the case of smoothing the whole evidence stream is always considered in the posterior probability computation. It should also be clear that the specific task of prediction can be obtained by asking for a time horizon T greater than the last time point considered for an observation.

The classical computation of the unreliability of the top event of a (D)FT is a special case of filtering, with an empty stream of observations (i.e., it is a filtering assuming $y_{0:0}$).

Smoothing may be, for instance, exploited in order to reconstruct the history of the system components for a kind of temporal diagnosis (e.g., given that the system has been observed failed at time t, to compute the probability of failure of basic components prior to t).

Different algorithms, either exact (i.e., computing the exact probability value that is required by the task) or approximate, can be exploited in order to implement the above tasks and some of them are available in Radyban (see [321] for more details).

13.4   A case study: The Hypothetical Sprinkler System

In order to illustrate the previously described approach, we take into consideration a slight variation of a simple case study introduced in [315]: the Hypothetical Sprinkler System (HSS).[1] HSS is a computer-aided sprinkler system composed of three sensors, two pumps, and one digital controller (see the block scheme in Figure 13.2).

Each pump has a support stream composed of valves and filters; each pump requires that the pump stream be operational in order to start (so if the stream is down, the pump is down as well). The sensors send signals to the digital controller, and when temperature readings at two of the sensors are above threshold, the controller activates the pump. HSS is operational if the controller is operational, at least two of the sensors are operational, and at least one of the pumps starts and operates. If a pump is activated on demand, then the filters and valves in the pump stream are in working condition.

The number of pumps is two, because one of the pumps is considered as a backup pump which runs if the primary pump fails. Since the backup pump is activated only if the primary pump fails, the former is considered to be a 'cold spare'. As said above, a system failure occurs if both pumps fail.

---

[1]We just consider the case where there is no distinction between the system being in demand or in stand-by mode, resulting in the first DFT model introduced in [315].

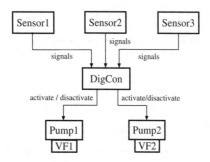

Figure 13.2 The block scheme of the HSS.

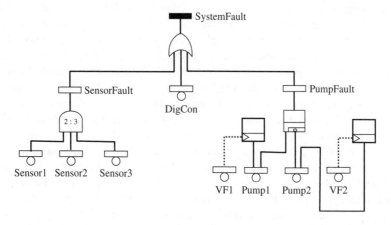

Figure 13.3 The dynamic fault tree (DFT) for the HSS system.

The component dependencies and the failure mode of the system can be modeled using the DFT in Figure 13.3.

In the DFT, the functional dependence of each pump on its associated valves and filters is captured by a PDEP gate with probability $p_d = 1$ (i.e., a FDEP gate); dashed arcs are used to identify the trigger event. A WSP gate with dormancy factor equal to 0 is used to model the cold spare relationship between the two pumps, with the circle-headed arrow to indicate the spare component (Pump2). The k:n gate over the sensors is actually a 2:3 gate and the whole system fault occurs when either the sensor subsystem fails or the digital controller (DigCon) fails or the pump subsystem fails.

The time to fail of any component in the system is considered to be a random variable ruled by the negative exponential distribution; Table 13.1 shows the failure rate of every component.

Figure 13.4 shows the canonical form DBN, for the HSS system, obtained from the DFT of Figure 13.3.

Table 13.1   The failure rates in the HSS example.

Component	Failure rate (λ)
Sensors	$10^{-4} \ h^{-1}$
Valves and filters (VF)	$10^{-5} \ h^{-1}$
Pumps	$10^{-6} \ h^{-1}$
Digital controller (DigCon)	$10^{-6} \ h^{-1}$

The nodes of the DBN derive from the translation of the basic events and of the gates (both static and dynamic) inside the DFT; we just show in Figure 13.4 the structure of the DBN and not the conditional probability tables of the nodes, which are however automatically constructed by RADYBAN as well (see [321] for the details of such a construction). Thick arcs in Figure 13.4 represent temporal arcs (i.e., arcs connecting the copies of the same variable at the different time slices), and nodes at the ulterior layer are shown with the '#' character at the end of the name.

It is worth remembering that the DBN formalism is a discrete-time model, in contrast with formalisms like CTMC (either explicitly specifies or generated by means of GSPN or SWN) which are continuous-time models. Indeed, one of the parameters that is possible to set in RADYBAN is the *discretization step* Δ: this parameter represents the amount of time separating the anterior and the ulterior layer of the DBN and is the minimum interval at which the measures of the model are computed. Of course, the smaller the discretization step, the closer the resulting model is to a continuous-time model. However, it is worth noting that a trade-off between computation time and result precision exists: if a looser approximation is sufficient, a quicker DBN inference can be obtained, by choosing a relatively large discretization step. In the following examples, we will always assume a discretization step of one hour.[2]

We have then performed different computations on the obtained model. First of all, the system unreliability (defined as the probability of the system fault at a given mission time) has been evaluated up to a mission time $t = 1000$; in Table 13.2 the results are visualized every 200 hours. Since this is the usual analysis in DFT, we have compared the results of RADYBAN with those obtained by other two tools: the DRPFTPROC tool [50, 379] (based on modularization [142] and SWN) and GALILEO [38] (based on modularization, *Binary decision diagrams (BDD)* [384] and CTMC). To perform such a computation in RADYBAN, we just used a filtering algorithm by querying the node SystemFault without providing any observation stream; in other words we have performed a standard prediction.

As we can see from Table 13.2, there is almost complete agreement among the values computed by the different tools for the system unreliability as a function of the time, even if RADYBAN uses a time-discrete algorithm.

---

[2]This means that if t is the time of the anterior layer, t' the time of the ulterior layer and C is a system component, $\mathbb{P}\left(C(\text{down at time } t')|C(\text{up at time t})\right) = 1 - e^{-\lambda}$.

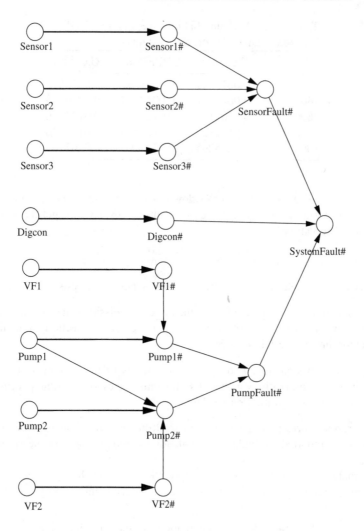

Figure 13.4 The dynamic Bayesian network (DBN) corresponding to the DFT in Figure 13.3.

As already mentioned, BNs also offer additional analysis capabilities with respect to Markov models and Petri nets. The most important one is the possibility of performing computation conditioned on the observation of some system parameters (i.e., a posteriori computation). Let us suppose that in the HSS system some sensors are monitorable and in particular Sensor2 and Sensor3. The status of such sensors can be gathered every 100 hours, but only one at a time. We decide to gather Sensor2 first (at time $t = 100$), then Sensor3 (at time $t = 200$) and so on until a failure is detected. The observations allow us to discover a fault in Sensor2 at time $t = 500$ and a fault in Sensor3 at time $t = 600$. Let S_i^t be

Table 13.2 The unreliability results obtained by RADYBAN, DRPFTPROC and GALILEO.

Time (h)	RADYBAN	DRPFTPROC	GALILEO
200	0.0013651	0.0013651	0.0013651
400	0.0049082	0.0049083	0.0049083
600	0.0104139	0.0104139	0.0104140
800	0.0176824	0.0176826	0.0176826
1000	0.0265295	0.0265295	0.0265295

the observation that the ith sensor is down at time t, and $\overline{S}_i^t$ be the observation that the ith sensor is up at time t (i.e., no fault in sensor i); we get the following stream of observations

$$\sigma_1 = \{\overline{S}_2^{100}; \overline{S}_3^{200}; \overline{S}_2^{300}; \overline{S}_3^{400}; S_2^{500}; S_3^{600}\}.$$

RADYBAN is then able to perform two different kinds of analysis:

- standard unreliability analysis of the system at mission time t conditioned on the stream gathered up to time t: this corresponds to executing a monitoring algorithm on the DBN model;

- given a mission time t, diagnosing the status of the system up to time t, given the observations gathered so far: this corresponds to the execution of a smoothing algorithm on the DBN model.

The results corresponding to stream σ_1 from mission time $t = 100$ to time $t = 1000$ are reported in Table 13.3, both for the monitoring and the smoothing analysis.

By analysing Table 13.3, we can interpret the different values provided by monitoring and smoothing. Until time $t = 300$ the monitoring unreliability is larger than

Table 13.3 Computation of the unreliability of HSS given the observation stream σ_1, under monitoring and under smoothing.

Time	Monitoring unreliability	Smoothing unreliability
100	0.0002001	0.0001011
200	0.0004016	0.0002046
300	0.0006043	0.0003103
400	0.0008083	0.0201133
500	0.0587333	0.5270140
600–1000	1.0000000	1.0000000

the smoothing unreliability, since the smoothing procedure takes into account the future information about the operativity of the sensors. For instance, at time $t = 200$, the monitoring unreliability is the probability of having a system fault, given that we know that Sensor2 was operational 100 hours before and that Sensor3 is currently operational. The smoothing unreliability considers, in addition to the above information, the evidence provided by the knowledge about the sensors status at future time instants with respect to $t = 200$. This means that, provided that we are doing the analysis at least at time $t = 600$ (i.e., the last time instant for which we have information), the entry of the third column of Table 13.3 corresponding to $t = 200$ represents the following measure: the probability that the whole system was down at time $t = 200$, knowing not only the sensor history until $t = 200$, but in addition also the remaining sensor history until the analysis time. Since we know that at time $t = 300$ and $t = 400$, Sensor2 and Sensor3 were still respectively operational, the value that we read from the table for such a measure is smaller than the corresponding value from the monitoring column. In other words, in the time interval [200,300], the smoothing procedure knows for sure that Sensor2 will not break down (we do not assume repair or maintenance here), while the monitoring procedure has to take into account also the possibility of Sensor2 breaking down.

On the contrary, at time instant $t = 400$, the smoothing unreliability becomes larger than the monitoring one, since we know that in the interval [400,500], the component Sensor2 will definitely break down. An even more consistent increase in the smoothing unreliability is evident at time $t = 500$, since we know that in the interval [500,600] the system will definitely fail, because of the failure of Sensor3. Of course, for the remaining time instants, both monitoring and smoothing determine an unreliability measure of 1, because of the certainty, gathered from the observations, of a system fault.

Another possibility of analysis offered by DBN consists in performing a diagnosis over the status of the components, given again a stream of observations. Let us consider gathering information about the whole system (i.e., by monitoring its global operativity) every 200 hours. Suppose we get the following stream of observations (where SF stands for SystemFault):

$$\sigma_2 = \{\overline{SF}^{200}; \overline{SF}^{400}; SF^{600}\}.$$

This means that the system has failed in the interval [400,600]. We can now ask for the probability of fault of the system components, given the evidence provided by σ_2. As before, we can perform monitoring as well as smoothing. Results are reported in Table 13.4, and in Table 13.5, respectively.

These results allow us to concentrate on the analysis of single components for diagnostic purposes. For instance, by analyzing smoothing results, we can notice that the DigCon component cannot be failed before mission time $t = 400$; this is consistent with the fact that a failure of such a component will cause a system failure (see the DFT of Figure 13.3) and that we have observed the system being operational until such a mission time. This is not reported in the monitoring

Table 13.4   Computation of the probability of fault of the HSS components, given the observation stream σ_2, under monitoring.

Time	Sensors	DigCon	Monitoring valves/filters	Pump1	Pump2
100	0.0099502	0.0001000	0.0009995	0.0010994	0.0009995
200	0.0190470	0	0.0019938	0.0021930	0.0019936
300	0.0288077	0.0000100	0.0029913	0.0032900	0.0029914
400	0.0363592	0	0.0039753	0.0043720	0.0039745
500	0.0459476	0.0000100	0.0049709	0.0054666	0.0049706
600	0.6450366	0.0361441	0.0096857	0.0106514	0.0100596
700	0.6485684	0.0362408	0.0106757	0.0117392	0.0110499
800	0.6520652	0.0363369	0.0116644	0.0128256	0.0120391
900	0.6555275	0.0364330	0.0126522	0.0139109	0.0130274
1000	0.6589551	0.0365291	0.0136390	0.0149949	0.0140148

Table 13.5   Computation of the probability of fault of the HSS components, given the observation stream σ_2, under smoothing.

Time	Sensors	DigCon	Smoothing valves/filters	Pump1	Pump2
100	0.0657449	0	0.0013842	0.0015225	0.0013841
200	0.1308356	0	0.0027672	0.0030436	0.0027669
300	0.1952788	0	0.0041483	0.0045631	0.0041483
400	0.2590806	0	0.0055294	0.0060811	0.0055283
500	0.4530237	0.0180732	0.0076176	0.0083774	0.0077854
600	0.6450363	0.0361446	0.0096859	0.0106516	0.0100598
700	0.6485685	0.0362407	0.0106757	0.0117392	0.0110499
800	0.6520653	0.0363368	0.0116644	0.0128256	0.0120391
900	0.6555274	0.0364328	0.0126522	0.0139109	0.0130274
1000	0.6589551	0.0365291	0.0136390	0.0149949	0.0140148

analysis, where the `DigCon` component can be assumed operational only at the instants when we gather the information about the system being up (i.e., $t = 200$ and $t = 400$); for the other time instant before $t = 400$, the monitoring analysis predicts a possibility of fault for such a component. Of course, since we do not assume component repair in the present example, once we gather the information that the system is failed ($t = 600$), then the results from monitoring and from smoothing will coincide.

Finally we would like to remark that the results shown in Tables 13.4 and 13.5 are just the marginal probability of fault of the various system components; DBN algorithms are able to provide even more detailed information, by computing the joint probability distribution of a set of query variables. This means that we can ask for the probability of a given scenario, where some components are operational and some component are not. Since the case study of Figure 13.2 has eight basic components, the joint distribution of the system components is composed by $2^8 =$ 256 entries for each mission time. As an example we can report that RADYBAN has computed that, at the time of the discovery of the system fault ($t = 600$), the probability that this fault is caused by the fault of the DigCon component, only (all the other components working) is $p = 0.0003001$. Since the system fault cannot be ascribed only to this component, the corresponding entry in Tables 13.4 and 13.5 is larger, corresponding to the marginalization of the joint distribution on the considered component (which in this case is $p' = 0.0361441$).

13.5   Conclusions

In this chapter, we have described how the use of a formalism like DBN can augment the analysis capabilities of a reliability tool, while still maintaining a compact modeling framework. We have pursued this goal by considering a tool where the system to be analyzed is modeled by means of a well-known formalism like DFT and then analyzed by means of a conversion into a DBN. This allows us to take into account all the dynamic dependencies that can in principle be addressed by a DFT model or by a state-space model. Standard inference algorithms for DBN, like monitoring and smoothing, can then be adopted to perform interesting

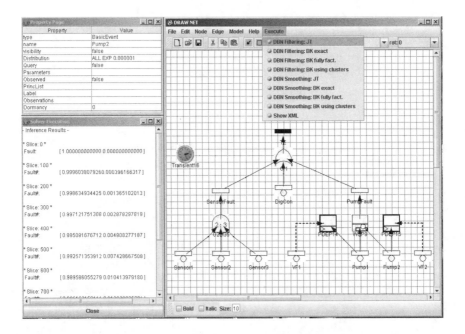

Figure 13.5   A screenshot of the RADYBAN tool.

reliability analyses. We have shown the suitability of the approach by considering a case-study taken from the literature (the Hypothetical Sprinkler System) and by running different reliability analysis with the RADYBAN tool. Our experimental results therefore demonstrate how DBNs can be safely resorted to, if a sophisticated quantitative analysis of the system is required.

Acknowledgment

This work has been partially supported by the Italian Ministry of Education under project FIRB-PERF and by the EU under Grant CRUTIAL IST-2004-27513.

14

Terrorism risk management

David C. Daniels
8260 Greensboro Drive, Suite 200, McLean, VA 22102, USA

Linwood D. Hudson
8260 Greensboro Drive, Suite 200, McLean, VA 22102, USA

Kathryn B. Laskey
Dept. of Systems Engineering and Operations Research, George Mason University, Fairfax, VA 22030-4444, USA

Suzanne M. Mahoney
Innovative Decisions, Inc., 1945 Old Gallows Road, Suite 215, Vienna, VA 22182, USA

Bryan S. Ware
8260 Greensboro Drive, Suite 200, McLean, VA 22102, USA

and

Edward J. Wright
Information Extraction and Transport, Inc., 1911 North Fort Myer Drive, Suite 600, Arlington, VA 22209, USA

Bayesian Networks: A Practical Guide to Applications   Edited by O. Pourret, P. Naïm, B. Marcot
© 2008 John Wiley & Sons, Ltd

14.1   Introduction

Recent events underscore the need for effective tools for managing the risks posed by terrorists. The US military defines antiterrorism as the defensive posture taken against terrorist threats. Antiterrorism includes fostering awareness of potential threats, deterring aggressors, developing security measures, planning for future events, interdicting an event in process, and ultimately mitigating and managing the consequences of an event. These activities are undertaken by government and private security services at military, civilian and commercial sites throughout the world. A key element of an effective antiterrorist strategy is evaluating individual sites or assets for terrorist risk. Assessing the threat of terrorist attack requires combining information from multiple disparate sources, most of which involve intrinsic and irreducible uncertainties.

Following the bombing of US Air Force servicemen in Khobar Towers in Saudi Arabia and the bombings of the US Embassies in Africa, investigations revealed inadequacies in existing methods for assessing terrorist risks and planning for future terrorist events. Interest in improved methods for assessing the risk of terrorism has grown dramatically in the wake of the World Trade Tower bombings of 2001. Because of its inherent uncertainty, terrorist risk management is a natural domain of application for Bayesian networks. This chapter surveys methodologies that have been applied to terrorism risk management, describes their strengths and weaknesses, and makes the case that Bayesian networks address many of the weaknesses of other popular methodologies. As a case study in the application of Bayesian networks to terrorism risk management, we describe the Site Profiler Installation Security Planner (ISP) suite of applications for risk managers and security planners to evaluate the risk of terrorist attack. Its patented [45] methodology employs knowledge-based Bayesian network construction to combine evidence from analytic models, simulations, historical data, and user judgements. Risk managers can use Site Profiler to manage portfolios of hundreds of threat/asset pairs.

14.1.1   The terrorism risk management challenge

Events such as the New York, Madrid, London, and Mumbai public transportation bombings vividly demonstrate the global reach of terrorism. Terrorists present an asymmetric threat for which neither domestic civil security forces nor the conventional military are well suited. Terrorists exploit the freedom of open, democratic societies to hide their planning, training, and attack preparations 'below the radar'. By attacking when least expected using unconventional means, terrorists exploit the weaknesses of conventional military forces organized and trained to fight clearly defined enemies in definitive engagements.

Although vulnerabilities cannot be eliminated, risks can be contained by identifying exploitable vulnerabilities, estimating the likelihood that these vulnerabilities will be targeted, evaluating the magnitude of the adverse consequences if they were exploited, and prioritizing risk mitigation efforts. Assessing the likelihood of an event and the severity of the consequences requires integrating disparate data sources. Information about terrorist intent and targeting preferences, usually the province of intelligence staff, is largely subjective and highly uncertain. Vulnerabilities and mitigation options, typically areas for physical security specialists, are often based on experience or best judgement. Estimating the consequences of an attack requires sophisticated models used by engineers and scientists. The antiterrorism planner at each site is responsible for assimilating all of this information for all assets and managing a dynamic risk portfolio of potentially thousands of threat-asset pairs. Although a limited number of experts may be able to understand and manage a given risk, no human can manage all of the components of thousands of risks simultaneously.

14.1.2   Methodologies for terrorism risk management

Risk is commonly defined as the possibility of suffering some type of harm or loss to individuals, organizations, or entire societies. Risk management is the discipline of identifying and implementing policies to protect against or mitigate the effects of risk. Degree of risk is typically quantified by multiplying the likelihood of a catastrophic event by a measure of the adverse effect if the event occurs. Adverse effect is typically measured in a standard unit such as monetary loss. Nonmonetary adverse consequences such as death, suffering, or aesthetic damage are commonly assigned a monetary value to facilitate comparison. Traditional risk assessment uses historical data to estimate correlations between observable variables catastrophic events, and to predict future events from observable variables. However, historical data on terrorist events is scarce, and an intelligent and continually adapting adversary is unlikely to repeat past behaviors. Thus, straightforward forecasts based on historical data are poor indicators of future terrorist events.

Prior to the attacks of 9/11 and the establishment of the Department of Homeland Security (DHS), antiterrorism planners in the US military employed manual procedures derived from conventional military doctrine or special forces targeting criteria and supported by paper and pencil tools. Since 2001, a proliferation of tools and methods has emerged, signifying both the intense interest in evaluating terrorism risk and the failure of any single tool to satisfy the need. Similar risk factors appear across the myriad methods, suggesting that they are important risk drivers. The unsatisfactory element in most of the models is how the factors are quantified and combined into a single risk metric. A brief review of some of the most prominent risk analysis methods follows.

14.1.2.1   Risk mnemonics

Mnemonics are simple tools to help a practitioner remember a more complicated framework. The oldest risk mnemonic is CARVER, which stands for Criticality, Accessibility, Recognizability, Vulnerability, Effect (on the populace), and Recoverability [468]. CARVER was developed by US special operations forces during the Viet Nam conflict to optimize offensive targeting of adversary installations. Originally, it provided a framework for subjective analysis by highly trained operators. It has since been applied to asset risk evaluations by scoring each CARVER factor on a ten point scale and adding the scores. This approach is non-specific to particular threats, labour-intensive, and not scaleable to many assets. CARVER has been extended and customized by adding or redefining terms in the acronym [59], [359].

DSHARPP is a subjective risk assessment process developed by the US military to identify assets at highest risk for terrorist attack. An installation planner assigns a score from 1 to 5 for each term – Demography, Susceptibility, History, Accessibility, Recognizability, Proximity, Population. Points are summed to rank potential targets on a scale ranging from 7 to 35 points [4]. Similar approaches include M/D-SHARPP, the Homeland Security Comprehensive Assessment Model (HLS-CAM) [332] (used at both the Democratic and Republican National Convention sites in 2004), and the Navy's Mission Dependency Index (MDI) [335].

The Force Protection Condition (FPCON) System is a military risk management approach that is based on compliance with a set of prescribed standards [235], the Force Protection Condition Measures. Five FPCON levels (Normal, Alpha, Bravo, Charlie, and Delta) represent an increasing level of terrorist threat, as determined by military intelligence. As the FPCON level increases, the installation employs FPCON Measures. Though these measures make it easy to develop consistent plans, they have been shown to be inadequate [462].

14.1.2.2   Algebraic expressions of risk

Many probabilistic approaches to risk use algebraic $=$ expressions to represent risk: e.g., Risk = Threat $\times$ Vulnerability $\times$ Consequence. Depending on their intended audience and usage, these methods sometimes make simplifying assumptions, such as that the probability of the Threat and/or Vulnerability is 1. They also typically give rise to static equations that are difficult to update as new information about terrorists' intent and capabilities is obtained. In addition, despite their apparent mathematical rigor, most of these models suffer from an inability to accurately quantify Threat, Vulnerability, and Consequence.

Direct or decomposed assessments of Threat, Vulnerability, and Consequence are commonly provided by experts, who rate the values on a scale. Ordinal values (e.g., Low/Medium/High) are typically converted to cardinal values for combination. It is our experience that most subject matter experts do not evaluate subjective quantities well on cardinal scales. If they are asked to rate a collection of assets on a scale of 1–5, the results will be at best ordinal. For instance, it is unclear whether an asset with a visibility rated a '2' is really twice as visible as another rated a '1' by subject matter experts, yet the mathematical treatment of the scores often assumes it is.

The Special Needs Jurisdiction Tool Kit (SNJTK) [466], was developed for DHS by the Office of Domestic Preparedness (ODP, formerly part of the US Department of Justice). Several variations share a similar approach. SNJTK is an asset-based risk approach that uses Critical Asset Factors (CAFs) to evaluate the risk of threat-asset scenarios. CAFs represent characteristics of assets that would result in significant negative impact to an organization if the asset were lost. Examples include Casualty Impact, Business Continuity, Economic Impact, National Strategic Importance, and Environmental Impact. Through expert elicitation, each CAF is assigned a weight for its relative importance to the organization; each asset is assigned a score for its impact on each CAF; and each threat-asset scenarios is assigned a CAF score indicating how much that scenario would impact the CAF. The SNJTK approach is similar to a Bayesian network in that expert judgement is used to construct new variables and assign them weights. SNJTK could in principle be updated as new threats emerge or as asset characteristics change. Since threat likelihood is not considered, the output of SNJTK is a conditional consequence, not a true risk metric; however, threat likelihood could be included in a relatively straightforward way. Because of the reliance on expert opinion each time an assessment is made, the methodology does not lend itself to large numbers of assets or to process automation. To date, models have only been developed and customized for transportation sectors.

The Texas Domestic Preparedness Assessment document [347] was produced under the ODP 2002 State Domestic Preparedness Program (SDPP) and was designed to help standardize asset vulnerability assessments. A limited number of high-risk assets are assessed for visibility, criticality to jurisdiction, impact outside the jurisdiction, potential threat element (PTE) access to target, target threat of hazard, site population, and collateral mass casualties. The assessor assigns each factor a rating on a 0–5 scale; ratings are summed to create the Vulnerability Assessment Rating.

The Critical Asset & Portfolio Risk Analysis (CAPRA) methodology [306] was developed at the University of Maryland and will be used by the Maryland Emergency Management Agency (MEMA) for data input into MEMA's Critical Asset Database. CAPRA is an asset-driven approach that measures risk with a parametric equation using variables estimated by subject matter experts. CAPRA employs a five phase expert evaluation process: (1) define mission-critical elements; (2) match critical elements to applicable hazards to form hazard scenarios; (3) estimate consequences and select high-consequence scenarios for vulnerability assessment; (4) solve for the probability of adversary success; (5) incorporate consequences to compute the conditional risk of each hazard scenario. Though informed by expert opinion, it is unclear how the CAPRA risk equation was derived or validated.

The CREATE Terrorism Modeling System (CTMS) [475], developed by the University of Southern California's DHS-funded Center for Risk and Economic Analysis of Terrorism Events (CREATE), is both a methodology and a software system. It is based on threat-asset scenarios built and evaluated by expert judgement, and supports the assessment of risks within a framework of economic analysis and structured decision-making. A threat assessment is used as a filter to select likely scenarios. The model defines threat as the likelihood of a successful terrorist attack. Threat factors considered are (1) criticality to the economy or government, (2) human occupation and vulnerability, (3) damage vulnerability, (4) symbolism worldwide and to the US population, and (5) existing protection measures. The highest threat scenarios are evaluated for human, financial, and symbolic consequences, which are rated on a logarithmic Terrorism Magnitude Scale (TMS).

The Risk Analysis & Management for Critical Assets Protection (RAMCAP) framework was developed by DHS to analyze and manage risk of terrorist attacks on critical infrastructure [467]. RAMCAP includes a seven-step process of data collection and analysis on a number of risk management dimensions, including assets, attack consequences, threats, and vulnerability. RAMCAP defines Risk as the product of vulnerability, consequence, and threat for a specified attack scenario on a particular asset with a defined asset failure mode. Owners of assets can use either expert elicitation or a simple event tree to estimate vulnerability and consequence. RAMCAP provides more of a risk framework than a risk metric. It does not specify how to collect the data for vulnerability and consequence, and does not detail how DHS will estimate threat. Instead, it provides suggestions of viable alternatives, such as a worked example showing how a multi-node event tree and an expert elicitation approach can be used to obtain the same vulnerability estimate in the case of a truck bomb attack against a refinery.

Beginning in 2006, DHS decided to move from a population-driven to a risk-based methodology for infrastructure protection grant funding under certain of their grants programs. The DHS Risk Management Division (RMD) developed the Risk Analysis Calculator (RASCAL) as a universal asset-based risk methodology, applicable across DHS grant programs [391, 464, 465]. RASCAL computes a sum of normalized asset-based and geographic-based risk terms. The asset-based term sums risks for each identified critical asset in an entity of interest (state, urban area, etc.). For each of 48 asset types, risk is considered for each of 14 threat scenarios, but for most assets only the 'most-likely, worst' scenario is used to estimate risk. The geographic-based component considers features that cannot be captured in the asset-based risk component. For both asset and geographic terms, risk is estimated as the product of consequence, threat, and vulnerability variables, which are obtained as sums of normalized observables collected from various public and private databases. Subjective data is avoided in RASCAL, so expert elicitation is not required. The validity and practicality of the RASCAL construction have been challenged, and DHS is moving away from the geographic plus asset risk formulation for the 2007 grant allocations.

The US Coast Guard (USCG) created the Port Security Risk Assessment Tool (PSRAT), and its successor, the Maritime Security Risk Assessment Model (MSRAM) for port risk assessments [134]. DHS has incorporated MSRAM scores into risk analysis for the Port Security and Transportation Security Grant Programs. MSRAM quantifies risk to an asset as the product of consequence, threat, and vulnerability. The risk variables for each asset are evaluated subjectively and scored by the relevant captain of the port (COTP). In order to minimize variances and reduce potential biases among the different COTPs, a national-level quality review is conducted. MSRAM is a port-specific risk methodology, but it could in principle be extended to other domains. Because the data is collected qualitatively by sector experts, however, it would be problematic to compare risk scores from one sector with those from another.

14.1.2.3   Fault trees

Fault tree analysis assumes a threat baseline (often assuming the probability of occurrence is unity) and uses decision paths to evaluate the probabilities and magnitudes of different outcomes. Fault trees are used in industrial safety applications to perform failure analysis of complex systems. Because probabilities are usually estimated from historical data, fault trees are of limited use when dealing with an intelligent, adaptive agent. Nevertheless, fault tree structures can provide insight into important risk factors.

The Operationally Critical Threat, Asset, and Vulnerability Evaluation (OCTAVE) [83] is a risk-based information systems security analysis tool originally designed for large organizations and sponsored by the US Department of Defense. OCTAVE uses expert-assessed event trees to identify threats and create a threat profile. The output of an OCTAVE assessment is a written summary of the team's findings; there are no quantifiable risk measures. Other fault tree analyses tailored for specific types of terrorism risk analysis include Sandia National Laboratory's RAM-D program for dam security, and Argonne National Labs' ERSM program for event response assessment [17, 410, 411].

14.1.2.4   Simulations

Several simulation packages have been developed to inform terrorism risk analysis. Most of these focus on the consequences of a terrorist attack, although some physical models can calculate asset vulnerabilities as well. Detailed simulations tend to be cumbersome to set up and use, and therefore do not lend themselves to easy updates when new information about terrorist intentions is received. In addition, most simulations are specific to certain types of assets or threat scenarios, and cannot incorporate the full range of terrorism scenarios.

The Critical Infrastructure Protection Decision Support System (CIP/DSS) national infrastructure interdependency model [463] was developed for DHS by Sandia National Laboratories, Los Alamos National Laboratory, and Argonne National Laboratory to simulate the effects of a disruption in one CI/KR sector on others. It is a highly-specific model of the economic interdependencies among 14 of the nation's CI/KR sectors. CIP/DSS does not evaluate the direct consequences of any particular terrorist attack, nor does it consider threats nor vulnerabilities. It is therefore not a risk model, but it could be used to help inform the evaluation of the broader consequences from a terrorist attack.

Geospatial Information Systems (GIS) (e.g., [154, 224]) are commonly used by the emergency management community to perform natural hazards risk analysis, and are being extended to incorporate man-made hazards as well. GIS systems can identify the areas most at risk from predictable area-wide catastrophes, and can locate damage profiles for large-scale terrorist attacks. However, GIS systems cannot evaluate the likelihood of terrorism threats because these do not correlate strongly with geography.

The US Environmental Protection Agency (EPA) has developed the Threat Ensemble Vulnerability Assessment (TEVA) Modeling program to study contamination threats to drinking water and wastewater systems [328]. TEVA is a physics-based software package that simulates water flow and quality behavior in pipes under various detection and response scenarios. It appears to be well-suited to evaluating alternative sensor placement schemes and mitigation strategies.

14.1.2.5   Other non-Bayesian tools

Automated Critical Asset Management System (ACAMS) [267, 398] is an asset inventory system developed by the City of Los Angeles and DHS for use by state and local governments in conducting risk assessments as part of an overall risk management program as mandated by HSPD-8 [76]. Although ACAMS is an asset catalog and not a risk methodology, it could serve as the asset database for other asset-based risk analysis algorithms.

The DoD employs dozens of expert teams to conduct risk assessment. Each team is composed of eight to ten experts in terrorist options, structural engineering, chemical weapons, law enforcement, and disaster response. During a site visit, the team interviews and observes, selects several likely targets and potential threats, and prepares a report and briefing that discusses the risks and mitigation options. The installation commander is then responsible for addressing these findings. Since the findings of the expert teams are based solely on their judgement and experience, the analytical integrity of their assessments is subjective and their results are not repeatable. While expert assessment is a powerful tool, it must be structured to produce repeatable, high confidence results.

14.1.3   The Site Profiler approach to terrorism risk management

Site Profiler was initially designed to provide tools for antiterrorism planning at the site level. Since initial deployment, the system has been extended for use as the analytical foundation for an asset portfolio risk management program at the municipality, state, or national level. Site Profiler has been deployed in a number of settings for antiterrorism planning at individual sites and for terrorism risk management across larger portfolios of assets. A generic application development environment enables new deployments to be constructed rapidly by tailoring the knowledge base and user interface to new sites and/or asset portfolios.

14.1.3.1   Requirements for a new terrorism risk management system

Our initial research identified a broad consensus among both experts and policy makers that incremental improvements to existing methods would not be sufficient. Rather, a truly revolutionary approach to terrorism risk management was needed. Our research identified the following requirements for Site Profiler.

1. *Individuality of risk scenarios*: A 'one size fits all' approach is unacceptable. Each asset, each threat, and each potential terrorist attack situation has its own unique characteristics. Site Profiler must be able to account for

the large number of factors that need to be considered in evaluating any given risk assessment problem.

2. *Intrinsic uncertainty*: Site Profiler must be capable of accounting for the intrinsic uncertainty in the identities of the terrorists, their capabilities, the factors that make an asset attractive, the method of attack, the consequences of an attack, and how all these factors combine to affect risk.

3. *Defensible methodology*: The methodology used to combine inputs and produce an overall risk assessment must be transparent and analytically defensible. It must be possible to construct a clear and credible rationale for the results of analysis.

4. *Flexibility*: The system must be capable of accepting and combining inputs from a wide variety of sources. These include expert judgement, direct user observation, results from models and simulations, and data from external sources such as facility databases.

5. *Modifiability, Maintainability, and Extensibility*: The system must be designed to be easily maintained and updated, and to adapt as the terrorist threat adapts.

6. *Customization*: The system must be easily customized to a range of facility types, threat types, and threat scenarios.

7. *Usability*: The user interface must be cognitively natural for analytically unsophisticated users, must economize on data entry, and must support a workflow that fits the thought processes and organizational environment of the end user population.

8. *Portfolio management*: Security managers must be able to assess risks for a single site, or to analyze and manage a large portfolio of risks simultaneously. The system must be capable of storing and managing data for large numbers of threats and assets.

9. *Tractability*: The system must be capable of producing accurate results in a timely fashion when assessing a large portfolio of risks.

14.1.3.2 Bayesian networks for analyzing terrorism risk

These requirements led us to consider Bayesian networks as the analytical methodology for quantitative assessment of risks. At the time Site Profiler was designed, we found no other applications of Bayesian networks to risk analysis. Recently, a number of applications have appeared (e.g., [190, 275]), but to our knowledge, Site Profiler is unique in using Bayesian networks to quantify the risk of terrorism. Other approaches we considered suffer from the lack of a coherent means to combine objective and subjective data, and from the inability to update a risk assessment as new threat information is received. Furthermore, analytical approaches tend to be based on a historical understanding of threat and risk, and do not account for subjective assessments of how intelligent adversaries will modify future attack strategies to be different from the historical pattern. Bayesian networks overcome these limitations. They are well suited to complex problems involving large numbers of interrelated uncertain variables. Unlike 'black box' technologies such as neural networks, the variables and parameters in a Bayesian network are cognitively meaningful and directly interpretable. Unlike traditional rule-based systems, Bayesian networks employ a logically coherent calculus for managing uncertainty and updating conclusions to reflect new evidence. Tractable algorithms exist for calculating and updating the evidential support for hypotheses of interest. Bayesian networks can combine inputs from diverse sources, including expert knowledge, historical data, new observations, and results from models and simulations.

The knowledge engineering process uncovered natural clusters of variables that matched our users' domain concepts. These clusters were used to define Bayesian network fragments [269, 270] to represent uncertainty about the attributes of and relationships among entities in the domain. For example, there was a natural grouping of variables that corresponded to characteristics of a valuable asset that must be protected from an attack. We created a fragment consisting of variables pertaining to the concept of an asset. Some uncertain variables pertained to more than one type of entity (e.g., the accessibility of fragment consisted of relational information regarding threat/asset pairs). In this case, we defined a relational entity type to represent the pairing, and a fragment to represent uncertain relational variables for the pairing. Each fragment was developed using information from a small number of sources, and could be tested as a unit independent of the other modules. Thus, our modular approach supported a manageable knowledge engineering and model evaluation process.

Site Profiler uses an object-oriented database to store the Bayesian network fragments, and to manage information about individual entities (e.g., individual assets). The database schema was based on the entities, attributes and relationships we identified in developing the Bayesian network fragments. We have reused the network fragments and database schema across applications with minor tailoring. For each Site Profiler application, the database is populated with data from a variety of sources, including managers' subjective and objective assessments, historical information from databases external to Site Profiler, analytic model results, and simulation results. Whenever possible, information is obtained directly from external sources without direct user intervention. For example, Site Profiler interfaces directly to simulations and executes them without user intervention as needed. For each type of entity, values for some attributes must be obtained directly from users. Our design kept the number of such attributes to a minimum. We developed an intuitive graphical user interface for users to enter the required information.

In addition to information about attributes of entities, the model included relational information regarding threat/asset pairs. Because the number of risks scales roughly as the number of assets times the number of threats, manual entry of relational information for each threat/asset combination was infeasible. We were able to model the relational aspects so that all relational information could be calculated from a complete characterization of threats and assets. Many of these calculations were computed from external simulations external invoked automatically. For example, a simulation calculates the accessibility of an asset to a threat, and a physics-based model predicts the consequences of a given explosive against a structure or a chemical weapon against a group of people.

14.1.3.3   Interacting with end users

To perform a risk assessment using Site Profiler, a risk manager specifies the scenarios to be evaluated, tasks the system to perform the evaluation, and views the results. Optionally, the user can drill down to obtain details and/or adjust inputs to see the impact on results. A threat scenario is defined as an attack by a particular threat against a specific asset. To specify scenarios, the user identifies a set of assets to be evaluated and a list of threats to be considered. For each threat/asset pair, the system retrieves the relevant Bayesian network fragments and combines them at run-time into a Bayesian network tailored to the scenario. In Site Profiler, this run-time Bayesian network is called the Risk Influence Network (RIN). Variables for which the database contains a value are entered as evidence to the RIN. For variables that correspond to the results of simulations and models, Site Profiler interfaces directly to the relevant simulations, executes them without user intervention, and applies the results as evidence to the RIN. Additional judgemental and/or observational evidence is obtained directly from users through the graphical user interface. The system propagates the evidence, computes probabilities for variables of interest, and displays the results to the user.

Because our users are not analysts, they needed a tool they could learn simply, use effectively, and trust. We wanted to avoid a black box into which the user feeds information and out of which an answer magically appears. Users invoke the model via a natural and understandable interface to describe their assets, specify characteristics of their installation, and select threats to consider. The system constructs RINs for each threat/asset combination, runs offline simulations and database queries as needed, applies evidence, and computes risks which are presented back to the user in tables formatted for understandability. At this high level view, a utility function is used to reduce the probability distributions to single indicators such as High, Medium, or Low, as shown in Figure 14.1. Users are then able to

Threat	Target	Plausible	Attractive	Susceptible	Consequences	Risk	Conf	Vectors	Model
Panel Truck Bomb	Operations	High	Moderate	High	Critical	1	‡	↘	◎
Panel Truck Bomb	Operations	High	Moderate	High	Critical	2	‡	↘	◎
Vehicle Bomb	Operations	High	Moderate	High	Critical	3	‡		
Hoaxes	Operations	Very High	Moderate	High	Moderate	4	‡		
Panel Truck Bomb	Dignitaries	High	Moderate	Moderate	Critical	5	‡		
Vehicle Bomb	Dignitaries	High	Moderate	Moderate	Critical	6	‡		
Panel Truck Bomb	Warehouse	High	Low	Low	Critical	7	‡		
Vehicle Bomb	Warehouse	High	Low	Very Low	Moderate	8	‡		

Figure 14.1 The risk table allows the user to view and sort by key network nodes.

drill down into the components of the risk by clicking on rows in the risk table and walking down the Bayesian network. This ultimately takes them to leaf nodes at which the information may have come directly from a question they answered or from the results of a model calculation. We present users with a graphical view of risks and probability distributions for each node in the RIN. Users can adjust inputs as necessary but can feel confident that they understand the underlying components of a given risk score.

The structure of the network and domain fragments facilitates the risk management process. Users can easily see the threat that is most plausible, or the asset that has the highest consequences, or a common element among many risk scenarios. Countermeasures, procedures, and other adjustments can be applied to the site baseline to address issues identified in the risk influence network.

14.2   The Risk Influence Network

The heart of the Site Profiler risk methodology is the Risk Influence Network (RIN). The RIN is a Bayesian network, constructed on the fly from a knowledge base of Bayesian network fragments, and used to assess the relative risk of an attack against a particular asset by a particular threat. Depending on the configuration of fragments in the network, the number of nodes in the RIN can range from 35 to 60. The nodes of the RIN contain information about the installation as a whole, the asset, the threat (tactic, weapon system, and terrorist organization), the threat/asset target pairing, and the attack event.

14.2.1   Knowledge representation

As work on Site Profiler progressed, it became clear that developing a single, all-encompassing Bayesian network would not be a viable solution. First, the size and complexity of the network would make maintenance and extensibility nearly impossible. Second, the network would have to be constructed from inputs provided by a diverse collection of individuals (e.g., intelligence analysts, physical security specialists, civil engineers, facility managers). Each individual is very knowledgeable about his or her own specialty, but may know little about the other specialties required for building the model. The process of identifying the relevant inputs to be obtained from each specialist, collecting the inputs, and assembling them into a single monolithic network would be an unmanageable task. In addition, once the network was constructed, testing and debugging the model would be nearly impossible. Finally, because the relevant features are not the same for all scenarios, the RIN structure itself differs from problem to problem. Because a traditional Bayesian network has a fixed number of variables related according to a fixed dependency structure, no single Bayesian network could represent our problem. Greater expressive power is required for complex problems like the one for which Site Profiler is designed. Site Profiler uses multi-entity Bayesian networks (MEBNs) [268], a language for specifying knowledge bases consisting of generic Bayesian network fragments. Each MEBN Fragment (MFrag) represents a relatively small, repeatable, conceptually meaningful part of a total model structure. Site Profiler uses knowledge-based model construction [489] to construct a Bayesian network to represent features pertaining to a given scenario. The constructed Bayesian network is called a Risk Influence Network, or RIN. The ability to represent a large number of Bayesian networks that vary in both structure and evidence, and to construct models on the fly for any scenario of interest, is key to the power of Site Profiler.

14.2.2   Knowledge base development

Following a network engineering process [294], we iteratively moved from initial concepts and definitions to a set of reusable MFrags that could be combined into an asset-threat specific RIN. The effort proceeded in six stages, as described below.

Stage 1: Concept definition.   All the categories of data needed to support the system were identified. This data fell cleanly into two categories: physical and domain data. Physical data includes information necessary to describe the state of a physical object, such as position, size, shape, and weight. Domain data represents abstract concepts such as attractiveness, risk, and plausibility. These two types of data suffice to develop a complete model of the terrorist realm.

Data elements were clustered according to the domain structure. Each cluster represents information regarding a type of entity, (e.g., a building, asset, or threat) or information regarding relationships between entities (e.g., accessibility of an asset to a threat). MFrags were defined for seven types of entity: assets (including seven sub-types), threats, tactics, weapon systems, targets (defined as threat/asset pairs), attacks, and attack consequences. There were also entity types that did not explicitly represent uncertainty. For example, Countermeasures, or risk mitigation actions, represent contextual factors affecting some of the likelihoods in our model. Countermeasures can be turned off and on to perform what-if analysis, but the model treats them as deterministic variables. The core knowledge representation consists of a set of Bayesian network fragments expressing information about attributes of and relationships among entities of these seven types, conditioned as appropriate on nonprobabilistic entity types.

Stage 2: Formal definition and analysis.   The next stage of developing the knowledge base was formal definition of the structure and probability distributions for the MFrags. Working with a combination of documents and experts, we drew an initial graph for the RIN. Nodes in the network included both evidence nodes and measures of aspects of the risk. Because the RIN was a new concept to our experts, initial definitions for many nodes (e.g., Accessibility, Recoverability) were informal and imprecise. To develop a computational model, we needed formal definitions for the allowable states of each node. To define state spaces, we worked with the experts to construct concrete examples for each state. These examples helped us to decide how many states were appropriate for each node. The examples also made it much easier to communicate the concepts to the experts who later reviewed the network. Concurrently with defining the state spaces, we identified inferentially interesting network fragments of five to a dozen nodes, revised their structure, and populated their conditional probability tables with rough guess values based on information we had obtained from domain experts and literature.

Stage 3: Subsection review with experts.   We reviewed subsections of the RIN with three different groups: threat experts, damage experts, and accessibility experts. Each review took two days. Most of the effort was spent communicating and revising the terminology. We used Netica to display the fragments, one fragment at a time. Rather than explicitly asking for probability distributions, we elicited relative strengths of influence. As noted above, we had already populated the probability tables with rough guess values. Elicitation was speeded up by presenting results from the rough guess model and allowing experts to make adjustments as appropriate. We entered distributions consistent with the judgements experts provided, and displayed the inferential results to the experts for additional feedback. The process concluded when experts were satisfied with the results.

There is a common view in the literature that elicitation of structure is relatively straightforward relative to the difficult problem of eliciting probabilities [135]. In our experience, the most difficult and time consuming part of the knowledge engineering process was establishing a common understanding of terminology and definitions. Once we had reached agreement on terminology and definitions, we found that obtaining consensus on the quantitative aspects of the model was far less contentious. We found that experts had little difficulty understanding and suggesting improvements to network fragments we presented to them.

One reason for the relative ease of specifying the probability distributions may have been that we made no attempt to elicit probabilities directly from experts. Instead, we adopted and developed an initial model based on a review of the literature, reviewed the model with experts, and obtained feedback. We asked experts for relative strengths of influence rather than for probabilities. We found this process to be an efficient and effective approach for rapid knowledge engineering.

In a purist view, our approach would be an unacceptable compromise: probabilities in a Bayesian network should reflect the knowledge of experts and not be contaminated by the judgements of knowledge engineers. Our elicitation method could give rise to anchoring bias [458], in which elicited values are biased toward an 'anchor', or initial value used as a starting point for elicitation. It is quite possible that a more thorough and extensive knowledge elicitation process would have resulted in changes to the RIN, and that the rough guesses we used to populate the initial model had an influence on the probabilities in the final model. However, if we had adopted the purist view, Site Profiler would not have been built. Even under the dubious assumption that we could have induced the experts to cooperate in a more extensive elicitation process, we would not have been able to build the system within reasonable budget and time constraints. Furthermore, it is not at all clear more extensive elicitation would yield a better model. Our expert reviewers felt much more comfortable critiquing our initial model and assessing the quality of the results than with providing direct judgements of structure and probabilities. The purpose of Site Profiler is to provide an analytically defensible risk analysis, a problem for which there is no established correct solution. The process we followed was the best we could do within available resources, and in our view represents an improvement over alternative methodologies. This view is supported by our expert evaluators and the clients who have purchased Site Profiler.

Stage 4: Scenario elicitation and revision.   After developing an initial set of fragments and reviewing them with experts, we held additional sessions for expert review of the overall model. In these sessions, we elicited evaluation scenarios from a cross-section of experts, analyzed the scenarios, and presented the results to the experts. Due to time constraints, our initial evaluation focused on a relatively small set of threat/asset combinations, and focused primarily on scenarios the experts considered exceptional. We discovered that we could present the constructed networks to users to give them a clearer picture of the factors influencing risk. They felt comfortable with evaluating the plausibility of the overall conclusions, and with evaluating the plausibility of the reasoning chains they examined while inspecting the constructed network. The experts were satisfied with the initial evaluation. In their judgement, the RIN appeared to order asset-threat pairs sensibly for the small set of scenarios we evaluated during initial validation of the model.

Our initial evaluation was conducted in 2000, and included a scenario involving a terrorist attack on the Pentagon by means of a mortar shot from the Potomac River. At the time, this was intended to represent an exceptional case to stretch the limits of the model, rather than as a realistic scenario that might reasonably be expected to occur. In our evaluation, the model results indicated that the Pentagon was vulnerable to terrorist attack.

In the years since the initial development, Site Profiler has been deployed in a range of settings. Each new deployment requires the collection of data specific to the problem setting. The test and evaluation process for new deployments involves developing and evaluating new test scenarios relevant to the new setting.

Stage 5: Implementation and operational revision.   After our initial evaluation of the RIN, we incorporated the model into an implemented system. This provided our experts with the full capabilities of the Site Profiler graphical user interface for examining model results and drilling down to explore the reasons for model conclusions. Figure 14.2 shows a view users can access through Site Profiler's RIN viewer. We call this view the uRIN. The uRIN shows the nodes of the Bayesian network, color coded according to the fragment type. Each node is labeled with a summary value chosen by means of a utility function, and an icon indicating whether its distribution is obtained via user input, database lookup, belief propagation, or analytical computation. If desired, users can see details such as belief bars and confidence measures.

The Bayesian network used to compute results, which we call the cRIN, has the same connectivity as the uRIN, but many of its arcs are oriented differently. We found it necessary to reorient some arcs when depicting the uRIN, in order to make the diagram understandable to users who are not quantitatively sophisticated. Our users found it natural to trace evidential flows from observable indicators to

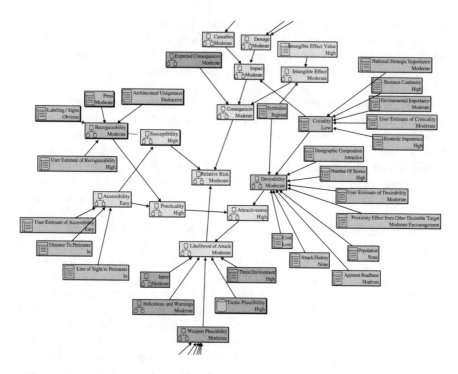

Figure 14.2 The RIN viewer allows users to mine through the uRIN in detail.

inferred measures, but found it confusing when arcs were oriented in the opposite direction from the flow of inference. Although the uRIN graph does not correspond to the dependence structure of the internal system model, it does reflect the experts' intuitive concept of the flow of evidence. Figure 14.3 shows a portion of the cRIN corresponding to a subset of the nodes shown in Figure 14.2. Many, but not all, of the arcs are oriented differently in the two graphs.

We are still using the set of MFrags developed for the initial implementation, with little modification. It is encouraging that our initial set of domain entities, attributes and relationships and features seems to capture essential structural features of the domain that are robust and stable over time. Of course, there is considerable variability in the information for specific assets and threats relevant to a given deployment. But the basic structure we identified in our initial development has remained remarkably consistent. This suggests that we have been successful in identifying essential structural characteristics relevant to assessing the risk of terrorist attacks on the kinds of assets included in our model.

14.3   Software implementation

Site Profiler uses an object-oriented database to manage the MFrags used to construct the RIN, the evidence to be applied to the constructed RIN, and the

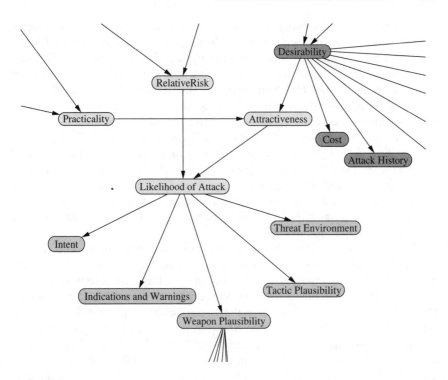

Figure 14.3 The cRIN has the same connectivity as the uRIN but some arcs are oriented differently.

probability distributions computed by the inference algorithm. We designed the database architecture to support knowledge based Bayesian network construction. The database contains domain objects that represent information about entities in the domain, including the current probability distribution of attributes that correspond to nodes in the RIN. The database also contains the information necessary to construct the RIN. Given a scenario to be evaluated, Site Profiler's knowledge-based model construction module uses the information represented in an object-oriented database to construct the RIN. After constructing the RIN, Site Profiler applies evidence where available, uses the inference algorithm to calculate probability distributions for nonevidence nodes, and stores the resulting probability distributions back into the database as values for their corresponding Bayesian attributes. These results can be viewed by the user through the GUI.

14.3.1 Bayesian attributes and objects

We defined a special type of attribute, called a Bayesian attribute, to represent attributes of a domain object that appear as nodes in the Bayesian network. For example, because the RIN has a node for criticality of an Asset, the Asset domain object has a Bayesian attribute to represent its criticality. Bayesian attributes store current belief values for nodes in the RIN. Belief values represent either evidence entered into the network, or propagated belief generated by queries against the network. Each Bayesian attribute has an associated Bayesian object. A Bayesian object contains the data necessary to represent a node in the RIN, such as the states of the node, the probability distribution, and the parents of the node. Together, the set of Bayesian objects associated with a domain object represents an MFrag corresponding to the subset of RIN nodes pertaining to that object.

Bayesian attributes and objects drive the knowledge-based RIN construction process. Bayesian objects are building blocks that define the structure and local distributions of the RIN. Although designed to be generic, our Bayesian objects are optimized for use with IET's Quiddity*Inference, which is the Bayesian inference engine used by Site Profiler. Bayesian attributes provide persistent storage for a snapshot in time of the RIN. In addition, the values of the Bayesian attributes can be displayed to users through the GUI.

Figure 14.4 depicts the RIN construction process. When a domain object containing Bayesian attributes comes into existence, the Bayesian objects associated with its attributes are also created. The collection of Bayesian objects, defining an instance of the domain object's MFrag, remain with the object during its life cycle. During RIN construction, all nodes in the MFrag for each applicable domain object are brought into the RIN as a group. When one domain object becomes associated with another, such as when an Asset and Threat form a Target, the MFrags are also associated with one another, based on the parent/child relationships identified by the network structure. These associations control how the MFrags are connected to construct an instance of the RIN. They also control the construction of the local probability distributions of the constructed Bayesian network. The conditional probability table for the Bayesian object is imported into the RIN, in a manner that may depend on the scenario. For example, the Asset domain object has a Bayesian attribute representing its accessibility. The distribution for the associated RIN

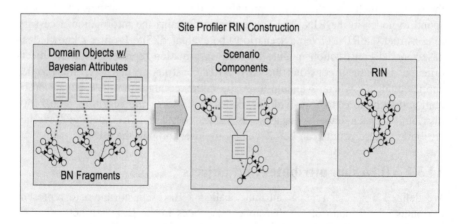

Figure 14.4 Fragments associated with domain objects combine to form the RIN.

node depends on the type of threat: an asset may have very different accessibility depending on whether it is attacked from the air, from a ground vehicle, or on foot In some cases, constructing a scenario-dependent local istribution may involve a simple look-up; in other cases, we used simulations or physics-based models to obtain local distributions.

14.3.2   RIN structure

The structure of the RIN is determined by the schema for our object-oriented database. The schema is organized around the domain model developed through the knowledge engineering process. The domain objects reflect concepts our experts use in reasoning about the domain. Bayesian attributes represent attributes of domain objects about which there is uncertainty. The Site Profiler domain objects combine in fundamental relationships to describe risk. Assets and Threats combine to form Targets. When Targets are created from Threat-Asset pairs, an instance of the RIN is created. The RIN is composed of network fragments from the Asset, the Threat, and other domain objects.

For Assets, the MFrag contains nodes that represent how critical the Asset is to the organization's mission, how desirable it is to an enemy, and how soft or accessible the Asset is. For Threats, the MFrag describes how plausible the tactic and weapon are, the likely intent of an actor to target the organization, and the asset types the actor is most likely to target. These risk elements combine to contribute to the key risk nodes associated with a Target: Likelihood of event, Susceptibility of an Asset to the event, the Consequences of the event, and ultimately, the Risk of the event. Countermeasures can act to mitigate any of the positive influencers of risk. These Bayesian attributes and their associated Bayesian objects represent the critical elements of risk for each threat-asset pair.

In traditional Bayesian network applications, a single, fixed Bayesian network is used for every problem and only the evidence varies from problem to problem. In contrast, the RIN is constructed from small collections of nodes (MFrag instances) that are attached and detached to form different networks. For instance, consider the case in which a single asset A may be attacked by two separate threats (T1 and T2). This results in two different instances of the RIN, one solving the risk to A from T1, and the other solving for the risk to A from T2. The network fragment structure for A is the same in both cases, and this structure is shared by both RIN instances. This reconfigurability of fragments not only allows the RIN to be flexible, but it reduces the input requirements from the user, because information about A can be entered once and reused in multiple scenarios.

The local distribution for a node may depend on characteristics of the entity to which it is related. In such cases, the structure is shared by RIN instances, but the local distribution is not. For example, the local distribution for accessibility of an Asset depends on characteristics of the threat. Thus, in the example above, the structure of the RIN relating A and T1 is the same as the structure of the RIN relating A and T2, but the local distribution for some of the nodes, including the accessibility node, is different in the two RIN instances. Additionally, Site Profiler includes seven subtypes of assets, each with its own MFrag that represents type-specific concepts. For instance, a Building Asset includes different factors from a Person Asset. Similarly, Site Profiler supports different threat types as well, and the number of nodes for these types varies. Although these fragments are different, they can still combine with other domain objects to form a RIN. When they involve different subtypes of threats and assets, different RIN instances will have different structures as well as different local distributions. Depending on the combinations of assets and threats, the number of nodes in a particular RIN instance can range from 35 to 60. The flexibility of the RIN and its fragments allows for a tailored risk analysis based on information appropriate to each scenario.

14.3.3   Evidence from other modules

Along with the database, RIN, and user interface modules of Site Profiler, we developed a 3D Modeling environment for building a site in 3D, an intelligent terrorist module that attempts to infiltrate the site in order to identify physical vulnerabilities, and analytic models for simulating weapons effects. These three modules provide evidence that can improve users' understanding of their risk. Integrating these results into the RIN was another requirement of Site Profiler.

Evidence in the Site Profiler architecture is supplied through the Risk Evidence Interface, which is an application programmer interface (API) for accessing various data sources. As shown in Figure 14.5, this allows the RIN to fuse information from the graphical user interface, models and simulations, an historical database, a corporate information system, or a real-time information source. This interface allows the RIN to consider new and existing evidence sources for evaluating risk contributors.

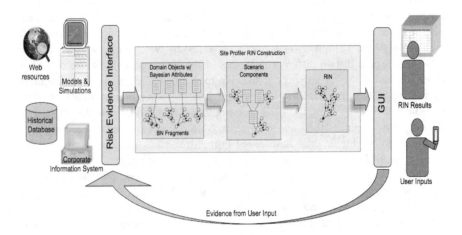

Figure 14.5 The RIN uses evidence from a wide range of sources to evaluate risks.

14.3.4   Confidence measures

In order to differentiate between the various types of RIN evidence, we developed a confidence model that recognizes the difference between subjective user evidence and objective analytical/historical evidence. Using credibility nodes, we apply softer evidence to the RIN when the data is subjective in nature. When applying analytic data, however, we only soften the evidence if the analytic model itself is less credible than other models. This not only allows for greater confidence in analytic versus user evidence, but also for recognition of the levels of fidelity in analytic models. This approach also works for our historical data, in that we vary the credibility of the data depending upon the reliability of its source. We use a very simple heuristic credibility adjustment. More sophisticated explicit credibility models (e.g., [503]) could be included in a future version of Site Profiler.

14.4   Site Profiler deployment

The preliminary validation of the Site Profiler RIN is encouraging. Our generic environment provides the ability to rapidly develop and deploy decision support systems employing knowledge based Bayesian network constructions across a wide range of application domains. In the past five years, we have deployed the Site Profiler system in military installations, seaports, municipalities, and states. Additional deployments are planned. Applications break into two broad categories: individual site assessments and risk assessments of a portfolio of assets.

14.4.1   Site assessments

Initially, Site Profiler deployments were focused on military installations and facilities within the United States and overseas. These sites represented a single, co-located collection of assets that were typically enclosed within a well-defined perimeter. This site-based model matched very well with the original philosophies of the RIN, especially in the areas of threat, attractiveness, accessibility, and consequences. The numbers of assets were constrained, as were the types of threats that were normally considered. The physical countermeasures in place were easy to identify and catalog, and the standard method of security represented an 'outside-in' approach to keeping threats at bay outside the walls of the site. Over time, this model evolved into also considering the areas surrounding the installation (referred to as a 'buffer zone', incorporating the risk posed to the site from the surrounding area as well as the risk posed to the surrounding community in the event of an incident (e.g., chemical release at the site). For each facility, Site Profiler would construct a collection of RINs for the relevant asset-threat combinations, from which installation security planners could analyse and make decisions. There was seldom a need to aggregate risk results to a higher level (e.g., across sites, cities, counties, states), and the number of scenarios was small and manageable (normally less than a hundred). In some instances, we made modifications to the basic model to handle special scenarios (such as drug trafficking through a seaport), but the underlying approach remained relevant.

As our customer base expanded and Site Profiler began to be used by states and local governments, several assumptions and design decisions had to be reconsidered. At first, our state and local customers used Site Profiler to assist in the assessment and management of the assets and risks within their jurisdictions. These users, mostly made up of law enforcement and counter-terrorism specialists, appreciated the structured assessment workflow and processes within Site Profiler. As these customers built up large data sets of assets and risks, spanning multiple counties and jurisdictions, it became clear that we needed to shift the focus of our applications. These larger data sets introduced the need to manage a portfolio of assets and the risks posed from a wide array of threats.

14.4.2   Asset portfolio risk management

Congress has mandated that DHS allocate some of its antiterrorism grant funds according to risk. In addition, HSPD-8 and other directives have called for state and local governments to begin risk management programs of their own. With this trend toward quantifying risk at the jurisdictional level, we have seen increased interest in using the Site Profiler application to measure and manage the risk across large groups of assets.

Managing the risk across a large portfolio of assets is a very different problem from that faced by a site security officer. Not only can the numbers be vastly different, from a single structure to a small handful on a military base to tens of thousands of critical assets in a large state, but the threats and assets themselves are different. In a single site implementation, the assets tend to be similar: similar construction, similar usage, similar environment. In addition, sites such as ports or military bases or other campuses tend to have some sort of perimeter barrier which filters out certain types of attack. Thus, the attack types of concern for assets at one site tend to be similar. For instance, at a port one might be concerned with ramming a ship into a terminal or pier, or an explosive attack on a docked ship from a small, fast-moving boat or a swimmer or diver; at a military base, one might be more concerned with a vehicle-borne improvised explosive device (VBIED) attacking a command center or barracks. In contrast, across a larger jurisdiction a risk manager must be concerned with a wide range of possible weapons and attack types, and the targets of these attacks could vary from food processing plants to high-rise office buildings to chemical plants to railroad tunnels. In addition, the assets can be located in densely populated urban areas or in more sparsely populated areas; they can have varying levels of defensive measures in place; and they can have different intrinsic abilities to withstand certain types of attacks.

The Bayesian network construction of the Site Profiler application has allowed us to use the same application to manage thousands of assets as we use to assess a handful. The MFrag representation allows our application to handles both the comparative and the combinatorial problems. The risk to a chemical facility from a VBIED attack can be directly compared with the risk to a tall building from an anthrax spore release in the ventilation system because the fundamental risk fragment is the same in both cases, with only the asset and threat fragments being different. The application recognizes this fundamental risk association and is able to evaluate, manage, and present the various combinations in a coherent and effective manner. Similarly, the RIN can easily be configured to calculate the risk to an asset from multiple attack types by swapping out the threat network fragments. Automating this across all the assets in a portfolio allows the straightforward calculation of all the combinatorics associated with a large asset portfolio and a significant collection of relevant threats. Site Profiler deployments have operated in environments with up to 5000 assets and up to 19 different threat types.

14.5 Conclusion

Site Profiler's knowledge based Bayesian network construction module is an essential component of a decision support system for assessing terrorist threats against military installations and civilian sites. Site Profiler uses a coherent, repeatable, and efficient methodology to give risk managers and security planners a complete picture of risks due to terrorism. By embedding the RIN in Site Profiler's intuitive interface, users can perform complex data analysis without the need to develop expertise in risk management or Bayesian reasoning. The network fragments provide flexibility in reconfiguring the network, validating with different expert communities, and tailoring to a variety of application types. The modular structure of Site Profiler offers a minimally intrusive pathway to updating the network to accommodate different risk algorithms or updated information.

The Bayesian network approach to thinking about risk has influenced our recommendations about risk Modeling to DHS. For instance, there has been a general reluctance to use expert elicitation to inform risk algorithms because of the perception that such subjective information could not be sufficiently quantified. Our experience constructing and validating the network fragments in the RIN has demonstrated that this is not necessarily the case. In the 2006 grant cycle, for the first time DHS has begun to incorporate quantified expert judgement into the eligibility risk equations. DHS has developed several risk methodologies, none of which takes a Bayesian network approach, but our experience with Site Profiler suggests that this would be a viable approach for an asset-based risk model.

Our experience with Bayesian networks with a nonexpert user community has yielded insights that may prove helpful with other Bayesian network applications. First, although there are clear advantages to using a Bayesian network approach, we have found that clients care more about price, ease of use, and the defensibility/intuitiveness of the results of a risk analysis application than about the technical details of the calculations. Although we believe the Bayesian network approach is superior to other risk approaches, we have not found it to be an effective selling point.

Second, if Bayesian networks are to become more widely used, tools will be needed to quickly and easily construct, populate, and validate network fragments.

The power of Bayesian networks – that they can incorporate both objective data and subjective expert judgement into a consistent data model – is also a formidable barrier to their use. It took six months to construct the RIN used in Site Profiler, primarily due to the difficulty of extracting useful data from experts. We often found ourselves building tools from scratch. Some decision support tools, such as implementations of the analytical hierarchy process [405], are simple to use and require judgements users find natural. However, tools designed to construct Bayesian networks tend to be cumbersome to use and to demand a high degree of technical sophistication. Addressing this issue would, we believe, encourage wider application of Bayesian networks.

Third, many casual observers and even some users in the terrorist risk management field do not take maximum advantage of the Bayesian network because they are asking the wrong questions of it. Many people want to ask questions about the future, such as 'When and where will the next terrorist attack occur?' Accurately forecasting specific plans of an intelligent, resourceful and adaptive adversary is something no model can do. The power of our Bayesian network comes from its ability to answer questions such as: 'What are the factors that make the risk high or low?' and 'Which of these proposed mitigation actions results in the greatest reduction to risk?' Bayesian networks are especially well suited to Modeling complex and uncertain relationships among the many factors that contribute to the risk of terrorism. We chose to apply Bayesian networks because they provide a powerful tool to help risk managers to make comparisons of relative risks, and to analyse the benefits of proposed mitigation actions. For this purpose, they are unmatched by any of the other methodologies we considered.

We have successfully implemented a Bayesian-network-based risk analysis toolkit and deployed it in antiterrorism environments at the state and local government levels. The system has many advantages over other risk models, and we believe there is great opportunity to expand its use. However, we have found significant barriers to more application development using Bayesian techniques, and we have found that the user community does not fully exploit the power of the construction. In our view, the most promising applications are those that rely on expert judgement and must adapt to new information on a regular basis.

Dedication

This chapter is dedicated to the memory of journalist Daniel Pearl, murdered by terrorists in Pakistan in February 2002, and to the pioneering research of his father Judea Pearl, inventor of the Bayesian network representation language and computational architecture. Daniel Pearl's spirit will live on in the work of those who apply his father's research to protecting the open society for which he gave his life.

15

Credit-rating of companies

Shigeru Mase

Department of Mathematical and Computing Sciences, Tokyo Institute of Technology, O-Okayama 2-12-1, W8-28, Meguro-Ku, Tokyo, 152-8550, Japan

15.1 Introduction

Credit-rating or appraisal of companies is the evaluation of desirability of companies as targets of investment. It provides important information not only to investors but also to companies themselves because it has a crucial influence on stock prices. They are published regularly and as circumstances demand by economists of credit-rating agencies, security firms and business journals. Credit-ratings are based primarily on much current financial data on performance of companies. But, frequently, unofficial or insider information, current market conditions and past data are also important. As such, credit-ratings are after all objective judgements of economic experts. Different economists may give different ratings on the same company.

In this chapter, we try to predict (classify) actual credit-ratings of experts based solely on several typical items of financial data of companies. The working data set used is of Japanese electric and electronic companies for the terms 2000 – 2003. It consists of 523 complete cases consisting of credit-rating data by experts and 15 official financial indexes. We will use the Bayesian network model as a classification tool. Applications of Bayesian networks are numerous in almost every field and, in particular, in finance, there are applications as a general inference tool [180], or a classification tool [30, 414, 426].

Bayesian Networks: A Practical Guide to Applications   Edited by O. Pourret, P. Naïm, B. Marcot
© 2008 John Wiley & Sons, Ltd

In the following, we construct and apply several Bayesian networks to our credit-rating problem. Some are constructed following financial experts' points of view. Also simple type of Bayesian networks, naive Bayesian networks [133, 161, 203], are examined. We perform the leave-one-out cross-validation test on each model to check its classification accuracy. As a conclusion, a naive Bayesian network with probability parameters adjusted so that a weighted mean squared errors of leave-one-out cross-validation prediction is minimized is shown to have the best performance. For comparison, several traditional classification methods were also applied.

15.2   Naive Bayesian classifiers

Let $X_1, X_2, \ldots, X_n$ be observable feature variables and let X_0 be the class variable to be predicted. It is easily seen that as either the number of nodes n or the number of involved links among nodes increases, the number of structural parameters which should be estimated increases rapidly. Hence it becomes difficult or impossible to get reliable estimates of parameters from a given limited amount of data.

One possible remedy to overcome this difficulty is the so-called naive Bayes assumption, which assumes that all the feature variables are conditionally independent on each other given the class variable [133, 161]. With this assumption, we have the following simple structural relation:

$$\mathbb{P}(x_0, x_1, x_2, \ldots, x_n) = \mathbb{P}(x_0) \times \prod_{i=1}^{n} \mathbb{P}(x_i \mid x_0). \qquad (15.1)$$

Figure 15.5 below (p. 271) is the graphical representation of a naive Bayesian network.

This network seems too simple to be a realistic model. Nevertheless, it is known to be useful for the classification task. One additional merit is that the calculation of the posterior marginal density of $\mathbb{P}(x_0)$ is quite easy. It should be thought of as an operational model rather than a descriptive model.

15.3   Example of actual credit-ratings systems

It is instructive to show an actual example of credit-ratings of companies. The most well-known and comprehensive example in Japan is called the *NEEDS-CASMA* (Corporate Appraisal System by Multivariate Statistical Analysis of Nikkei Economic Electronic Databank System) of Nihon Keizai Shimbun Inc. which is the most influential financial press in Japan.

The NEEDS-CASMA system covers almost all companies (2275 companies in the 2005 ratings) the shares of which are listed on Japanese stock exchanges. The employed procedure in 2005 is as follows:

1. Fifty economic journalists of Nihon Keizai Shimbun listed 10 excellent and 10 risky companies each.

2. From their opinions, excellent and risky company groups were selected. The total was about 80 companies.

3. Sixteen financial indexes which discriminate two groups best were selected.

4. Applying factor analysis, these 16 indexes were factored into four major factors. They can be interpreted as 'profit performance', 'financial security', 'growth potential' and ' firm size'.

5. Factor loadings of each factor for 2275 companies were calculated and, then, scaled so that the highest is 100 and the mean is 50.

6. Applying discriminant analysis to factor loading data, a discriminant function was selected which can discriminate distributions of the excellent and the risky group best.

7. From the thus obtained discriminant function, scores of each companies were calculated.

8. Scores were finally scaled so that the highest is 1000 and the mean is 500.

The NEEDS-CASMA system can be explained schematically by the network in Figure 15.1 and relevant financial indexes are shown in Table 15.1.

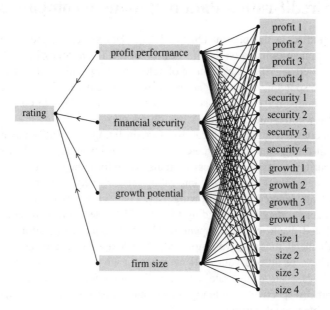

Figure 15.1 NEEDS-CASMA system network. Reproduced with permission from Inderscience.

Table 15.1 Sixteen financial indexes employed in NEEDS-CASMA system.

Profit Performance	profit 1	ROE (return on equity)
	profit 2	ROA (return on assets)
	profit 3	operating income margin
	profit 4	business interests per employee
Financial Security	security 1	business liquidity ratio
	security 2	ratio of funded debt to assets
	security 3	ratio of interest expenses to interest-bearing debt
	security 4	equity capital ratio
Growth Potential	growth 1	increase in overall sales
	growth 2	increase in equity capital
	growth 3	increase in gross capital
	growth 4	increase of employees
Firm Size	size 1	total capital employed
	size 2	number of employees
	size 3	amount of sales
	size 4	operating cash flow

15.4   Credit-rating data of Japanese companies

The data studied in the present chapter is the data set of Japanese electric and electronic companies listed on the first section of the Tokyo Stock Exchange during the period 2000 – 2003. The number of relevant companies is about 200. Since some of corresponding financial data need two consecutive years to be calculated and several items of data are missing, the data set consists of 523 complete cases. One company corresponds to three cases on average.

Since a convenient and complete list of credit-rating data was not available, the actual credit-rating data were gathered mainly from investment analyst reports of several securities firms and business journals. Therefore, they are never consistent. These data were finally categorized into ratings ec (= 'risky', 'ordinary' and 'excellent') which are denoted in the following by 1, 2 and 3 respectively for convenience.

Also financial data corresponding to these 523 cases were gathered. They are all official and disclosed by companies. Although there exist various (over 100) kinds of financial indexes, we employed 15 indexes which are most commonly referred. They are all real-valued. If necessary, each of the indexes are classified into three rates, 'bad', 'moderate' and 'good', which are denoted in the following by 1, 2 and 3 respectively. Actually, these rates simply divide the range of each index into three equal portions.

Table 15.2 shows 15 indexes employed in the present study. They can be grouped into four groups. Clearly there are functional relationships among these variables which are never independent.

Table 15.2   Fifteen financial indexes employed in the present study.

Profit Performance	p_1 = ROE (return on equity)
p_2 = ROA (return on assets)	p_3 = operating income margin
Financial Security	
s_1 = equity capital ratio	s_2 = ratio of funded debt to assets
s_3 = D/E ratio	s_4 = interest coverage ratio
s_5 = free cash flow	s_6 = liquidity ratio
Growth Potential	g_1 = increase in overall sales
g_2 = increase in operating profit	g_3 = increase in ordinary profit
Efficiency and Productivity	e_1 = overall sales per employee
e_2 = ordinary sales per employee	e_3 = total assets turnover

15.5   Numerical experiments

In this section, we apply Bayesian networks to predict (classify) credit-rating data. The used data is that of Section 15.4. Fifteen feature variables are discretized as 1, 2 and 3. In the following, we explain the employed classification methods BN1, BN2, nBN, nBN* and several classical methods and compare their performances. Performances of each method are examined using the leave-one-out cross-validation. We exclude one case from the data in turn, construct a predictor from the remaining 522 cases, and predict the ec value for the excluded case. Thus we can get 523 prediction results for which 'true' values are known. Results are summarized in Tables 15.3 and 15.4. Table 15.3 displays classification errors (= true value of ec – estimated value of ec) for all 523 cases.

Classifications are done by applying the junction tree algorithm for the first two networks whereas the last two are based only on Equation (15.1). For checking predictive performance of four networks, the leave-one-out cross-validation technique is used. Summaries are shown in Tables 15.3 and Table 15.4.

Table 15.3   Overall performances of all the methods for all cases.

method	classification errors: number of cases				
	−2	−1	0	1	2
qda	2 (0.4%)	21 (4%)	255 (49%)	233 (45%)	12 (2%)
lda	0	46 (9%)	423 (81%)	54 (10%)	0
nnet	0	51 (10%)	412 (79%)	60 (12%)	0
lm	7 (1.3%)	103 (20%)	286 (55%)	123 (24%)	4 (0.8%)
BN1	0	109 (21%)	329 (63%)	85 (16%)	0
BN2	0	96 (18%)	366 (70%)	61 (12%)	0
nBN	1 (0.2%)	75 (14%)	353 (68%)	94 (18%)	0
nBN*	0	47 (9%)	429 (82%)	47 (9%)	0

Table 15.4 Errors of all the methods according to the true ec. Numbers of cases and their percentages.

method	true $ec = 1$ (59 cases)				
	-2	-1	0	1	2
qda	1 (2%)	22 (37%)	36 (61%)		
lda	0	20 (34%)	39 (66%)		
nnet	0	28 (47%)	31 (53%)		
lm	7 (12%)	41 (69%)	11 (19%)		
BN1	0	40 (68%)	19 (32%)		
BN2	0	59 (100%)	0		
nBN	1 (2%)	10 (17%)	48 (81%)		
nBN*	0	5 (8%)	54 (92%)		

method	true $ec = 2$ (342 cases)				
	-2	-1	0	1	2
qdq		34 (10%)	270 (79%)	38 (11%)	
lda		26 (8%)	298 (87%)	18 (5%)	
nnet		23 (7%)	299 (87%)	20 (6%)	
lm		62 (18%)	237 (69%)	43 (13%)	
BN1		69 (20%)	241 (71%)	32 (9%)	
BN2		37 (11%)	305 (89%)	0	
nBN		65 (19%)	210 (61%)	67 (20%)	
nBN*		42 (12%)	261 (77%)	39 (11%)	

method	true $ec = 3$ (122 cases)				
	-2	-1	0	1	2
qda			74 (61%)	48 (39%)	0
lda			86 (70%)	36 (30%)	0
nnet			82 (67%)	40 (33%)	0
lm			38 (31%)	80 (66%)	4 (3%)
BN1			69 (57%)	53 (43%)	0
BN2			0	61 (50%)	61 (50%)
nBN			95 (78%)	27 (22%)	0
nBN*			114 (93%)	8 (7%)	0

15.5.1 BN1: Clustering by principal component analysis

One of the current rating methods of economists is to summarize feature variables first through factor analysis or principal component analysis. An obvious way to do this with Bayesian networks is to include these summary variables as new intermediate nodes of the Bayesian network, along with feature variable nodes and the class variable node. In the present experiment, we applied principal component analysis to each of the feature variable groups 'profit performance' (p), 'growth potential' (g) 'corporate efficiency and productivity' (e) and 'financial security' (s).

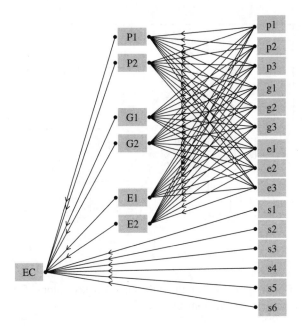

Figure 15.2   Bayesian network BN1. Reproduced with permission from Inderscience.

It was found that, for categories p, g and e, only the first two principal components are sufficient. However, we could not reduce the dimension for s and they were left as is. Other feature variables are connected to respective principal components nodes (Figure 15.2). For example, P_1 and P_2 are two principal components to p_1, p_2 and p_3 and so on in Figure 15.2. Values of principal component nodes were discretized into 1 (bad), 2 (moderate) and 3 (good) as values in original feature variable nodes. The structure parameters were estimated using MLEs.

15.5.2   BN2: Clustering by measuring conditional dependencies

Next we tried to cluster feature variables according to conditional dependencies among them given ec. We built a hierarchical cluster tree which arranges strongly conditionally dependent variables into the same cluster. We used the Matusita distance rather than the conditional mutual information. The Matusita distance [385], between two discrete probability distributions ϕ and ψ, is

$$M(\phi, \psi) = \left\{ \sum_x \left| \sqrt{\phi(x)} - \sqrt{\psi(x)} \right|^2 \right\}^{1/2}. \tag{15.2}$$

It is a true metric and can be used to measure a degree of (conditional) dependence. In order to measure the strength of the conditional dependency between X_1 and X_2 given X_3, we let $\phi(x) = \mathbb{P}(x_1, x_2 \; given \; x_3)$ and $\psi(x) = \mathbb{P}(x_1|x_3) \mathbb{P}(x_2|x_3)$ in

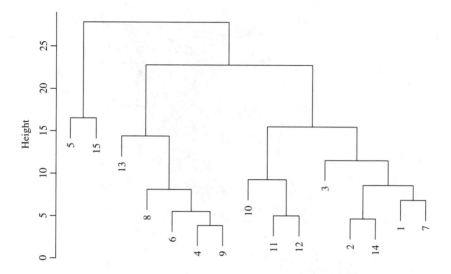

Figure 15.3   Dendrogram of conditional dependencies among feature variables. Reproduced with permission from Inderscience.

Equation (15.2). The smaller the distance, the weaker the conditional dependency. In particular, zero distance implies conditional independence. As the hierarchical cluster tree brings dependent variables together according to the degree of the dependency, the reciprocals of the Matusita distance were used. Figure 15.3 shows the hierarchical clustering where the y-axis refers to reciprocal distances.

As a result, we found five clusters:

$$\{p_1, p_2, p_3, e_2, s_4\}, \ \{s_1, s_3, s_5, s_6, e_1\}, \ \{g_1, g_2, g_3\}, \ \{s_2\}, \ \{e_3\}$$

when we cut the cluster tree at height 15. The hierarchical clustering using the conditional mutual information gave a slightly different network BN2 from those preferred by economists, see Figure 15.4.

15.5.3   nBN: Naive Bayesian network

A straightforward way to build a Bayesian network appropriate for a given data set is to learn a structure from the data using mutual and conditional mutual information or any other method, see, e.g., [92, 106, 204]. However, such Bayesian networks as classifiers are often reported not to increase the prediction accuracy, see [161]. Moreover, they are likely to become densely connected complex networks. More edges in the network increase the number of parameters and require even more data for reliable estimation.

The naive Bayesian classifier shown in Figure 15.5, having the simplest network structure, assumes the conditional independence of feature variables given the class variable even though most real-world data sets do not support such a property.

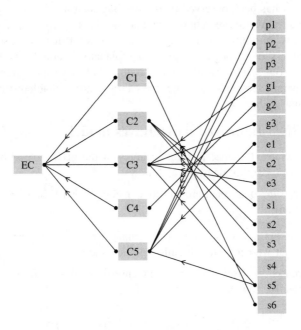

Figure 15.4   Bayesian network BN2. Reproduced with permission from Inderscience.

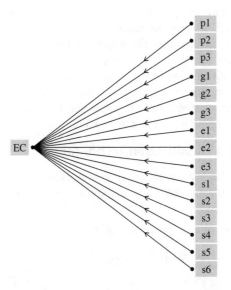

Figure 15.5   Naive Bayesian network nBN. Reproduced with permission from Inderscience.

Nevertheless, it has been observed that naive Bayesian networks frequently show considerable accuracy in prediction. This may not be so surprising if we take note of the usefulness of multiple linear regression models for many data sets for which linear dependency of an object variable on explanatory variables is dubious. One merit of naive Bayesian network models is that they have the minimum number of structure parameters and, therefore, we can get more reliable estimates from a limited amount of data.

15.5.4   nBN*: Improved naive Bayes model

We tried to increase the classification accuracy of the nBN method by modifying its conditional probability densities as follows:

- initialize (un)conditional probability densities by MLE (maximum likelihood estimation);

- iterate an optimization procedure by changing conditional probability densities so that we get better classification accuracy.

For this purpose, we used the simulated annealing tool of R to minimize the following objective function:

$$obj(\boldsymbol{\theta}) = \sum_{i=1}^{N} (|ec_i - 2| + 1)^2 (\widehat{ec}_i - ec_i)^2 \qquad (15.3)$$

where $\boldsymbol{\theta}$ stands for the logarithm of all the conditional density parameters except $\mathbb{P}(ec)$ and $\widehat{ec}_i$ is the leave-one-out prediction of the classifier for the ith case based on the current system parameters $\boldsymbol{\theta}$. The weight $(|ec_i - 2| + 1)^2 = 1, 4$ was added so that classification errors when true ec values are either 1 or 3 should be exaggerated. We did not alter the parameter for the density $p(ec)$ since it is desirable that classifications with no evidence on feature variables should be based on the true sample distribution of the class variable.

The simulated annealing procedure was iterated 10 000 times and the parameters $\boldsymbol{\theta}$ which gave the smallest value of the object function was selected. This may be by no means optimal but nearly optimal hopefully.

15.5.5   Classical classification techniques

In order to examine the performance of Bayesian classifications, the main subject of this research, we also apply the following typical classical classification techniques for comparison:

1. quadratic discriminant analysis (QDA);

2. linear discriminant analysis (LDA);

3. neural network (NNET);

4. multiple linear regression (LM).

Since all these methods suit real-valued data, we used raw feature variables.

15.6   Performance comparison of classifiers

We performed the leave-one-out cross-validation test for all the methods. The classification results are summarized in Tables 15.3 and Table 15.4 and displayed in Figure 15.6 for nBN*. We list brief summaries of the performance of each tested methods in the following:

BN1: Although it gives no errors $= \pm2$, rates of error $= \pm1$ is fairly high, in particular, for the case ec $= 1$.

BN2: It has a similar overall performance to BN1. But rates of error $= \pm1$ is high. In particular, it always gives $\widehat{ec} = 1,2$ when $ec = 3$. It has a tendency to classify many cases as $\widehat{ec} = 2$ which leads to a superficial overall accuracy.

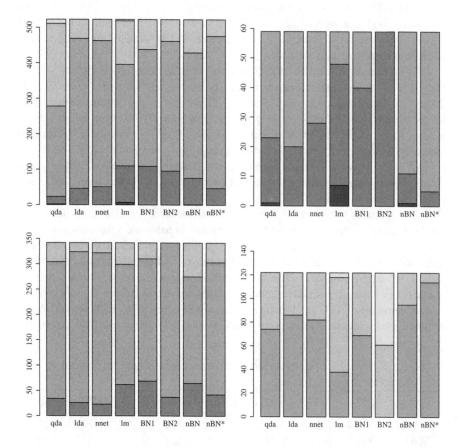

Figure 15.6   Bar plots of classification error percentages of nBN*. All cases (top left), cases $ec = 1$ (top right), $ec = 2$ (bottom left) and $ec = 3$ (bottom right). Shades of bars become darker as error varies from 2 to -2. Reproduced with permission from Inderscience.

nBN: It has a moderate overall accuracy. In particular, those for cases with the true $ec = 1, 3$ are fine. On the other hand, the accuracy for the cases with the true value of ec is 2 is not so impressive.

nBN*:   It has the best overall classification accuracy and it is also the best for cases with the true $ec = 1,3$. The classification accuracy for cases whose true value of $ec = 2$ is slightly lower.

Classical methods: Both NNET and LDA have fairly good overall accuracies whereas both LM and QDA do not. In particular, LM and QDA made several disastrous misclassifications, that is, $\widehat{ec} = 1$ when $ec = 3$ and $\widehat{ec} = 3$ when $ec = 1$.

It is seen from Table 15.3, that both LDA and NNET give very close figures to that of nBN* at least for the overall classification accuracy. BN1, BN2 and nBN have moderate overall classification performances. The relative sparsity of data compared with the abundance of parameters to be estimated for these models seems to lessen their performance.

It is important to note that the 'conservative' classifier which classifies every case as $\widehat{ec} = 2$ has overall performance 0, 59 (11.28 %), 342 (65.39%), 122 (23.33 %), 0 corresponding to Table 15.3. This misleading feature comes from the fact that 65.39% of cases belong to the class $ec = 2$. This type of problem is refereed to as 'credit card fraud detection' in [88]. The main interest of credit card companies is to predict a relatively minor group, i.e., potential deceitful applicants, as much as possible. In terms of investors, it is desirable to have a classifier which has very good classification performances for the cases with the true ec is either 1 (risky) or 3 (excellent), see Table 15.4.

Of the methods tested, nBN* has the best performance for cases $ec = 1, 3$. Note that this feature is naturally expected since the relevant parameters were chosen so that they gave a heavy penalty for misclassifications when $ec = 1, 3$. Also this causes its performance for $ec = 2$ relatively worse than those of LDA, NNET and BN2. However, the good performances of lda, nnet and BN2 for the case $ec = 2$ comes from the fact that they have a conservative tendency to classify cases $ec = 1, 3$ as $\widehat{ec} = 2$.

The superiority of nBN* can be seen also from Table 15.5 which shows weighted mean squared errors corresponding to (15.3), mean absolute errors and

Table 15.5   Weighted mean squared errors, mean absolute errors and mean squared errors for each method.

	QDA	LDA	NNET	LM	BN1	BN2	nBN	nBN*
WMSE	0.704	0.512	0.602	1.46	0.706	2.85	0.566	0.254
MAE	0.275	0.191	0.212	0.474	0.321	0.533	0.327	0.180
MSE	0.279	0.191	0.212	0.516	0.321	0.767	0.331	0.180

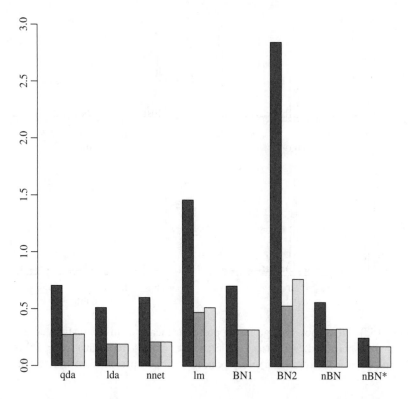

Figure 15.7 Bar plots of weighted mean squared errors (left), mean absolute errors (middle) and mean squared errors (right) for each method. Reproduced with permission from Inderscience.

mean squared errors. nBN* is also the best in the sense of both mean absolute errors and mean squared errors, see Figure 15.7.

Note that all the preceding models classify a given company to either 1, 2 or 3 according to the mode of the posterior density. But, sometimes, this type of rating may be misleading. For example, consider the case where the posterior density is 0.01, 0.44 and 0.45 for $ec = 1$, 2, 3, respectively. The MAP estimate is $ec = 3$ but $ec = 2$ is also very probable. Therefore, a better summary of results may be the posterior mean $1 \times 0.01 + 2 \times 0.44 + 3 \times 0.45 = 2.24$. The scaled score $50 \times$ [posterior mean $- 1$] $= 62$ that varies from 0 to 100 may be easier to interpret.

Figure 15.8 shows the histograms of posterior means of ec for the nBN* classification. Kernel-type density estimate curves are superposed. Means have sharp peaks at $ec = 1$, 2 and 3 as expected. The superiority of the nBN* method can be clearly seen from histograms drawn separately for each true ec-value.

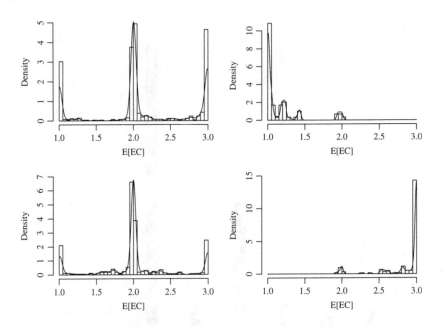

Figure 15.8 Frequency histograms of posterior means of nBN* (estimated density curves are superposed). All cases (top left), cases $ec = 1$ (top right), $ec = 2$ (bottom left) and $ec = 3$ (bottom right). Reproduced with permission from Inderscience.

15.7   Conclusion

Our analysis shows the usefulness of naive Bayesian networks, in particular, that of nBN*, for credit-rating. An important feature of Bayesian network models is that their structures are more transparent than that of, say, neural networks. This allows efficient interactive communication between modelers and application domain users. Further, for discrete variables, Bayesian networks assume no distributional restriction. Also Bayesian network models may work even if evidence for several feature nodes is missing, which happen sometimes. We can also calculate the posterior density of the class variable, although it may cause more probabilistic ambiguity.

It should be noted that nBN* as well as all our Bayesian network based methods used only partial information, i.e., discretized value of original data. This feature makes them 'robust' even if feature variables are erroneous. Also this discretization makes it unnecessary to preprocess feature variables, say, by logarithmic transformations, which is often necessary for other methods since feature variables have often extremely different scales.

Credit-ratings may be eventually objective judgements of financial specialists and our experiments tried to imitate existing experts ratings as faithful as possible. Nevertheless, the fine performance of nBN* suggests that judgements of economists are fairly subjective based in principle on typical official financial data. Hence it is likely that the classification based on nBN* trained for a certain period can also be useful for future periods unless economic fundamentals changes. Moreover, although no comparison were made so far, it may also be applicable to ratings of industry sectors other than electric and electronic companies.

Acknowledgment

The contents of this chapter is mainly taken from [496] and all computations were done using the R system [377] and its packages.

16

Classification of Chilean wines

Manuel A. Duarte-Mermoud, Nicolás H. Beltrán

Electrical Engineering Department, University of Chile Av. Tupper 2007, Casilla 412-3, Santiago, Chile

and

Sergio H. Vergara

Consultant for ALIMAR S.A. Mar del Plata 2111, Santiago, Chile

16.1   Introduction

During the last two decades several papers have been written on wine classification using different approaches and methodologies. Previous work [5, 79, 145] used for wine classification the concentrations of some chemical compounds present in the wine obtained through liquid or gas chromatograms. In the chromatograms the concentration of a substance is given by the area under the peaks of the curve, each peak representing a specific compound.

In [43] and [78], a different approach is presented where it is not necessary to quantify the concentration and to identify each peak (compound) but the whole information contained in the chromatograms is used and analyzed. The database used in [42] and [43] is composed of 111 wine chromatograms from high performance liquid chromatography (HPLC) for phenolic compounds, distributed as 27 samples of Cabernet Sauvignon, 35 samples of Merlot and 49 samples of Carménère. The chromatograms last 90 minutes and contain 6751 points. Due to the high dimensionality of the input data a dimension reduction process is first attempted using resampling and feature extraction techniques, including the fast

Bayesian Networks: A Practical Guide to Applications   Edited by O. Pourret, P. Naïm, B. Marcot
© 2008 John Wiley & Sons, Ltd

Fourier transform (FFT) [482], discrete wavelet transform (DWT) [199], Fisher transform (FT) [482] and typical profiles (TP) [42, 78]. Then several classification methods were analyzed and compared, including linear discriminant analysis (LDA) [482], quadratic discriminant analysis (QDA) [2], K-nearest neighbors (KNN) [482] and probabilistic neural networks (PNN) [2]. The best performance was attained when using the wavelet extraction with a classifier based on PNN. With a cross-validation process of the type leave-one-out (LOO), an average percentage of correct classification of 92.5% was obtained.

On the other hand, scientists and engineers are electronically reproducing some of the human senses with significant advances in seeing, hearing, tasting and smelling. Detecting aromas or odorants is, at the date of printing this book, being done with commercially available systems with applications for quality assurance of food and drugs [13, 65], environmental monitoring, military uses, security, medical diagnosis [284] or safety. Today there is a growing market for electronic noses (e-nose) for different applications.

Traditionally, human panels backed by sophisticated gas chromatography or mass-spectroscopy techniques quantify odors, making these methods expensive and time-consuming. The beauty of using an e-nose is that the information is obtained, for all practical purposes, on-line [176].

An e-nose is a system where an aroma sample is pumped into a small chamber housing the electronic sensor or an array of them. Then the sensor is exposed to the aroma sample giving an electrical response by using the transduction sensor properties, originated in the interaction of both the volatile organic compound (VOC) and the sensor active material. The response is recorded during the acquisition time of the signal processing subsystem of the e-nose. The stage ends with a sensor cleaning step for resetting the active material for a new measurement.

Recently, an e-nose based on metal oxide semiconductor thin-film sensors has been used to characterize and classify four types of Spanish red wines of the same grape variety [175]. Principal component analysis (PCA) and probabilistic neuronal networks (PNN) were used with satisfactory results. In [378] is studied the effect of ethanol, the major constituent of the head-space of alcoholized beverages, which generate a strong signal on the sensor arrays used in e-noses, impairing aroma discrimination.

Results of aromatic classification of three wines of the same variety but different years are presented in [507]. The input data for classification is obtained from an e-nose based on six sensors of conducting polymers [413]. Thus each pattern generated by the e-nose has six points. For classification purposes a multilayer perceptron (MLP) trained with the backpropagation algorithm (BP) [200] and a time delay neural networks (TDNN) trained with the Levenberg–Marquadt algorithm [200], were used. The database contained 5400 patterns, divided into sets for training (50%), validation (25%) and test (25%). It was shown that by incorporating a temporal processing the classification rate improved, i.e., the TDNN had better performance than the MLP. Other wine classification results using e-noses are reported in [171] and [412].

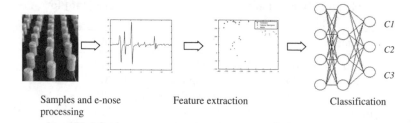

Samples and e-nose   Feature extraction   Classification
processing

Figure 16.1 Block diagram of the proposed methodology for wine classification.

The first stage of the proposed methodology is concerned with the dimension reduction of the patterns preserving the original information. This is done using feature extraction methods like principal component analysis (PCA) [48] and the discrete wavelet transform (DWT) [199]. Once the dimension of the input data has been reduced the information is introduced to a classification stage where BN techniques are used [331, 500]. A technique based on radial basis function neural networks (RBFNN) [48, 184] is used for comparison purposes. Finally, a classifier based on support vector machines (SVM) [109, 317, 470, 471] was also studied and compared. Figure 16.1 shows a block diagram of the used methodology.

16.2 Experimental setup

In this section we describe the experimental setup used to perform the study on wine classification. All the studies were carried out at room temperature and for each sample 10 aroma profiles were obtained from the electronic nose.

16.2.1 Electronic nose

The most important operating parameters of the e-nose are the temperature of the SAW detector in °C (sensor), the temperature of the GC column in °C (column), the temperature of the six positions valve in °C (valve), the temperature of the input gas in °C (inlet) the temperature of the trap in °C (trap), the slope the temperature ramp in °C/s (ramp), the time duration of the analysis in seconds (acquisition time) and the rate at which the information is registered in seconds (sampling period). This set of parameters defines the method under which the instrument operates. After a series of tests and experiments it was determined that the best values of the parameters for our study are those shown in Table 16.1.

To obtain an aroma profile, 40 ml of each wine sample were introduced into a 60 ml vial with septa cap avoiding sample exposure to oxygen in the air. The measurements were done immediately after the bottle was opened, maintaining the room temperature at 20 °C. Figure 16.2 shows a photograph of the e-nose during the measurement of a wine sample.

Table 16.1   Operation parameters for the e-nose. See [146] for more details.

Parameter	Value	Units	Parameter	Value	Units
Sensor	60	°C	Trap	300	°C
Column	40	°C	Ramp	10	°C/s
Valve	140	°C	Acquisition time	20	s
Inlet	175	°C	Sampling rate	0.01	s

Figure 16.2 Electronic nose model Fast GC Analyzer 7100 from Electronic Sensor Technology (EST).

16.2.2   Database

The database used in the study is formed by 100 commercial samples of Chilean wines of the type Cabernet Sauvignon, Merlot and Carménère. These wines belong to 1997–2003 vintages coming from different valleys of the central part of Chile. The distribution of the samples is shown in Table 16.2 while a complete description of the database is contained in [432].

The information from each sample was obtained by setting the e-nose parameters to the values given in Table 16.1. Ten runs were carried out for each one of the 100 wine samples generating a total of 1000 profiles (chromatograms).

16.2.3   Data preprocessing

A typical wine chromatogram obtained from the e-nose is a 12 s measurement with 0.01 s sampling period, containing 1200 points in total. From preliminary

Table 16.2   Distribution of wine samples.

Class	Type	Number	Percentage
1	Cabernet Sauvignon	36	36%
2	Merlot	44	44%
3	Carménère	20	20%

classification tests it was determined that using a sampling period of 0.02 s, similar results were obtained as in the case of a 0.01 s sampling period, in terms of the information content (Nyquist frequency). Therefore, for classification purposes chromatograms of 12 s composed of 600 points were considered. Another important factor on the signal preprocessing is the normalization. The amplitude of the profiles is variable around zero with positive and negative values. To minimize errors coming from measurement uncertainties a scale factor was applied to normalize the amplitude in the interval $[-1, 1]$. To this extent the maximum amplitude was used to normalize the signal according to the relationship $x_i' = x_i/x_{max}$ where x_{max} is the maximum amplitude of all profiles. With this procedure typical normalized profiles are shown in Figure 16.3 for Cabernet Sauvignon, Merlot and Carménère samples.

16.2.4   Methodology

In order to classify the profiles described in Section 16.2.2, Bayesian networks were used and compared with two other classification techniques: radial basis function neural networks (RBFNN) and support vector machines (SVM). As already mentioned, due to high data dimensionality previous to the classification process, a feature extraction procedure of the original data was performed, using principal component analysis (PCA) and wavelet analysis (WA).

Once the data dimension was reduced, the total database of 1000 profiles (360 Cabernet Sauvignon or Class 1, 440 Merlot or Class 2 and 200 Carménère or Class 3) was divided into two sets; one for training-validation (containing the 90% of the samples) and the remainder for test (containing the 10% of the samples). The sample distribution is the following.

- *Training-validation set:* 900 profiles corresponding to 90 wine samples, distributed as 330 profiles Cabernet Sauvignon (33 samples), 390 profiles Merlot (39 samples) and 180 profiles Carménère (18 samples).

- *Test set:* 100 profiles corresponding to 10 wine samples, distributed as 30 profiles Cabernet Sauvignon (3 samples), 50 profiles Merlot (5 samples) and 20 profiles Carménère (2 samples).

The samples for each set were randomly selected and proportionally to the number of samples contained in each class of the original data.

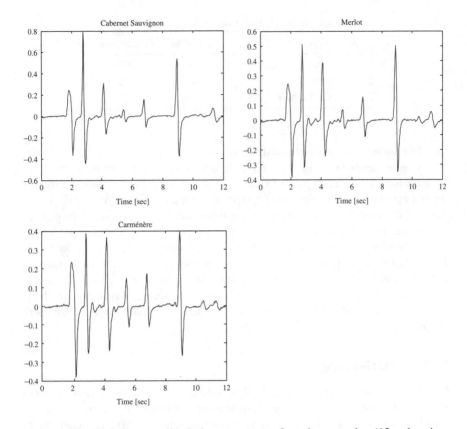

Figure 16.3 Typical normalized chromatograms for wine samples (12 s duration, sampling period 0.02 s and 600 points.

As a measure of the behavior of the method and to obtain the optimal values of the parameters for each method, cross-validation was used [166, 396, 419]. The database was divided into n sets, using $n - 1$ for training and the remainder for validation. The process is repeated n times so that all n sets are used once for validation.

Since 10 profiles were measured for each wine sample, the size of cross-validation sets will be 10 and therefore the training-validation base will be divided into 90 subsets of 10 elements, each one representing one wine sample. Thus, for each method the training is done using 890 profiles and one simulation for validation having 10 elements. The process is repeated 90 times so that each subset of 10 elements is used once to validate the method. The measure of the behavior is the average and the standard deviation of the percentage of correct classification in validation

Finally, once the cross-validation stage is done and the optimal parameters for each method are determined, a simulation with the test set is carried out for performance evaluation of each method when unknown samples are presented to the trained classification system. The behavior is measured again in terms of the average and the standard deviation of the percentage of correct classification using the test set. It is important to notice that the test set is never used in the training stage and therefore it will be completely unknown to the classifier.

16.3   Feature extraction methods

The main goal of feature extraction techniques is to reduce the dimension of data input to make the data analysis simpler. These techniques are usually based on transformations from the original data space into a new space of lower dimension. In this study we will use wavelet analysis (WA) and principal components analysis (PCA).

16.3.1   Feature extraction using wavelet analysis

Wavelet analysis is a mathematical tool of great importance due to its multiple applications and it is an interesting alternative to the popular Fourier transform (FT) [395]. In particular, the wavelet transform (WT) is a good tool in analyzing nonstationary signals since it uses variable size windows, i.e., small windows (detail windows) for a fine analysis of high-frequency signals and large windows (approximation windows) for a coarse analysis of low-frequency signals.

Wavelet analysis is based on a function $\psi(t)$ called the *mother wavelet function*, through which by means of shifting and scaling the signal, the signal can be decomposed. Scaling corresponds to a simple signal compression or signal stretching given by a scale factor $a = 2^{-m}$. The shifting is equivalent to a temporal displacement of the signal and it is determined by a shift constant denoted as $b = n2^{-m}$, where n denotes the number of points and m is the decomposition level. A criterion to determine the optimal decomposition level is to choose the level having minimum entropy [230].

The mother wavelet $\psi(t)$ chosen in this study is the orthonormal *Haar* basis. This choice is due to the simpliciy of the Haar wavelet and because there are no free design parameters to be chosen by the user, as suggested in [230] and [354].

In this study, different decomposition levels are analyzed by choosing the values 2, 3, 4 and 5, considering the approximation coefficients for the classification, because they contain most of the energy signal [230, 395]. Since each chromatogram has 600 points, the first decomposition level generates a curve with 300 points with the initial profile approximation coefficients. At the second level it has 150 points and so on. Figure 16.4 shows the profiles obtained after a wavelet analysis for decomposition levels 2, 3, 4 and 5, corresponding to a Cabernet Sauvignon sample. Note that the profile shape is preserved but having fewer points. For example, the profile of the fifth decomposition level contains only 19 points $(600/2^5 \simeq 19)$ showing clearly the compression effects exhibited by the wavelet transform [354, 395, 493].

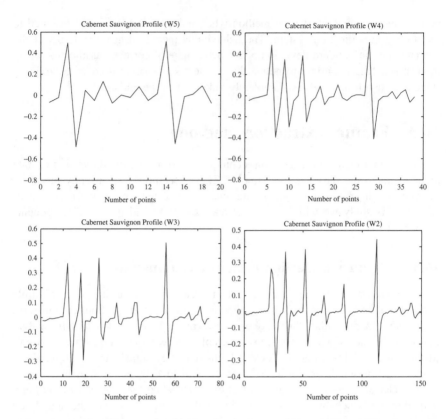

Figure 16.4   Cabernet Sauvignon profiles obtained after a wavelet analysis for decomposition levels 5, 4, 3 and 2, containing 19, 38, 75 and 150 points, respectively.

16.3.2   Feature extraction using principal components analysis

In this method the main idea is to transform the original feature space P into a space P' in which the data is not correlated, i.e., the variance of the data is a maximum. This is achieved by computing the eigenvalues and the eigenvectors of the of covariance matrix of the initial data and selecting those eigenvectors that have the largest eigenvalues. These components represent the axes of the new transformed space. By projecting the initial data onto these axes the largest data variance is obtained [48].

In our case the profiles can be seen as characteristic vectors belonging to R^{600} and the database as a matrix of 600×1000, where the 1000 columns correspond to each profile and the 600 rows to the points (that are going to be reduced). Considering the training-validation set we have a matrix of 600×900 (900 profiles (columns) of 600 points (rows)); then the covariance matrix of the training-validation set is

$$\Sigma_X = X * X^T \tag{16.1}$$

with X the training-validation matrix and Σ_X the covariance matrix of X of 600 × 600. Then computing the eigenvalues and the eigenvectors of Σ_X and selecting the eigenvectors with the largest eigenvalues, the principal components transformation matrix will be determined. One way of choosing the eigenvalues (and the associated eigenvectors) is considering the contribution to the global variance [104, 396] of each eigenvalue, γ_i, as:

$$\gamma_i = \frac{\lambda_i}{\sum\limits_{j=1}^{N} \lambda_j} \tag{16.2}$$

where $N = 600$ is the total number of eigenvalues of the covariance matrix Σ_X. It should be remarked that the parameter associates to each eigenvalue (and each eigenvector or principal component) a factor of relative importance considering its contribution to the total variance. When computing the eigenvalues of the covariance matrix Σ_X, these are ordered in ascending order [104, 396], thus the last components are those contributing most to the information (in terms of the covariance), whereas the first can be considered as noise and therefore disregarded. Figure 16.5 shows the contribution of the last 25 eigenvalues of the training-validation covariance matrix and it is observed that these retain practically all

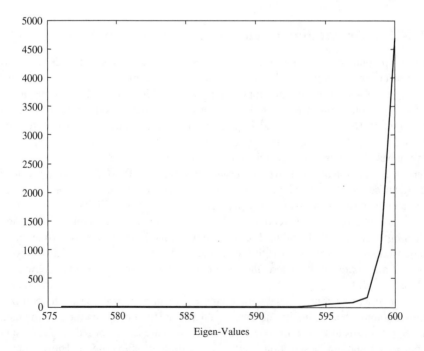

Figure 16.5   Contribution of the last 25 eigenvalues of the training-validation covariance matrix.

the information in terms of the covariance. When computing the contribution of the last 20 eigenvalues to the global covariance using Equation (16.2), the contribution to the total information is 99.87% and the last 10 eigenvalues contribute 99.46%. Therefore we will choose the matrix transformation composed by the 20 eigenvectors associated to the last 20 eigenvalues, generating a 600×20 matrix (the 600 rows represent the initial characteristics or points and the 20 columns the eigenvectors or new characteristics).

16.4   Classification results

Classification and pattern recognition techniques can be classified into three groups. In the statistical methods patterns are classified according to a model or statistical distribution of the characteristics. Neural methods perform the classification by means of a network formed by basic units (neurons) responding to an input stimulus or pattern. Finally, in the structural methods the patterns are classified based on a measure of structural similarity. In this section the results using the two feature extraction methods (PCA and Wavelets) together with BN as a classification method are presented. For comparison purposes results using RBFNN and SVM are also discussed.

16.4.1   Classification results using Bayesian networks

In this particular case we will use BN to classify Chilean wines in order to show its power when compared with other classification techniques.

The software used in this application is Weka [56, 500], which is the name of a large collection of machine learning algorithms developed by the University of Waikato (New Zealand) and implemented in Java. These are applied to data through the interface provided or to be incorporated inside another application. Weka has also tools to perform data transformation, classification, regression, clustering, association and visualization tasks. Its license is GPL (GNU Public License) and can be used freely. Since Weka is programmed on Java, it is independent of the architecture and runs over any platform where Java is available.

In a Bayesian network, learning is done in two steps; first structure learning and then probability table learning. For this study we have chosen a structure learning based on local score metrics [56] with the TAN (tree augmented Naïve Bayes) search algorithm [500], where the optimum tree is found using the Chow–Liu algorithm [96].

For structure learning, a network structure B_S can be considered as an optimization problem where a quality measure of a network structure given the training data $Q(B_S|D)$ needs to be maximized. The quality measure can be based on a Bayesian approach, minimum description length, information or other criteria. Those metrics have the practical property that the score of the whole network can be decomposed as the sum (or product) of the score of the individual nodes. This procedure allows for local scoring and thus local search methods. A more detailed explanation about the metrics can be found in [56] or [500].

Once the network structure has been learned, we choose how to adjust probability table learning. In this case the SimpleEstimator class produces direct estimates of the conditional probabilities and with the BMAEstimator, we get estimates for the conditional probability tables based on Bayes model averaging of all network structures that are substructures of the network structure learned [55]. The structure of the BBN used in this study is given in Figure 16.6. The upper most node is associated with the wine class and the other nodes correspond to the 38 components of the feature vector in both cases (wavelet and PCA extraction). Initially all the node probabilities were set to 0.5.

16.4.1.1   Results with wavelet extraction

In this study a wavelet transform was applied to the original 600 points signals with mother Haar wavelet function, using a decomposition level 4 to reduce the dimension of the vectors to 38 points. Then the preprocessed information (38 dimensional vectors) was introduced to the BBN classifier. The results are presented in Table 16.3 for five different local score metrics; minimum description

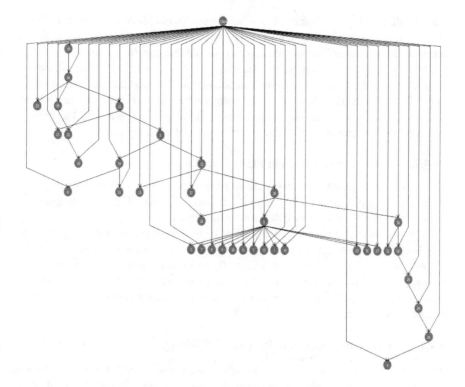

Figure 16.6 Structure of the BN used in the study.

Table 16.3   Classification results using wavelet+BBN for different metrics.

Local score metric	% of correct classification	Standard deviation	% of correct classification	Standard deviation
	Training-validation set		Test set	
MDL	90	0.3015	85	0.3589
Bayes	94	0.2387	91	0.2876
BDeu	93	0.2564	85	0.3589
Entropy	95	0.2190	89	0.3145
AIC	96	0.1969	88	0.3266

Table 16.4   Confusion matrix for the best case (91%) when using wavelet+BBN.

	Cabernet	Merlot	Carménère
Cabernet	33 (91.66 %)	3 (8.33 %)	0 (0 %)
Merlot	3 (6.82 %)	41 (93.18 %)	0 (0 %)
Carménère	2 (10 %)	1 (5 %)	17 (85 %)

length (MDL), Bayesian metric, Bayesian–Dirichlet likelihood equivalent uniform (BDeu), entropy metric and Akaike information criterion (AIC). The best result in the test set is obtained for the case when the Bayesian metric is used achieving 91.0% correct classification with a standard deviation of 0.2876. The confusion matrix for the best case (Bayesian metric) obtained in the test set is shown in Table 16.3. Three Merlot samples are misclassified as Cabernet Sauvignon and vice versa. Only two Cabernet Sauvignon samples are confused with Carménère and one Merlot sample is erroneously classified as Carménère.

16.4.1.2   Results obtained using PCA extraction

To compare this result with that obtained using the wavelet transform (in terms of data dimension), 38 principal components were chosen. These 38 components represent 99.997% of the original variance information.

The results are presented in Table 16.5 for the five local score metrics studied. The best case in the test set is obtained for BDeu and entropy metrics with only 60.0% correct classification, far from the 91.0% obtained when wavelet extraction is used. The confusion matrix for the best case obtained in the test set (BDeu metric) is shown in Table 16.6. From this matrix it can be seen that the classifier makes a lot of errors.

16.4.2   Classification results using RBFNN

Artificial neural networks (ANN) are mathematical models inspired by brain functioning and have the ability to perform determined tasks. Taking the brain as a model, ANN use a large number of interconnection of basic units called *neurons*.

Table 16.5   Classification results using PCA+BBN for different metrics.

Local score metric	% of correct classification	Standard deviation	% of correct classification	Standard deviation
	Training-validation set		Test set	
MDL	71	0.4560	59	0.4943
Bayes	71	0.4560	56	0.4989
Bdeu	71	0.4560	60	0.4924
Entropy	74	0.4408	60	0.4924
AIC	71	0.4560	59	0.4943

Table 16.6   Confusion matrix for Bdeu and Entropy metrics (60%) when using PCA+BBN.

	Cabernet	Merlot	Carménère
Cabernet	16 (44.44 %)	6 (16.66 %)	14 (38.88 %)
Merlot	7 (15.91 %)	31 (70.45 %)	6 (13.63 %)
Carménère	6 (30 %)	1 (5 %)	13 (65 %)

Radial basis function neural networks (RBFNN) constitute the main alternative to the multi-layer perceptron (MLP) for data interpolation and pattern classification problems. They use functions with symmetry around a center c in the n-dimensional space of the input patterns instead of using a linear activation function. The output of the neuron in a RBFNN is given by the general equation

$$y_k(x) = \sum_{j=0}^{M} w_{kj}\phi_j(x), \tag{16.3}$$

where $\phi_j(.)$ are the RBF and w_{kj} are the weights in the output layer. In a classification problem, the objective is modeling the a posteriori probability densities $\mathbb{P}(C_k|x)$ for each of the k classes. These probability densities can be obtained through the Bayes theorem [396], using the a priori probability densities of each class $\mathbb{P}(C_k)$. It can be proved that the a posteriori probabilities of each class can be expressed as

$$\mathbb{P}(C_k|x) = \frac{\sum_{j=1}^{M} \mathbb{P}(j|C_k)\, p(x|j)\mathbb{P}(C_k)}{\sum_{j'=1}^{M} p(x/j')\mathbb{P}(j')} \frac{P(j)}{P(j)} = \sum_{j=1}^{M} w_{kj}\phi_j(x) \tag{16.4}$$

where $\mathbb{P}(C_k|x)$ is the a posteriori probability density of each class given a pattern x.

The previous expression can be interpreted as a RBFNN where the normalized basis functions are given by

$$\Phi_j(x) = \frac{p(x|j)\mathbb{P}(j)}{\displaystyle\sum_{j'=1}^{M} p(x|j')\mathbb{P}(j')} = \mathbb{P}(j|x) \tag{16.5}$$

and the weights of the second layer are given by

$$w_{kj} = \frac{\mathbb{P}(j|C_k)\,\mathbb{P}(C_k)}{\mathbb{P}(j)} = \mathbb{P}(C_k|j). \tag{16.6}$$

In this way the results of the basis functions $\phi_j(x) = P(j|x)$ can be seen as the a posteriori probabilities indicating the presence of specific characteristics in the input space. Similarly, the weights $w_{kj} = \mathbb{P}(C_k|j)$ can be interpreted as the a posteriori probabilities of the members of a class given specific input characteristics. That is the reason why it is natural to apply RBFNN to pattern classification problems [396].

In this study basis functions $\phi(x)$ of Gaussian type were chosen for each neuron with

$$\phi(x) = \exp\left(\frac{||x - c||^2}{2\sigma^2}\right) \tag{16.7}$$

where c determines the center and the parameter σ determines the size of the receptive field. σ is also known as the *spread* and $1/\sigma$ defines the *selectivity* of the neuron. A small σ implies a high selectivity whereas a large value of σ makes the neuron less selective. Then for a RBFNN it is necessary to define the spread σ and the centers c, of the neurons forming the receptive fields of the network. The usual way is to set the center c at each one of the training patterns of the problem. Thus, if we have p training patterns the network has p neurons centered at each pattern. This strategy guarantee zero error in the training set and the freedom to choose σ that generates a controlled spatial overlapping to guarantee a good generalization. Depending on the computational implementation utilized, σ can be equal for all neurons or have different values for each unit.

The next step is to choose the weight vector $w \in R^m$. To this extent the RBFNN is evaluated at the p training patterns

$$\phi_{ji} = \phi\left(||x_i - x_j||\right) \ \forall i, j = 1, 2, ..., m \tag{16.8}$$

where the symbol $||.||$ corresponds to the Euclidean norm of a vector. We define the matrix Φ, composed by all the ϕ_{ji}, as the *interpolation matrix* of the problem [316], from which the weights can be obtained through the relationship

$$\Phi w = T \tag{16.9}$$

where $w \in R^m$ is the weight vector and T is the objective vector (target) containing the desired outputs. Then if Φ is nonsingular the weights are obtained as

$$w = \Phi^{-1} T. \tag{16.10}$$

Michelli's theorem [316] guarantees that if all vectors x_i used to compute Φ are all different, then Φ will be nonsingular. For all simulations the neurons were located at each training pattern [184]. Thus when cross-validation is carried out the network has 890 neurons corresponding to each profile. Recall that the NN has two layers; the first has radial basis activation functions and the second linear activation functions. Simulations were carried out making cross-validation with the training-validation set for different values of the selectivity σ, and computing the performance. The same was done for the test set.

16.4.2.1   Results using wavelet extraction

To study the performance of the classifier based on RBFNN, different wavelet decomposition levels and different values of σ were analyzed. The results are presented in Table 16.7 for the training-validation set. From Table 16.7, the best classification rate (76.8%) is obtained for a Wavelet decomposition level 5 and for a selectivity 0.02 ($\sigma = 50$). The results obtained for the test set are shown in Table 16.8, where the best result reaches 88.0%.

Simulations were performed with Matlab 6.0 using the Neural Network Toolbox, the Wavelet Analysis Toolbox and the Signal Processing Toolbox. Table 16.9 shows the average processing time considering three runs.

Table 16.7   Average percentage of correct classification in validation using wavelet+RBFNN.

Wavelet decomposition level	Selectivity = 0.1		Selectivity = 0.02		Selectivity = 0.01	
	% correct classification in validation	Standard deviation	% correct classification in validation	Standard deviation	% correct classification in validation	Standard deviation
5 (19 points)	75.6	0.33879	76.8	0.3289	76.5	0.3198
4 (38 points)	75.6	0.3388	76.5	0.3385	72.7	0.3732
3 (75 points)	74.8	0.34027	76.6	0.3464	75.4	0.3315
2 (150 points)	72.3	0.3594	75.4	0.3538	74.1	0.335

Table 16.8   Average percentage of correct classification for the test set using wavelet+RBFNN.

Wavelet decomposition level	Selectivity = 0.1 % correct classification in validation	Selectivity = 0.02 % correct classification in validation	Selectivity = 0.01 % correct classification in validation
5	88	82	83
4	77	78	79
3	71	79	78
2	63	60	69

Table 16.9   Average processing time for the
simulations of wavelet+RBFNN.

Wavelet level	t[s]	Standard deviation
5	553.49	9.651
4	635.32	11.477
3	1129.19	10.544
2	1236.44	9.279

16.4.2.2   Results using PCA extraction

For this method 20 principal components containing 99.86% of the total information
of the training-validation data were considered. Different values of the selectivity
were chosen in the interval $[2^{-9}, 10]$. The results are presented in Table 16.10,
where the best results are 71.4% in validation and 76.0% in test.

Simulations were carried out using Matlab 6.0, the Neural Network Toolbox and
the Signal Processing Toolbox. The average processing times for each simulation
over three runs are shown in Table 16.11.

Table 16.10   Classification results using PCA+RBFNN,
obtained in validation and test sets, employing 20 principal
components.

Selectivity	% correct classification in validation	standard deviation	% correct classification in test
10	39.8	0.4913	30.0
1	36.6	0.4845	50.0
0.1	60.8	0.4402	52.0
0.02	35.3	0.3557	63.0
0.01	53.5	0.3846	67.0
0.0078125	61.3	0.3641	65.0
0.00390625	66.1	0.3853	76.0
0.00195313	71.4	0.3776	60.0

Table 16.11   Average processing time employed in
simulations of PCA+RBFNN.

Number of principal components	t[s]	Standard deviation
5	891.82	4.517
10	905.64	4.498
15	968.37	4.209
20	998.22	4.008

16.4.3   Classification results using SVM

Support vector machines (SVM) is a technique introduced by Vapnik and his collaborators [470] as a powerful classification and regression method. The main idea is that SVM minimizes the empirical risk (defined as the error in the training set) and minimizes the generalization error. The main advantage of the SVM applied to classification models is that it supplies a classifier with a minimum Vapnik–Chervonenkis dimension [470], which implies a small error probability in generalization. Another characteristic is that SVM allows us to classify nonlinearly separable data, since it does a mapping of the input space onto the characteristic space of higher dimension, where data is indeed linearly separable by a hyperplane, introducing the concept of optimal hyperplane.

For a classifier based on SVM it is necessary first to choose a kernel to carry out the mapping of the input data space. In [74] and [221] it is suggested to use a radial basis function (RBF) type of kernel defined as

$$K(x, x') = \exp\left(-\frac{1}{2\sigma^2}||x - x'||^2\right). \qquad (16.11)$$

The parameter σ is determined by the user, but the number of RBF and their centers are automatically determined by the number of support vectors and their values. The choice of the value of the regularization parameter C that penalizes the training errors is another parameter to be chosen [471].

16.4.3.1   Results using wavelet extraction

Several tests with different values of C and σ, for different decomposition levels were carried out. Tables 16.11 and 16.13 show the results obtained with the SVM classifier, using an RBF kernel, for different values of C and σ for a wavelet decomposition level 5. After running a series of simulations, the values $2^{13} = 8192$ and $2^{-3.2} \simeq 0.1$ were selected for C and σ respectively. Table 16.12 shows the results fixing σ at 0.1 and varying C for a decomposition level 5, whereas Table 16.13 summarizes the results fixing C at 8192 and varying σ for a decomposition level 5. From a series of test experiments, it was observed that the best classification rates were obtained for decomposition level 5 in test.

An analysis of the results in Tables 16.12 and 16.13 gives that the best classification rate in validation was 84.8%, whereas in test this rate was 90%, highlighting the good generalization property of the SVM.

In this case the simulations were performed using Matlab 6.0, with the OSU Support Vector Machines Toolbox version 2.33 [89], the Wavelet Analysis Toolbox and the Signal Processing Toolbox. Table 16.14 shows the average processing time employed in the simulations, without including wavelet decomposition.

16.4.3.2   Results using PCA Extraction

In this case 20 principal components were considered containing 99.864% of the total information of the training-validation database (in terms of the variance). The

Table 16.12   Classification results as a function of C, for decomposition level 5 and $\sigma = 0.1$, using Wavelet+SVM in validation and test.

C	σ	% correct classification in validation	Standard deviation	% correct classification in Test
128	0.1	76.6	0.38336	80.0
256	0.1	77.3	0.36896	83.0
512	0.1	79.5	0.34312	84.0
1024	0.1	81.3	0.32333	88.0
2048	0.1	82.2	0.31293	90.0
4096	0.1	83.3	0.30354	89.0
8192	0.1	84.0	0.30011	89.0
16384	0.1	84.8	0.29575	88.0
32768	0.1	84.8	0.28961	84.0

Table 16.13   Classification results as a function of σ, for $C = 2^{13} = 8192$ and decomposition level 5, using Wavelet+SVM, in validation and test.

C	σ	% correct classification in validation	Standard deviation	% correct classification in Test
8192	0.5	81.8	0.30645	82.0
8192	0.25	84.6	0.29536	87.0
8192	0.125	83.6	0.3044	90.0.
8192	0.0625	83.4	0.30691	88.0
8192	0.03125	81.4	0.31357	90.0
8192	0.015625	81.3	0.3254	86.0
8192	0.0078125	81.1	0.3296	85.0
8192	0.00390625	77.4	0.35776	82.0
8192	0.00195313	75.0	0.37842	77.0
8192	0.00097656	74.3	0.40086	72.0
8192	0.00048828	69.7	0.41717	64.0
8192	0.00024414	53.8	0.46678	39.0

Table 16.14   Average Processing time for three runs using Wavelet+RBFNN for different decomposition levels.

Wavelet decomposition level	t[s]	Standard deviation
5	33.61	3.145
4	35.32	3.737
3	55.93	3.951
2	66.14	4.279

performance was measured as a function of the parameters C and σ of the RBF kernel. Fixing one of them at its best value ($C = 8192$ and $\sigma = 0.1$) and varying the other around the best value, the results obtained are shown in Tables 16.15 and 16.16 for each case.

The best classification results (83.5% and 51.%) are reached for $C = 8192$ and $\sigma = 0.00781$ in validation and $C = 8192$ and $\sigma = 0.00391$ in test. Using PCA

Table 16.15   Classification results obtained in the validation and test sets using PCA+SVM, for $C = 8192$ and different values of σ.

C	σ	% correct classification in validation	Standard deviation	% correct classification in Test
8192	50E-01	81.3	0.3349	49
8192	2.50E−01	81.3	0.3428	40
8192	1.25E−01	80.1	0.3554	44
8192	6.25E−02	80.4	0.3499	54
8192	3.13E−02	82.0	0.3369	54
8192	1.56E−02	83.2	0.3193	55
8192	7.81E−03	83.5	0.3200	55
8192	3.91E−03	82.4	0.3274	59
8192	1.95E−03	75.8	0.3659	52
8192	9.77E−04	77.4	0.3646	47
8192	4.88E−04	80.1	0.3426	48
8192	2.44E−04	81.7	0.3373	42
8192	1.22E−04	80.4	0.3515	40

Table 16.16   Classification results obtained in the validation and test sets using PCA+SVM, for $\sigma = 0.1$ and different values of C.

C	σ	% correct classification in validation	Standard deviation	% correct classification in Test
128	0.1	83.5	0.3254	58
256	0.1	82.4	0.3402	54
512	0.1	80.8	0.3517	49
1024	0.1	79.8	0.3586	43
2048	0.1	80.0	0.3619	42
4096	0.1	80.3	0.351	46
8192	0.1	80.4	0.3528	47
16384	0.1	79.7	0.3537	46
32768	0.1	79.7	0.3537	46

Table 16.17   Average processing time employed in simulations of PCA+SVM for different pattern size.

Number of principal components		Standard deviation
5	1121.24	5.175
10	1129.46	8.117
15	1130.73	9.937
20	1149.61	9.808

extraction leads to much lower classification rates as compared with wavelet extraction. The simulations were performed using the software Matlab 6.0, and the Signal Processing Toolbox. Table 16.17 shows the average processing time considering three runs for each simulation when using PCA+SVM. These times include the computation of the principal components.

16.5   Conclusions

Chilean wine classification of the varieties Cabernet Sauvignon, Merlot and Carménère, from different vintages and valleys, was successfully performed based on aroma information (gas chromatograms) measured by an electronic nose, by using Bayesian networks.

Two feature extraction techniques were analyzed to reduce the original data dimension: principal component analysis and wavelet analysis. Then three classification techniques; Bayesian networks, radial basis function neural networks (RBF) and support vector machines were studied and compared. For all six combinations the performance was measured as the average percentage of correct classification in the validation set as well as in the test set, using cross-validation. The best parameters for each method were obtained from a cross-validation process with the training-validation set.

The results show that BN with the Bayesian local score metric gave the best performance (91.0% of correct classification in the test set) when wavelet extraction with decomposition level 4 is used.

The second best classification rate (90.0% in the test set) was obtained using wavelet extraction with decomposition level 5 together with SVM with a RBF type of kernel with parameters $C = 8192$ and $\sigma = 0.03125$. The best result of RBFNN in the test set was 88.0%, reached using wavelet extraction with decomposition level 5 and a selectivity of 0.1.

For all three classifiers the performance notoriously decreased when PCA was used as extraction method.

The results obtained in this study are promising and the first on Chilean wines using gas chromatograms supplied by an e-nose. They provide the basis for future work on the classification of Chilean wines.

Acknowledgment

The results presented in this work were supported by CONYCIT- Chile, under grant FONDEF D01-1016, 'Chilean Red Wine Classification by means of Intelligent Instrumentation'.

17

Pavement and bridge management

Nii O. Attoh-Okine and Olufikayo Aderinlewo

Department of Civil and Environmental Engineering University of Delaware, Newark DE 19716, USA

17.1 Introduction

Infrastructure management systems are designed to provide information and useful data analysis so that engineers and decision makers can make more consistent, cost-effective, and defensible decisions related to preservation of the infrastructure network. This chapter will focus on pavement and bridge management. In pavement management systems, the problem setting often involves a large number of uncertain, interrelated quantities, attributes, and alternatives based on information of highly varied quality.

Pavement management is generally described and developed at two levels: the network level (overall road network, with no detailed technical analysis) and the project level (when work/maintenance is to be done on a specific road section). There is also a third level, known as the project selection level, which ties the network and project levels together. The primary differences between network- and project-level decision-making tools include the degree or extent to which the decision is being made and the type and amount of data required. Generally, network-level decisions are concerned with programmatic and policy issues for an entire network, whereas project-level decisions address the engineering and

Bayesian Networks: A Practical Guide to Applications Edited by O. Pourret, P. Naïm, B. Marcot
© 2008 John Wiley & Sons, Ltd

economic aspect of pavement management [26]. Traditional decision support techniques used in pavement management systems include a decision tree and linear programming methods.

Bridge management includes all activities related to planning for design, construction, maintenance, and rehabilitation. Bridge management systems are a relatively new approach developed after the successful implementation and application of pavement management. The essential elements of bridge management systems are as follows:

- data collection on bridge inventory and bridge element conditions;

- information management systems including a database and other data storage;

- analysis schemes for determining bridge condition and predicting bridge performance;

- decision criteria for ranking bridge projects for maintenance, rehabilitation, and repair (MR&R);

- strategies for implementing bridge MR&R.

Therefore bridge management systems also function as a decision support system. Bridge management involves a variety of activities that are carried out at different levels; these activities are primarily aimed at answering the following questions [112]:

- What, if anything, needs to be done to the bridge now?

- Are correct maintenance procedures being carried out effectively?

17.2   Pavement management decisions

Pavement management decision making has traditionally suffered from a lack of suitable analytical tools dealing with vagueness and ambiguity of the human decision process. The main objective in solving a decision problem in pavement management is the computation of an optimal strategy. This is achieved through a two-step process: computation of maximum expected values of the utilities and computation of an optimal strategy that gives the maximum expected value.

Attoh-Okine developed valuation-based systems and networks [427] in pavement management decision making [23]. Figure 17.1 shows the graphical representation of valuation-based systems. The valuation networks provide a compact representation of the decision making process emphasizing the qualitative features of decision making. The authors compared the decision tree analysis with that of the valuation-based systems. The weakness of the decision trees includes a proper method of addressing information constraints, combinatorial explosiveness and the preprocessing of probabilities that may be required prior to the tree representation.

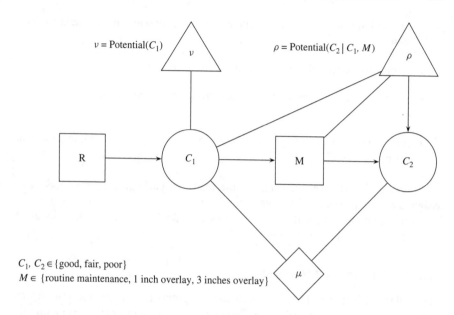

Figure 17.1 Valuation network for pavement management system decision making.

A brute force computation of the desired conditionals from joint distribution for all the variables is needed.

Figure 17.1 (VBS) consists of two decision nodes (R and M), and two random variables, C_1 (pavement condition at year 1) and C_2 (pavement condition at year 2). There are two 'potentials'[1]: ν is a potential for C_1, ρ is a potential $\{C_1, C_2, M\}$ and one utility valuation in the network. The valuation network can be used to compute the maximum expected utility value and an optimal strategy given by the solution M.

17.2.1   Pavement maintenance

In 1993, Attoh-Okine [24] proposed the use of influence diagrams in addressing uncertainties in flexible pavement maintenance decisions at project level. The Bayesian influence diagram is proposed as a decision-analytic framework for reasoning about flexible pavement maintenance; the influence diagram connects performance-related factors, historically related factors of the pavement, policy-related factors in maintenance, and environmental and cost-related issues. The following points are highlighted:

- identifying the key points of the pavement maintenance decision problem;

---

[1]As defined in [86]: If $C_1, C_2, \ldots, C_m$ are subsets of a set of variables $X_1, X_2, \ldots, X_n$, and the joint probability distribution of $X_1, X_2, \ldots, X_n$ can be written as a product of m functions $\Pi \, \Phi_i(C_i)$ $(i = i, \ldots, m)$, then the Φ_i functions are called *potentials* of the joint probability distribution.

- establishing the guidelines for quality decision making in pavement maintenance;

- quantifying uncertainties in pavement maintenance;

- assessing the judgment probabilities from pavement engineers and policy makers;

- estimating the value of information during flexible pavement maintenance at the project level.

Figure 17.2 is the model used. In the analysis, each probabilistic node was assigned an equal chance. The data input from the model comes from a pavement inventory database. The author did not present a final solution for this model; it was used for illustration purposes.

17.2.2   Pavement thickness

Attoh-Okine and Roddis [28] used influence diagrams to address uncertainties of asphalt layer thickness determination. The thickness of pavement layers is an important parameter in pavement management systems. The thickness data are used for pavement condition assessment, performance prediction, selection of maintenance

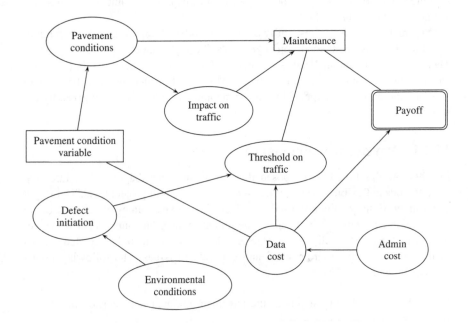

Figure 17.2   Bayesian influence diagram of a pavement maintenance decision model. Reproduced with permission from ASCE.

strategies and rehabilitation treatments, basic quality assessment, and input to overlay design. Negative consequences of both underestimating and overestimating
actual thickness may be seen in terms of increased cost and reduced service life.
For direct overlay projects, underestimation of existing pavement thickness will
result in a conservative overlay design with excessive cost. An overestimate will
result in a design that will not achieve the desired service life.

Layer thickness may be determined from historical records such as pavement
network databases, direct testing such as core samples, or nondestructive testing
such as ground- penetrating radar. The suitability of these procedures depends on
factors including the intended use of thickness data, the nature of the test method,
traffic patterns, and the cost of obtaining the data. Historical records normally
assume the same pavement thickness for pavement segment lengths of several
kilometers, coring is an inherently point-based method, and radar provides a continuous measure of thickness for the entire length of the pavement segment under
study. The nature of the test method affects the accuracy of the thickness value,
with the direct core being the most accurate, and historical records least accurate.

The influence diagram for pavement thickness assessment was represented at
three levels – graphical, dependence, and numeric. The graphical and dependence
level have qualitative (symbolic) knowledge and information flow. The numeric
level has quantitative knowledge. The influence diagram is shown in Figure 17.3.

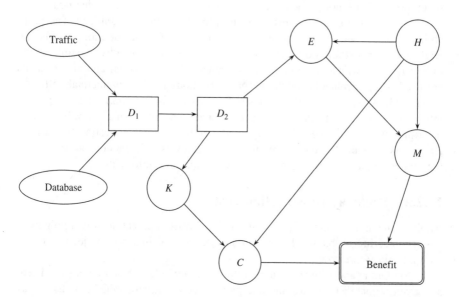

Figure 17.3 Pavement thickness layer modeling (D_1: application of thickness (decision node), D_2: thickness assessing methodology (decision node), E: estimated
thickness (probabilistic node), H: actual thickness (probabilistic node); M: benefit
of the estimated thickness (probabilistic node)).

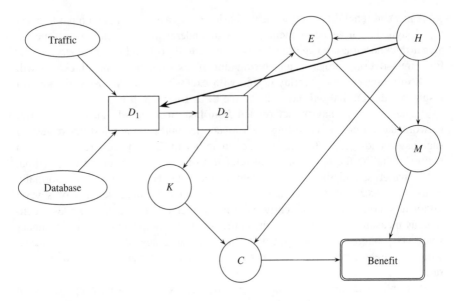

Figure 17.4 Value of Information analysis.

The influence diagram consists of 10 nodes connected by a direct arc. There are seven probabilistic nodes, two decision nodes, and one expected value node.

Expert elicitation procedures were used to obtain the probabilities. The probabilities were based on interviews with Department of Transportation and Federal Highway Administration officials. Optimization was performed by maximizing the expected value which represents the net benefit. Node removal and arc reversals are used in the evaluation of the probabilistic node [423]. Commercial software, India [123], was used for the analysis. Sensitivity analysis and value of perfect information were performed, Figure 17.4. The sensitivity analysis was used to determine which variables 'drive' the model and answers the question 'What matters in this decision?' The value of information analysis is the model that determines how much money one is willing to pay to obtain more information.

17.2.3 Highway construction cost

Attoh-Okine and Ahmad [25] present a risk analysis approach to solving highway cost engineering problems. The economic analysis for highway projects is based on uncertain future events.

Attoh-Okine and Ahmad used an influence diagram for the cost analysis. There are seven probabilistic nodes and one payoff node. The probabilistic nodes are as follows, Figure 17.5:

- Environmental regulation cost;

- Highway design cost;

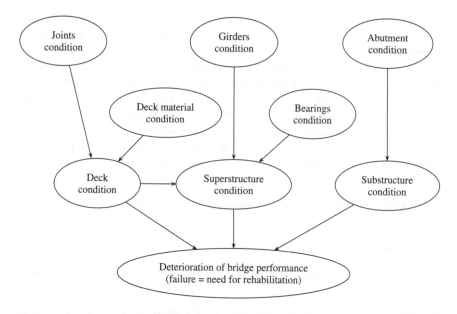

Figure 17.5 Global representation of the bridge deterioration problem. Reproduced with permission from Thomas Thelford Publishers.

- Directed labor cost;

- Material cost;

- Other labor cost;

and the utility node is highway construction cost. The authors presented steps on how to solve the problem.

17.3   Bridge management

Sloth et al. [431] proposed condition indicators within the framework of a Bayesian probabilistic network as a basis for bridge management decision making. The proposed method develops condition indicators for individual bridge components and is formulated in terms of time-dependent conditional probabilities. The authors show causal relationships between uncertain variables, including concrete mix, exposure conditions, and reinforcement in concrete bridges.

Bridge deterioration is a major component in making bridge management decisions. Attoh-Okine and Bowers [27] developed a Bayesian network model for bridge deterioration. The model is based on bridge elements including deck, superstructure and substructure conditions. LeBeau and Wadia-Fascetti [274] used a fault tree analysis to model bridge deterioration. A belief network representation of bridge deterioration was specified at three levels: graphical, dependence, and numeric. Figure 17.5 is a Bayesian network representation. The overall bridge deterioration is conditionally dependent on the deck condition, superstructure condition, and substructure condition. The deck condition is conditionally dependent on the condition of joints and deck material. The superstructure condition is conditionally dependent on the conditions of girders and bearings, and the substructure condition is conditionally dependent on the condition of abutments.

The probability updating method used for the analysis is based on the one proposed by Jensen in [232]. The algorithm does not work directly with the network, but on a junction tree, which is created by variables clustered into the tree. Various inferences can be made from the network:

- Predictive inference – given a particular model, what might happen?

- Posterior computation – given evidence, what is the model that explains the outcomes?

- Most likely composite hypothesis – given evidence, what is the most likely explanation?

The input (probabilities) used is obtained from experts. A series of questions to elicit probabilities was obtained from bridge engineers and inspectors with varying levels of experience. To eliminate the subjectivity, each expert's responses were weighted according to his or her own past experience, and the results were then reviewed by a selected expert.

Attoh-Okine and Bowers [27] used Netica software for the analysis. The authors investigated various scenarios. It was concluded that the Bayesian network is more appropriate than fault tree analysis, which is more appropriate for catastrophic failure. A Bayesian network is more appropriate for both catastrophic and noncatastrophic events. Development of the belief network model presented in [27] was limited to only a small amount of bridge element deterioration data available. The predictions made from this model are limited by the knowledge and experience of the experts involved.

Wadia-Fascetti [477] introduced a bidirectional model to bridge load rating based on a methodology following codes. [477] addresses both a prognostic and a diagnostic component which identifies the deficiencies. The author used a Bayesian network for the development of the model.

17.4   Bridge approach embankment – case study

A bridge approach embankment to be constructed in Delaware is presented. The bridge was 18 feet (5.49 m) thick, and it was required to be constructed in three layers, two of the layers constituting the top and bottom layers comprised of tire shreds and the middle layer consisting of normal soil. The top layer measured 10 feet (3.05 m) thick, the middle layer measured 3 feet (0.915 m) thick, and the bottom layer was five feet (1.525 m) thick. The reason for using tire shreds, otherwise called tire-derived aggregates (TDA), was its light weight and the resulting low lateral pressures it produces. Both of these are desirable qualities in this case, since the foundation soil is weak and low lateral pressure on the abutment wall is required.

The model that was developed using the Analytica decision analysis software program to simulate the critical characteristics of the embankment, namely, the temperature, leachate concentrations, horizontal pressures, and settlements, is as shown in Figure 17.6. These factors greatly influence its performance; hence, they are represented by four nodes, the outputs of which were compared with the analysis results of data obtained from the sensors in the field.

On the other hand, the input variables upon which the embankment characteristics are based were analyzed under the embankment simulation module. In the case of the temperature characteristic, the following variables were considered: the heat conductivity of the tire shred layer (Hct), the conductivity of the soil (Hcs), layer increment (Li), heat liberated (Hl), rate of transfer (Rot), and temperature per hour ($Tphr$) flowing through the embankment. Figure 17.8 below shows the submodel developed for simulating the temperature response.

Next, the submodel to analyze leachate characteristics of the embankment was developed based on the observed maximum leachate response of scrap tire fills as studied by Humphrey [222] at four major locations in the US and as documented in [22]. Some of the parameters tested for include pH, cadmium, manganese, sulphate aluminum, iron, chloride and zinc.

The major input variables considered in simulating the embankment settlement response included the surcharge load, normal stresses (3c3) at the point where the settlement is of interest, coefficient of compressibility of the embankment layer, and other variables like the layer thickness and the embankment height. Figure 17.7 shows the submodel used for simulating the settlement response.

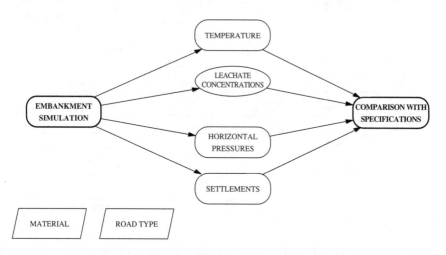

Figure 17.6 Model for embankment simulation.

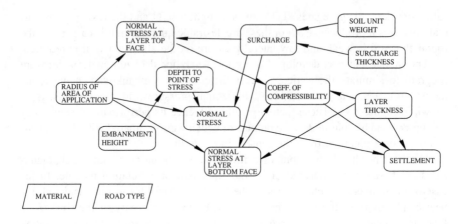

Figure 17.7 Diagram illustrating the settlement model for the embankment.

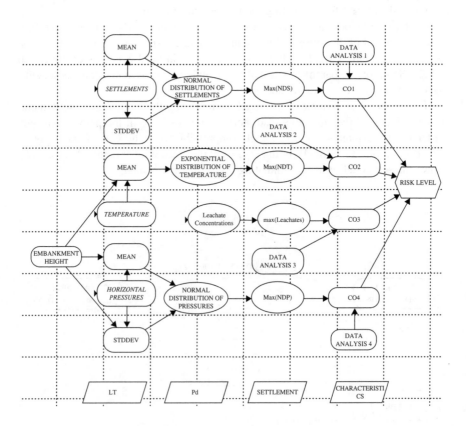

Figure 17.8 Diagram illustrating the output model.

Finally, the horizontal pressures exerted by the embankment on the abutment wall of the bridge were simulated as passive pressure during the hot season and active pressure during the cold season. The general variables considered in constructing the model for the horizontal pressures are friction angle, layer increment, surcharge, type of climate, coefficient of passive earth pressure, and coefficient of active earth pressure.

The simulation process was carried out by analyzing the output submodel as shown in Figure 17.8 under the 'comparison with specifications' module. The variables labeled MEAN and STDDEV give the average and standard deviation of the random values of the corresponding characteristics. The distribution nodes convert the mean and standard deviations calculated into probability distributions that best represent them, while the nodes labeled 'max' run a query through each set of distributions to identify the maximum probability density.

The index nodes LT, Pd, SETTLEMENT and CHARACTERISTICS specify the labels for the depths, probability density, settlements, and leachate properties respectively. The nodes labeled CO1, CO2, CO3 and CO4 are the maximum probability density of each simulated characteristic with that measured in the field, while the RISK LEVEL node gives the condition of the embankment using a 'SAFE' or 'FAIL' criterion.

The submodel then compares the maximum probability density of the embankment materials corresponding to the particular characteristic of settlements, temperature or horizontal pressure with that obtained from analysis of data obtained from the field. The output specifies a failure if the latter exceeds the former. In the case of the leachates, the maximum concentrations computed for the embankment based on the literature are used as the limiting criteria.

Based on the model analysis and the results obtained, it was observed that a greater proportion or density of the embankment was at lower temperature at the surface than further within it due to the low conductivity of scrap tire and soil. This means more sensors or highly sensitive sensors will be required at the surface to capture the temperature, while only a few will be required within it. For the embankment considered, the overall maximum probability density obtained for the temperatures is considerably high at 0.9703. This indicates that the possibility that temperature variations within an embankment of similar geometry on site could pose any reasonable risk is unlikely.

Some of the leachate substances such as lead, zinc, cadmium, selenium, chloride and sulphates whose characteristics were simulated using the model showed little possibility of their maximum probability of concentrations being exceeded by their corresponding values in the actual embankment. This is based on the fact that the simulated responses were very minimal.

In terms of the settlements, it was observed that at any particular depth, the relative settlements of points within the embankment measured at 0.1 m intervals decrease the further into the embankment the settlement is observed. On the other hand, the overall settlement varies nonuniformly from one layer to the other. In addition, a smaller proportion or density of the embankment experiences higher settlements near the surface of the embankment than further within. Since the relative settlements are highest near the surface of the embankment and the corresponding density of the tire shred materials is lowest here, very sensitive sensors will be needed to capture the information at these locations. For the embankment considered, the overall maximum probability density obtained for the settlements is 0.492.

In terms of the horizontal pressures, it was observed that in both cases, the pressures increase linearly with depth through the embankment. The active pressures are much lower than the passive pressures. Hence, the horizontal pressures under a hot climate should be considered critical when designing the abutment wall, since they are greater in magnitude than the horizontal pressures under a cold climate. For the embankment considered, the overall maximum probability density obtained for the horizontal pressures is 0.00344. Overall, the characteristics of the embankment which require critical assessment are the settlement and horizontal pressures because their corresponding maximum probability densities simulated by the model are considerably low at 0.492 and 0.0034 respectively.

Bayesian networks can thus be a tool in the planning and designing stages of road embankments. It will aid engineers in developing various risk scenarios, thereby reducing the amount of time for field trials.

17.5   Conclusion

The purpose of this chapter is to demonstrate how graphical probabilistic models and influence diagrams can be applied to pavement and bridge infrastructure. Although some of the ideas have been implemented, some of the topics are still academic exercises. The chapter illustrates both the formulation and method of analysis. The graphical model is appropriate for these kinds of problems, since it is capable of capturing both qualitative and quantitative knowledge, which are the major characteristics of pavement and bridge management decision making. The correct physical understanding of the inputs and the engineering interpretation of the results for decision making is the key in using Bayesian networks for bridge and pavement management.

18

Decision support on complex industrial process operation

Galia Weidl
IADM, University of Stuttgart, Pfaffenwaldring 57, 70550 Stuttgart, Germany

Anders L. Madsen
HUGIN Expert A/S, Gasværksvej 5, 9000 Aalborg, Denmark

and

Erik Dahlquist
Department of Public Technology, Mälardalen University, S-721 78 Västerås, Sweden

18.1 Introduction

In order to determine the origin of a process disturbance or fault, the need for a quick and flexible guidance tool for decision support at higher automation level has emerged. This includes analysis of process conditions and advice on cost-efficient actions. The technology of probabilistic graphical models has turned out to be the right choice out of several alternatives, when high diagnostics capabilities, explanation of conclusions for transparency in reasoning, and trustworthy decision support are expected by the users (process engineers, operators and maintenance crew).

Bayesian Networks: A Practical Guide to Applications   Edited by O. Pourret, P. Naïm, B. Marcot
© 2008 John Wiley & Sons, Ltd

Due to the existence of a number of first level diagnostic tools, the aim has been to provide decision support on process operation. The framework of probabilistic graphical models includes Bayesian networks and influence diagrams. It has been found to be an efficient and flexible tool in overall-level process operation analysis, since not all conditions are measurable or computable in real time, and the combinatorial reasoning procedure is subject to uncertainty.

In this chapter, we focus on key aspects of process monitoring and root cause analysis for complex industrial processes. If only a classification of the failure type is required, neural networks or statistical classifiers may be more adequate. However, if decision support is needed, Bayesian networks support reasoning under uncertainty while influence diagrams support decision-making under uncertainty. In a pre-study, we have considered alternative approaches for diagnosis of industrial processes, e.g., the neuro-fuzzy approach [334], however, these techniques do not give the same advantages as probabilistic graphical models as discussed in Section 18.3.

A probabilistic approach to fault diagnostics in combination with multivariate data analysis was suggested in [278] and [279]. Arroyo-Figueroa and Sucar [19] have been using temporal (dynamic) Bayesian networks for diagnosis and prediction of failures in industrial plants. For decision support (DS), we have combined an algorithm for decision theoretic trouble-shooting with time critical decision support [205, 238]. We have applied influence diagrams [487] for advice on the urgency of competitive actions for the same root cause. Model reusability, simple construction and modification of generic model-fragments, reduction of the overall complexity of the network for better communication and explanations, were other selection criteria in favor of probabilistic graphical models [253].

This chapter contributes with a description of an application based on a combined object-oriented methodology, which meets the system requirements in industrial process control. The application development has been closely related with the integration of the methodology into the ABB Industrial IT platform. The ABB Industrial IT platform is an automation and information platform that integrates diverse IT applications for process control and management. Monitoring and root cause analysis of the digester operating conditions in a pulp plant has been selected as the real world application for testing purposes. ABB has managed the reuse issues through a generic approach including object-oriented Bayesian networks modeling (reflecting the process hierarchy) supported by the Hugin tool.

18.2   A methodology for Root Cause Analysis

Industrial systems grow in their complexity. A sophisticated industrial process can generate output from hundreds or thousands of sensors, which should be monitored continuously. In the case of deviations from normal process conditions, the relevant information should be singled out, the cause of the failure should be found and appropriate corrective actions should be taken.

The huge amount of data and the continuity of the process demands a high level of automation of operation and maintenance control. But not all operations can be completely automated. Often it is necessary to let a human operator manually control the process in critical situations. This poses a formidable challenge on the concentration and capability of the human being and on the efficiency of his decisions.

To support the operator in the task of disturbance analysis, an adaptive system for root cause analysis (RCA) and decision support will collect the data from the sensors and transfer it into structured and relevant information. Simultaneously, the process overview should be maintained and relevant explanations provided with advice on corrective actions. Then the operator can make educated decisions, based both on artificial intelligence and human experience. This should help avoid unplanned production interruption or at least ensure that the lost production is minimal.

Early abnormality warnings, RCA and process safety are different aspects of condition monitoring and diagnostics for industrial process operation under uncertainty. One can define as the problem domain the overall plant performance and output quality, any of its processes, process sections, equipment or basic process assets, within which the actual failure is located, fault, abnormality or in general, any deviation from desired prerequisites. Moreover, the problem domain may include the operating and maintenance conditions of the process.

The generic mechanism of disturbance (or failure) build up includes a root cause activation, which causes abnormal changes in the process conditions. The latter represents effects or symptoms of abnormality. Abnormal changes in process conditions are registered by sensors and soft sensors. If not identified and corrected, these abnormal conditions can enable events causing an observed failure.

18.2.1   Decision support methodologies

In general, a decision support system for industrial process operation should satisfy the following requirements listed in [113]: early detection and diagnosis; isolability; robustness, multiple fault identifiability; explanation facility; adaptability; reasonable storage and computational requirements. These requirements are satisfied by the methodology and systems described in this chapter.

In a pre-study we have considered neural networks, fuzzy logic, neuro-fuzzy systems, and Bayesian neural networks as alternative methods for root cause analysis and decision support. In this section we discuss briefly the advantages and disadvantages of these methods.

Typical examples for suitable applications of neural networks include automatic classification, pattern recognition, pattern completion, determining similarities between patterns or data. Neural networks can provide refined classifications of complicated patterns, which can be specified by a large number of parameters, e.g., several hundred variables. The main weaknesses of neural networks for RCA is their inability to integrate domain knowledge in the connections between input and output patterns and that users perceive neural networks as black boxes. Neural networks do not explain the reasoning nor do their internal quantitative parameters have a real-world interpretation. Thus, neural networks are not an appropriate choice for decision support tasks, which would require background information on the underlying mechanism of undesired events or failure build-up.

The strength of fuzzy logic lies in the knowledge representation via fuzzy if-then rules, operating with linguistic variables. They are well suited for applications where the knowledge is incorporated via direct input–output correlations. Fuzzy logic has been implemented with success in consumer electronics, i.e., systems of relatively small size and complexity as compared to large industrial processes. Most applications of fuzzy logic are in the area of control engineering. The construction of fuzzy systems requires delicate tuning of membership degrees to reach satisfactory and robust system performance. In complex fuzzy systems, manual optimization of membership degrees is practically impossible. Therefore one needs automated learning algorithms for the specification of membership degrees.

Neuro-fuzzy systems combine the advantages of the fuzzy logic knowledge representation with the adaptive learning capability of neural networks. This compensates for the drawbacks of the individual approaches – the black box behavior of neural networks and the membership determination problems for fuzzy logic. The resulting model allows interpretation of inference conclusions and one can incorporate prior knowledge on the problem domain. Neuro-fuzzy systems still have some problems with the interpretation of their solutions [334]. Some of them can indicate an error, but cannot fully diagnose the situation. The fuzzy variables should not be used in long chains of inference calculations, since a small change in one value might have great influence on the final result. Therefore, they are not the first choice of technology for decision support, where explanation of the underlying causal mechanism of failure build-up is expected from the user.

A hybrid combination of Bayesian statistics and neural networks has proved to provide good solutions for classification tasks in adaptive learning systems [216]. They have been used to explain which inputs mean most for the outputs, but not for causal explanation of inference conclusions to the user.

More traditional methods for fault diagnostics of industrial processes include quantitative and qualitative model-based fault diagnostics and principal component analysis. The immense number of possible combinatorial configurations of causes and consequences as well as the inability to resolve ambiguities are the weak points of qualitative models which rule them out for large scale industrial fault diagnostics. In deterministic qualitative models, the problems appear from the system's incapability to handle contradicting impacts due to different variables. That is, the impact might be a combination of increasing and decreasing impact due to different root cause variables on the same measurable variable. Moreover, in deterministic qualitative models, the variables are divided into qualitative states with fixed thresholds, which are difficult to define and do not allow any flexibility in the signal classification. The last is making the system rigid to natural changes in normal process behavior. A solution to this problem has been provided by fuzzy sets, where the thresholds allow smooth transition between the states [334].

We find that Bayesian networks are in particular suitable for root cause analysis and decision support on complex process operation. Bayesian networks are compact and intuitive graphical models supporting stochastic modeling of chain causality structures in the problem domain. Bayesian networks may be constructed based on qualitative knowledge, experience and expertise on the causal structure of industrial processes. The observed events (abnormalities or failures) in the domain are well-defined and can be represented efficiently in the models. Furthermore, probabilistic inference on the state of the system is possible under few or no observations. The reasoning process and the results of inference are simple to explain given the model and observations made. Finally, the decision-making is time critical and alternative corrective actions for the same root cause are utility (cost) dependent. Thus, influence diagrams are suitable to estimate the optimal corrective action for utility maximization.

Based on the evaluation of the previously discussed methods and their suitability for RCA and DS, we have chosen to use Bayesian networks and influence diagrams.

18.2.2   Root cause analysis using Bayesian networks

The developed RCA system uses Bayesian inference. Its system architecture is given in Figure 18.1.

The system architecture for RCA is built on the following main modules:

- A database for gathering evidence obtained from various sources of relevant information.

- Creation of Bayesian networks (off-line) as knowledge bases for RCA.

- Signal preprocessing and classification of variables into discrete states. The DCS-signals classification is adaptive to changes in operation mode (e.g.,

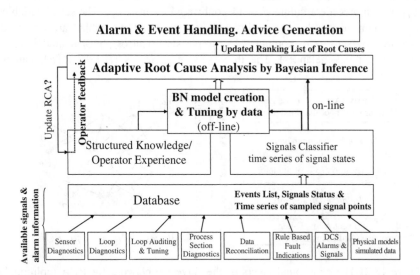

Figure 18.1 System architecture for RCA. Reproduced with permission from Elsevier.

changes in production rate or grade). The combination of signal levels and trends allows for estimating the risk of abnormality. This provides an early warning of abnormality in the process operation, which allows for performing a corrective action before any actual failure has occurred.

- The adaptive signal classification into states provides evidence for inference.

- Bayesian probabilistic inference supports the root cause analysis (on-line).

- Presentation of inference results and acquisition of operator feedback for update, explanation facility, advice as optimal sequence of corrective actions and adaptivity to process changes [486, 488].

18.2.3   Model development methodology

The main task of root cause analysis is to identify the possible root causes of process disturbances. A causal representation of the process disturbance mechanism produces a chain of events and transitions, which is of interest for root cause analysis under uncertainty and for the purpose of decision support on corrective actions, as shown in Figure 18.2, column 1.

A Bayesian network model for root cause analysis should reflect the causal chain of dependency relations as shown in Figure 18.2, column 2. The causal dependency relations are between the three symbolic layers of random variables, i.e.,

$$\{H_i\}, \{S_j\}, \{F\}, \tag{18.1}$$

where $i = 1, \ldots, n$ and $j = 1, \ldots, m$. In (18.1), the set of root causes $\{H_i\}$ contains all possible failure sources or conditions, which can enable different events $\{S_j\}$, which precede a failure F or its confirming events $\{S_{c_k}\}$. The set of variables $\{S_j\}$ contain also early abnormality effects and symptoms, which are observed, measured by sensors or computed by simple statistical or physical models (e.g., mass and energy balances). The three sets of variables $\{H_i\}, \{S_j\}, \{F\}$, can be

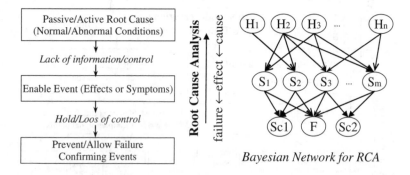

Figure 18.2   The conceptual layers of the Bayesian network for RCA (column 1) and the corresponding variables in each layer of the Bayesian network. Reproduced with permission from Elsevier.

viewed as three conceptual layers of variables in the Bayesian network (i.e., root cause → effect → failure), see Figure 18.2.

Several root causes can share the same effects or symptoms, which is expressed by multiple dependency relations between the layers. This qualitative structure is the starting point of all Bayesian network models used for RCA and DS. Each model is constructed from process knowledge. To this initial structure are added more dependency relations extracted by analysis of data from normal and abnormal process operation.

The root causes of abnormality may originate from various levels in the plant hierarchy:

- basic assets (e.g., sensors, actuators (valves, pumps), screens inside digester);

- equipment units (e.g., heat exchangers, evaporators, pulp screens, etc.);

- maintenance conditions;

- process operation conditions;

- plant conditions.

A number of data-driven packages for diagnostics of basic assets exist and form the base of available information for RCA, see Figure 18.1. Therefore, the Bayesian network is using a hybrid cross-combination of their outputs as evidence of the uncertainty or reliability of input information for RCA. The value added by root cause analysis is at process and plant level, especially since it can provide reusable design of process sections within the same process industry. The reusable design can be ensured by the use of object-oriented Bayesian networks (OOBNs). Moreover, the domain uncertainty and the need of adaptive treatment (within the same process section or at different levels of plant hierarchy) require a number of modeling assumptions and specifically developed OOBNs, which are used as building blocks in large and complex models.

A model of a problem domain, where all process variables are interconnected, can have very complex dependency relations. A large connectivity of the Bayesian network graph often results in huge conditional probability tables and/or complex inference. Therefore, the structure and/or parameter learning task become infeasible as there is simply not enough data from faulty operation. Even if enough data were available, processing huge conditional probability tables (CPTs) would be too slow for on-line inference. Instead of a pure data-driven approach, we have chosen to model the qualitative Bayesian network structure based on process knowledge. If there was data from faulty operation, we could have combined it with structural learning and/or parameter learning. For complex processes, this may imply complex multi-variable dependency relations. This means that the resulting CPTs can still be huge and become the source of inefficient probability calculations. Even if the probability calculation with huge CPTs were efficient, the knowledge acquisition would become a tedious time consuming task. To ensure feasibility of inference, we apply a number of simplifying assumptions. These also simplify the knowledge acquisition task. Moreover, the dependence properties of the problem domain allow for using a number of different modeling techniques to simplify the model. Notice that these modeling techniques are not approximations, if their simplifying assumptions are reflected by the underlying structure of the problem domain.

The use of OOBN models facilitates the construction of large and complex domains, and allows simple modification of Bayesian network fragments. We use OOBNs to model industrial systems and processes, which are often composed of collections of identical or almost identical components. Models of systems often contain repetitive pattern structures (e.g., models of sensors, actuators, process assets). Such patterns are network fragments. In particular, we use OOBNs to model (DCS and computed) signal uncertainty and signal level-trend classifications as small-standardized model classes (a.k.a. fragments) within the problem domain model.

We also use OOBNs for top-down/bottom-up RCA of industrial systems in order to simplify both the construction and the usage of models. This allows different levels of modeling abstraction in the plant and process hierarchy. A repeated change of hierarchy is needed partly due to the fact that process engineers, operators and maintenance crew discuss systems in terms of process hierarchies and partly due to mental overload with details of a complex system in simultaneous causal analysis of disturbances. It also proves to be useful for explanation and visualization of analysis conclusions, as well as to gain confidence in the suggested sequence of actions. The modeling and system requirements are all met by the developed methodology. This is further supported by the integration of the methodology and the Hugin tool into the ABB Industrial IT platform.

18.2.4   Discussion of modeling methodology

The development of the methodology incorporated the following system requirements and modeling issues:

- root cause analysis of industrial processes with adaptation to process operation/grade changes, aging and wear;

- reusable system design for various process applications;

- reusable modeling of repetitive structures, e.g., sensors, control loops, assets such as pumps and valves;

- risk assessment of disturbances by analysis of signal level-trend, adaptive to changes in process operation mode;

- ease of communication and explanations of conclusions at different process levels.

In addition, the Bayesian network modeling techniques used for RCA exploit the underlying structure of the problem domain. These are summarized below:

- modeling of root causes under the single fault assumption, where the single fault assumption is enforced by mutually exclusive and exhaustive states of the root cause variables.

- modeling of multiple root causes, where the single fault assumption is enforced by constraints on all root cause variables.

- noisy OR/AND, parents divorcing, modeling of expert disagreement.

- discretization of continuous signals on soft intervals

- mixture models.

- conditional independence of the effect or symptom variables given their common cause variable.

This reduces the complexity of the RCA model structure and its CPTs. For more details, see [485] and [488]. The issue of model reusability in different process sections or at different hierarchy levels is naturally realized by use of OOBNs.

18.3 Pulp and paper application

18.3.1 Application domain

For the proof of concept and to demonstrate the capabilities of the framework of Bayesian networks, a number of pulp and paper application examples have been developed. Next, to demonstrate a real world application, the monitoring and root cause analysis of the digester operating conditions in a pulp plant has been chosen, see Figure 18.3. The pulp mill is producing high yield unbleached pulp. It includes a continuous digester, a fiber line with washing and screening, a chemical recovery as well as utilities, e.g., a bark boiler. The pulp itself is obtained as a result of cooking of the wood chip in the digester. The preparation of the chips before digestion can also be crucial for eventual deviation from normal process conditions. Therefore, both chip preparation and actual digesting conditions are monitored. The RCA system is connected to the mills distributed control system (e.g., ABB's DCS), where from it reads sensor values and actuator settings. By monitoring sensor readings and control actions, disturbances are detected at an early stage. This makes it possible to adjust problems before they become serious.

The process is usually operated at several normal operation modes dependent on production rate, process load, etc. During normal operation modes, the variations of process variables should be within specified limits. Faulty change of operation mode, faulty process operation as well as equipment faults can be the root causes of abnormal process deviations. This can cause degradation of process output (e.g., quality) or failure in process assets, when exploited under improper conditions.

The above application has been used for testing the system performance in a simulated scenario with historical data from a real pulp plant. The structure of one of the developed Bayesian networks is shown in Figure 18.4. More details are provided in [488]. The application is being implemented in a Swedish pulp mill, but results have not yet been achieved when this is written.

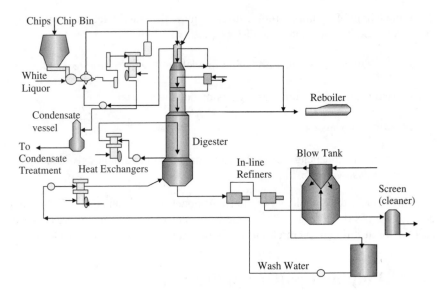

Figure 18.3 Digester fiber-line. Case-study: Monitoring of the digester operating conditions. Reproduced with permission from Elsevier.

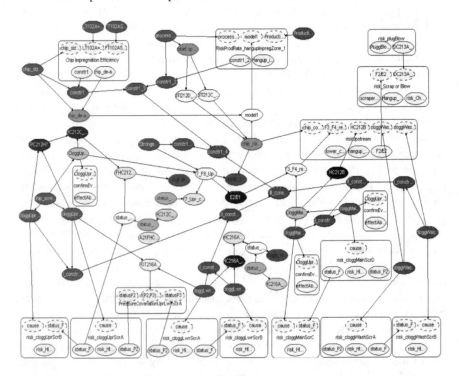

Figure 18.4 An OOBN version of the RCA for digester hang-up. Reproduced with permission from Elsevier.

18.3.2   Information, knowledge and data sources

One can rarely find an industrial process with an extensive database including a large number of cases from faulty operation with all variables needed to perform Bayesian network learning. One can use training simulators to simulate data on some faulty operation and their parameter configurations, but they do not cover all root causes, effects and failures listed by experts during knowledge acquisition. Instead, the RCA developer often has to rely on other approaches for initial estimations of probability distributions. The knowledge acquisition on the parameters of the Bayesian network model may use a mixture of different acquisition strategies for different fragments of the network. This becomes essential for large Bayesian networks modeling root causes of failures in large and complex industrial processes. For RCA on industrial processes, the knowledge acquisition is a mixture of:

- subjective estimates from process experts.

- Scientific theories for knowledge acquisition like fluid dynamics (for variables calculated from physical models, in the simplest case: mass, flow and energy balances, using DCS signals as inputs).

- Statistical model fragments based on a large number of database cases (applicable only for few root causes, which are the most frequent source of a problem).

- Subjective estimates on probability distributions. Striving from the beginning to estimate the Bayesian network parameters might not be efficient. There might be different sources of uncertainty:

  — The process behavior might be drifting over time as in the case of pulp plant commissioning when the production rate is sequentially increased over time and the mill is evaluating different suppliers of wood chips both for quality and economical reasons. All these natural process variations require model parameters that are able to adapt and follow the changes.

  — It might also be relevant for multivariable correlations, which are not measurable and difficult to estimate or calculated indirectly from DCS signals. In additions, even if these multivariate correlations are established for one piece of equipment, they are not generally applicable to complex equipment of the same producer and type, even when used at the same process stage. Different equipment units often exhibit individual behavior with different faults distributions. Even more variety is observed when the same type of screen is placed in sequence or in different process sections.

- Mathematical formulas describing the underlying physics and control strategy. For many processes, such simulations might take too long time to calculate on-line. Then, the simulations can be performed off-line for a properly reduced number of parent configurations and combined to estimate the probability distributions. The number of estimated parent configurations could be reduced by the design of experiment methodology, while exploiting symmetry properties of the matrix.

18.3.3   Model development

A case study on digester process operation has been the source of typical repetitive structures incorporated in the developed OOBN models. The general applicability of this methodology has been proved by its easy migration to a case study of pump operation problems in evaporation process.

References [485, 487, 488] present the learning algorithm and a number of OOBN models, combinations of which we have also used in the RCA of digester operation. For example, a generic Bayesian network model for adaptive signal classification is used as a generic building block in the RCA models. Such generic building blocks are typical examples of repetitive patterns in the Bayesian network model development. It is natural to represent them as OOBNs, e.g., the OOBN for adaptive signal classification provides information on the degree of reliability of sensor readings and reduces the degree of sensor uncertainty. In the OOBN for RCA of control loops the associated asset is the loop actuator. It models control loops of general interest to the process industry, e.g., pressure, flow, tank level and temperature control. The model output *class signal* shows whether the control loop is providing the target value for a process variable or its *set points* are wrong for the operation mode, alternatively its assets (sensors and actuators) are malfunctioning. A malfunctioning actuator (i.e., valve or pump) is a root cause from the set of basic assets. The OOBN for basic process assets is obtained from the OOBN for control loops by a simple extension incorporating root causes due to related equipment (e.g., screen status) and other basic components (e.g., pumps and tanks). On the other hand, this model allows under present evidence to reason about the performance of a selected diagnostic agent by pointing to that agent as a root cause that needs treatment. The OOBN model for risk assessment of process abnormality indicates improper operation conditions, recognized in changed level-trend pattern.

The OOBN models incorporate the causality steps of the basic mechanism of a failure build-up, e.g., an OOBN model of an event (e.g., pump plug), which is enabled by abnormal process conditions (e.g., high flow concentration) and is confirmed by another event (e.g., low pump capacity). Another type of OOBN evaluates simultaneously several events (e.g., various pump problems), which can cause process faults or failures (e.g., if a pump fails and it is indispensable for process operation, the plant should be shut down).

18.3.4   Validation

A detailed description of the validation for the above described methodology, BN models and application has been published in [488]. Here, we summarize only a few main features. The qualitative testing of the BN models included: (1) domain expert validation of the (in)dependency relations of the Bayesian network models; (2) design tests have been used to determine whether the Bayesian network is modeling correctly the expert and causal relations obtained from data analysis and to provide consequent revision of consistency in knowledge, experience and real plant data; (3) historical data analysis was used to set the allowed normal variations for measured signals, to examine expert opinions and to adjust the Bayesian network structure.

In addition, statistical tests were performed while exploiting the knowledge represented in the Bayesian network. The Bayesian network models have been evaluated based on simulated data and by off-line testing and calibration with real process data. The model-testing experiment proceeds in a sequence of steps. In the first step we sample the data to be used in the subsequent steps of the experiment. The second step of the testing phase is to process each of the sampled cases and determine the output from the model on the case. The third and last step is to determine the performance of the model based on the generated statistics.

We consider two different scenarios. The first scenario considers situations where no faults have occurred whereas the second scenario considers situations where a single fault has occurred. This experiment is used to generate statistics on error detection and correct RCA of all possible faults. The experimental results are described in [488].

The validation study indicated that the system actually works as designed, which has been a crucial ingredient in the proof of system concept and performance. Thus, this methodology represents an improvement (as automation technology) over existing techniques for manual root cause analysis of nonmeasurable process disturbances and can be considered as a powerful complement to industrial process control.

The sequence of repair actions can be considered as an improvement over existing policies, since it can provide on-line assistance to the operators and thus, it can save process operation time. Because of the wide applicability of this methodology, we expect that the results described will be of interest to other system designers who are considering similar problems.

This application demonstrates that fast and flexible disturbance analysis (RCA and DS) is feasible in industrial process control. It need not be a time-consuming task, if a computerized troubleshooting system is deployed. Thus, it has the potential of reducing substantially the production losses due to unplanned process break-downs.

18.4   The ABB Industrial IT platform

The ABB Industrial IT platform is an automation and information platform that integrates diverse standardizations of global processes and has a greater return on process assets. The developed RCA methodology has been intended as an advice functionality on process level in pulp and paper mills and as a part of the ABB's Smart Enterprise concept. The RCA application is connected to the mills distributed control system, wherefrom it reads sensor values and actuator settings. By monitoring sensor readings and control actions, disturbances are detected at an early stage. This makes it possible to adjust problems before they become serious.

The object-oriented Bayesian network framework fits naturally into the ABB Industrial IT environment, which utilizes aspect objects as containers of different applications communicating via the aspect integration platform in order to allow overall process optimization. ABB is utilizing for RCA the advanced Hugin tool from Hugin Expert A/S for implementing decision support into the Industrial IT platform.

The application development has been closely related with its integration on the Industrial IT platform and has required the development of special modeling conventions, such as variable names as well as conventional classes for measured/computed/observed, diagnosed or status variables.

In addition, history handler (for the filtered computation of signal trends) and state handler (for classification of raw data into states of evidence) have been developed and linked with the models through the Hugin application programming interface (API). The history and state handler have also been essential for the tests of the models on historical data and to simulate and evaluate the performance of the RCA system in an Industrial IT environment.

Thus, the infrastructure for applying this methodology in different domains is ready for immediate use, i.e., any new application of Bayesian networks is automatically integrated on the ABB Industrial IT platform.

The requirement of customization has also been addressed. Most measurements are real continuous data. The classification level (e.g., low, high) of the signals is customizable, i.e., they can be changed by the user, see Figure 18.5 and Figure 18.6. The extension of system functionality will include an automated classification of the signal limits as described in [488]. This will be of advantage for a higher level of automated system configuration.

18.5   Conclusion

The benefits of Bayesian networks relate in the first place to the advantages of the developed RCA methodology, which meets the requirements for operator decision support. As a consequence, any suitable RCA application will also benefit from the methodology based on Bayesian network.

With the growing size of domain applications, it might be preferable to have a simpler classifier (e.g., based on PCA or neural networks) for a binary decision on whether a fault is present or not. Once abnormality is detected, the RCA system will find the root cause and explain its underlying mechanism.

In any real process application, RCA needs adaptation to incorporate the ongoing changes in process behavior. Sequential learning is performed on the actual root causes and evidence for that particularly observed case. This is based on feedback from DCS and on operator/maintenance reports. A detailed discussion on

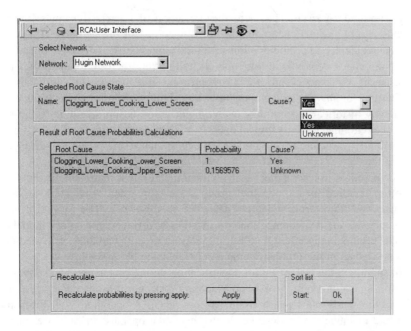

Figure 18.5 Source: ABB, RCA. HMI for presentation of the most probable root causes, acquisition of user feedback and following update of most probable root causes. Reproduced with permission from Elsevier.

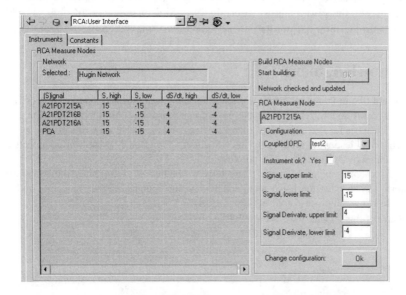

Figure 18.6 Source: ABB, RCA. GUI for configuration of the measurement instruments (sensors) and its status. Reproduced with permission from Elsevier.

adaptation and on issues related to the prediction of process dynamics is given in [488].

The scalability and modification of an existing system is feasible due to the OOBN-approach deployed in our tools. It will require the development of a new Bayesian network model only for a new process. This overall process model will use OOBNs for all standard components with default probabilities, which can adapt either with historical process data or by *in situ* cases during operation to the new process environment and its typical operation. This scalability is feasible as long as the qualitative structure of the process remains similar to the previously developed application.

For practical use we expect that six man-months is a reasonable estimate on the effort required for an industrial group to develop a new application by using the proposed framework and tools described in this work. The estimated work effort includes the development and verification of a Bayesian network model for a problem domain described by 200–500 relevant process variables, provided data and knowledge on the problem domain are available or acquirable.

In summary, the task of failure identification during production breakdown, its isolation and elimination is a troubleshooting task. On the other hand, the task of detecting early abnormality is a task for adaptive operation with predictive RCA and maintenance on demand. It also gives the operator an early warning of possible future problems, so that counter-actions may be taken before it is too late. Therefore, these two tasks have different probability-cost function. We combine both tasks under the notion of abnormality supervision. It aims at predicting both process disturbances and unplanned production stops, and to minimize production losses. Thus, the priority is to determine an efficient sequence of actions, which will ensure the minimal production losses and will maximize the company profit.

Acknowledgment

We would like to thank ABB for the support on the development of this technology, P. O. Gelin, J. Jaliff and B. Oldberg for the project management, J. Morel, S. Israelson and all ABB colleagues for sharing with us their knowledge and experience on process engineering, operation and testing, as well as for the stimulating discussions on application aspects and integration into the Industrial IT environment.

This chapter summarizes the characteristic features of a combined object oriented methodology to meet the listed requirements and incorporates various modeling and cost issues in industrial process control [486, 487, 488].

Parts of this chapter have been published in [486].

19

Predicting probability of default for large corporates

Esben Ejsing, Pernille Vastrup

Nykredit, Kalvebod Brygge 1-3, 1780 København V, Denmark

and

Anders L. Madsen

HUGIN Expert A/S, Gasværksvej 5, 9000 Aalborg, Denmark

19.1 Introduction

19.1.1 The Basel II Accord

Management of uncertainty is an important and significant part of doing any type of business in a modern economy. Uncertainty, as an inherent part of doing business, cannot be ignored or completely eliminated. Efficient handling of uncertainty is especially important in the financial services industry (e.g., banking and finance, investment, insurance, real estate, etc.). The new Capital Accord of the Basel Committee on Banking Supervision (Basel II) stresses the importance of handling risk.

The need for advanced tools for handling uncertainty in the financial services industry, and in particular risk management, is of the utmost relevance and importance. From January 2007 the Basel II Accord is a requirement for over 30 000 banks and financial institutions in over 100 countries. In order to comply with Basel

Bayesian Networks: A Practical Guide to Applications Edited by O. Pourret, P. Naïm, B. Marcot
© 2008 John Wiley & Sons, Ltd

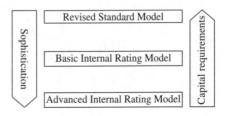

Figure 19.1 The advanced internal rating model for credit risk management under the Basel II Accord gives diversified capital requirements.

II, banks must have deliberate and transparent risk management initiatives in place by the end of 2006. In addition to the interest in advanced solutions originating from Basel II, there is an increased demand for reducing costs and cutting losses in all business areas.

Different models can be used to calculate the capital requirements for risk. The standard approach is the simplest approach, and can be characterized as the method used today to calculate capital requirements. Under the intermediate approach, the financial institution calculates the rating/probability of default of each customer on its own, and an external ratings authority provides an assessment of security collateralization. The advanced model is when both the customer rating and the security collateralization value of a given default situation is calculated on the basis of statistical models.

Under Basel II, banks are required to hold a capital charge based on an estimate of their risk exposure. The capital adequacy of credit, market, and operational risks can be adjusted by means of risk management. The quality of the capital charge assessment will depend on the measurement approach selected, with the simplest approach requiring a higher capital charge than the more advanced approaches. In the more advanced approaches there will be lower capital requirements for some customers, and larger for others. The estimates of the different parameters should be used in the daily business, for example for pricing and rating. This induces a capital incentive to use sophisticated internal risk modeling techniques with a prospect of benefiting from a more risk-sensitive and significantly reduced capital requirement (see Figure 19.1). The common interest of decision makers, supervisors and the like is to identify, assess, monitor, and control or mitigate risk.

19.1.2 Nykredit

Nykredit is one of Denmark's leading financial groups with activities ranging from mortgage banking and banking, to insurance and estate agency services. Nykredit is in particular a leading player in the Danish market for property financing. Nykredit's primary activity is providing mortgage credit products to customers in the private, commercial, agricultural, and subsidized housing sectors. In addition, Nykredit offers a wide range of financial services, including banking, insurance, asset management and real estate services. Nykredit has almost 500 000 private customers and 80 000 business customers, and in 2004 had recorded profits before tax of DKK 4.4 billion.

The Nykredit Group is Denmark's largest mortgage provider having lent DKK 622 billion, and having a market share of 40.6% of total Danish mortgage lending at year-end 2004. Total Group lending (banking and mortgage banking) amounted to DKK 645 billion, equal to a market share of 32.8% at year-end 2004. The Nykredit Group is therefore the second largest lender and the largest mortgage provider in Denmark, as well as one of the largest private bond issuers in Europe.[1]

Nykredit decided early on that it would calculate capital requirements based on the advanced model. Nykredit would employ statistical models both to rate customers, and to assess capital charges. Employing statistical models in connection with customer credit ratings, and subsequently in connection with capital allocation, provides an unprecedented degree of homogeneity in the customer rating process compared to the overall guidelines and general attitudes that characterize the credit process today.

The decision to adopt the advanced approach was made based partly on the fact that, as the second largest credit institution in Denmark, Nykredit has a particular obligation to *strive for the advanced model*. But also because Nykredit expects to be able to benefit from the risk assessment capability of doing so, both in connection with the direct capital assessment, and in connection with the overall evaluation of Nykredit as an institution and the evaluation of its loan portfolio by external ratings authorities.

At Nykredit, Group Risk Management is responsible for risk models, risk analysis, risk control and risk reports. It consists of a Risk Secretariat, a Risk Analysis Department and a Capital Management Department. The development of credit and other models is located in the Risk Analysis Department, whereas calculating and managing capital is the domain of the Capital Management Department. The Risk Secretariat makes sure that Nykredit is compliant with regulatory guidelines, etc.

In the mid-1990s Nykredit began to develop credit-scoring models in the private customer area. Initially, the models were developed to rate the credit risk of customers wanting to secure a mortgage loan on 80–100% of their property value. Because Danish mortgage finance legislation imposes a mortgage lending limit of 80% on the value of residential property, this type of loan must be secured through a bank. The implementation of credit scoring on home loans of this kind was an absolute success, and work to develop credit-scoring models for other bank products continued. In 2001 Nykredit introduced credit scoring on mortgage loans, thus ending the development of scorecards for private customers.

Therefore, when the new proposals to the Basel Accord saw the light of day at the beginning of the millennium, there was never any doubt that Nykredit would use the expertise accrued from mortgage lending as the launching pad for developing models for the other lending areas within the organizaton, namely the commercial, agricultural and subsidized housing areas.

---

[1]Parts of this and the above paragraph have been taken from [344].

19.1.3   Collaboration with Hugin Expert A/S

The Risk Analysis Department has traditionally collaborated with universities and external consultants. In this case Nykredit contacted Aalborg University to assist with project assessment. Initially, the collaboration took place as a part Virtual Data Mining – an EU supported research project on the application of alternative methods of data analysis. This project was the link to the university, and subsequently to Hugin Expert A/S (Hugin).

A working group consisting of members from the university and the Risk Analysis Department was established to look at the possibilities of using Bayesian networks to model the credit risk of commercial customers. Relatively quickly – six months later – a model prototype was developed from a Hugin model that was then tested, both statistically and by credit experts. Later the model was modified and expanded, and today it is used in all areas, and Nykredit's cooperation with Hugin is ongoing. For instance, when developing a more advanced model for retail customers, Hugin was consulted on how to use different features of the technology to optimize the calculation.

19.2   Model construction

19.2.1   Probability of default – PD models

When handling uncertainty in the financial sector a default event is defined. The event can be described as a situation where the borrower fails to meet his/her obligations. PD models most often calculate a probability of default within a year. The use of statistical default models gives a homogeneous and objective credit assessment and makes it possible to perform simulation and stress testing at portfolio level.

19.2.2   Data sources

A key requirement in the choice of technology and tool for implementing an advanced internal rating model is that the technology and tool should support the fusion of historical data and expert knowledge. The two main sources of data and/or knowledge for constructing a model are expert knowledge or models and historical data. Bayesian networks are well suited for combining different sources of information into a single model.

A wide range of data sources may be used in model development. For instance, in Figure 19.2 the potential data sources include financial statements, provision information and arrears data. These data sources should be combined to form a single united source of information for the model development as illustrated in Figure 19.2.

Both external and internal data can be used to form a model. Internal data consists of basic information on each corporate, whereas external data consists mainly of information about accounts. Some of the external data consists of bankruptcy

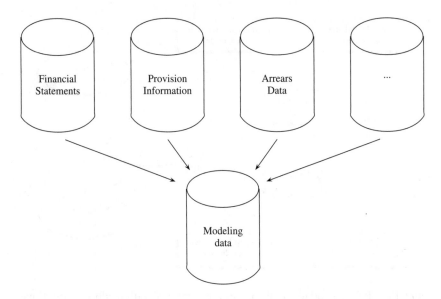

Figure 19.2   Multiple data sources are available and should be combined into a single data set of model development.

information and standard financial key ratios, and some of the internal data consists of arrears, provision and size of exposure. Both the internal and external data sources include expert knowledge as well as benchmark information about the corporate.

Figure 19.3 illustrates how different sources of information are combined to produce a single data source of information used for the model development. Once the models enter production, they start to produce additional information or data for use in model revisions and the development of additional models.

19.2.3   Computing probability of default

As a Bayesian network is an (efficient) encoding of a joint probability distribution, it can be used for computing the posterior probability of events and hypotheses given observations. That is, a Bayesian network can, for instance, be used for computing the probability of default for a given corporate given observations on the corporate. In the Nykredit application the probability of default is the single event of interest. The remaining variables are either hidden (i.e., never observed) or observed. The observations will be indicators of the event that a corporate will default within the next year. In general, the construction of a Bayesian network can be in general a labor-intensive knowledge elicitation task. For this reason Bayesian networks with restricted structures are often used. In addition, historical data and expert knowledge on the relation between variables are often exploited in the assessment of the parameters of a model.

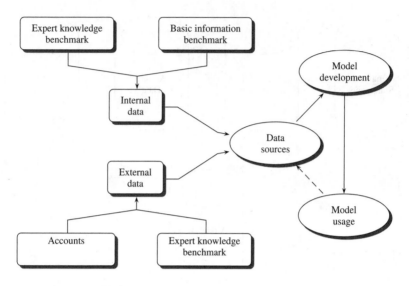

Figure 19.3 Internal and external sources are combined with feedback from Bayes-Credit in order to obtain the data source for model development and refinement.

The naïve Bayes model [249] (also known as naïve Bayesian network) is a particular simple and commonly used model. The naïve Bayes model is a popular choice of restricted model due to its simplicity, its computational efficiency and its performance. The model has complexity linear in its size. A naïve Bayes model consists of a hypothesis variable and a set of feature variables. Given observations on a subset of the features, the posterior probability distribution of the hypothesis variable is computed. Figure 19.4 illustrates the structure of a naïve Bayes model where Health is the hypothesis variable and the remaining variables are feature variables representing financial factors. The financial factors are indicators of the health state of a corporate. The structure encodes pairwise conditional independence between each pair of financial factors given the Health variable.

Figure 19.4 A naïve Bayes structure where financial factors are assumed to be pairwise independent given the health state.

The naïve Bayes model produces a particularly simple factorization of the joint probability distribution over its variables on the following form:

$$\mathbb{P}(H, F_1, \ldots, F_n) = \mathbb{P}(H) \prod_{i=1}^{n} \mathbb{P}(F_i \,|\, \text{pa}(F_i)) = \mathbb{P}(H) \prod_{i=1}^{n} \mathbb{P}(F_i \,|\, H),$$

where H is Health and F_i is a financial factor. The factorization consists of the marginal distribution over Health and one conditional probability distribution given Health for each financial factor F_i. In addition to identifying the variables and the structure of the model, it is necessary to identify the states of each variable and to quantify the model by specifying the conditional probability distribution of each variable in the model given its parent as defined by the graphical structure of the model. In general, variables may be discrete or continuous. The initial models in BayesCredit contained only discrete variables. This implied that some variables in the model represented discretized entities. For instance, capital equity was represented as a variable with six states where each state represents an interval of values (e.g., from DKK 678 millions to DKK 1 170 millions). The current set of models in BayesCredit includes discrete Bayesian networks and conditional linear Gaussian Bayesian networks with non-naïve Bayesian network structure.

19.2.4   Model development and production cycle

Figure 19.5 illustrates the model development cycle and its relation to the production cycle. The networks used when calculating PDs are developed, maintained and tested at a desktop PC using Hugin Developer. The use of the networks in Nykredit's production environment is solely on the mainframe. When a network is developed it is stored in a file on the desktop PC using the Hugin network specification language, i.e., the NET-format. By an in-house developed application the net-file is uploaded to the mainframe where it is stored in a DB2 table. This in-house developed application assigns a unique identifier and handles the versioning of the more than 40 networks used. Thereby the use of the networks in different scenarios is clear. The individual network is accessed by methods in the Hugin C API. Nykredit's application complex on the mainframe contains logic to identify the above-mentioned naming convention. A system component on the mainframe retrieves the relevant data used as arguments in the network. Methods in the Hugin C API are used for retrieving the value of the variables defined as output. All calculations are stored in DB2 tables along with information on which network is used, and input and output to and from the network. The results of the calculations are used in many of the front end systems in Nykredit.

19.3   BayesCredit

BayesCredit is a decision support system based on Bayesian networks. The system consists of a set of models for computing the probability of default for a large

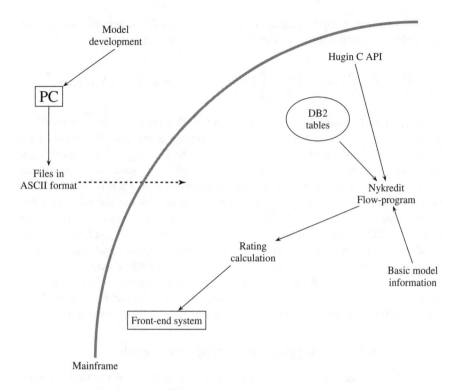

Figure 19.5 The model development and production cycle.

corporate. Using a single model for all types of corporates is unrealistic due to the diversity of the corporate portfolio. With this in mind, corporates have been divided into several segments, and today BayesCredit consists of three models, one for each of the segments industry, commerce and property.

Subsequently, models have been developed for calculating the credit risk of agricultural customers. These models have also been segmented according to the type of agricultural production of the customer (pig, cattle or crop production).

The construction of the BayesCredit model for computing the probability of default consists of a number of steps such as selecting the set of variables, identifying the states of each variable, specifying the structure of the model, and estimating or specifying of the (conditional) probability distributions as defined by the structure of the graph of the Bayesian network. Notice that the structure of the graph of the Bayesian network is more or less given by the choice of the naïve Bayes model.

19.3.1 Variable selection

The model consists of a health variable and a set of variables specifying financial factors. The financial factors are assumed to be independent given the health state of the company. This implies a naïve Bayes model. The naïve Bayes model encodes conditional pairwise independence between any pair of financial factors given the health variable. The purpose of the model is to predict the financial health of a corporate when the values of a number of financial factors are known. The financial health is represented using a variable **Health** with states corresponding to how healthy a corporate is. For instance, the states may be 'bad' and 'good'. In practice **Health** has a larger number of interval states reflecting the financial health of a company $0 - x_1\%, x_1 - x_2\%, \ldots, x_{(n-1)} - x_n\%$. Each interval corresponds to a quantile such that the state $0-x_1\%$ is the $x_1\%$ of the corporates with lowest health. The remaining states are defined in a similar way. The quantile intervals are finer grained for the lowest quantiles because the differentiation of corporates is particular important for the lower quantiles. Unobserved factors are given the 'average' value 0. The health of a corporate is computed by the sum of the transformed financial factors.

For each corporate the value of the heath variable is computed and stored in such a way that it can be used at the next analysis of the corporate. Thus, the health state of the company captures and represents the state of the corporate at the point where the analysis is performed. This health state becomes the previous health at the next analysis.

The variables representing financial factors are selected by single factor analysis on factors selected by credit experts, adviser knowledge, etc. The financial factors should have a high correlation with the health of a corporate in order to give good predictions of default. In addition to the financial factors, earlier arrears were found to have a high discriminatory power between defaulted and nondefaulted corporates. For example, Figure 19.6 shows the default rate given different values for earlier arrears across the considered segments.

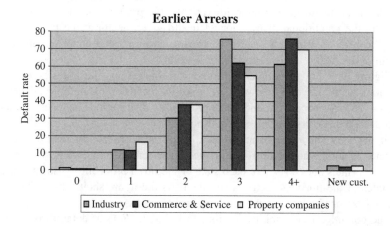

Figure 19.6 The default rate given earlier arrears across segments Industry, Commerce & Service, and Property companies.

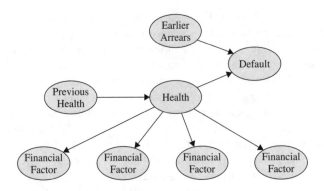

Figure 19.7 The general structure of the model.

19.3.2   Model structure

Based on the variable selection the model consists of variables representing financial factors, a variable representing the health of a corporate, a variable representing the health of a corporate at the previous time-step, a variable representing earlier arrears and the variable Default with two states 'no' and 'yes' indicating whether a corporate will or will not default within the next year.

The structure of each model is based on the naïve Bayes model as illustrated in Figure 19.7.

19.3.3   Assessment of conditional probability distributions

All conditional probability distributions have to be assessed before a fully specified model is obtained. This includes the conditional probability distribution of each financial factor given the health variable, the prior distribution on the health variable, the conditional distribution of health given health at the previous time-step, the prior distribution on earlier arrears, and the conditional distribution of default given health and earlier arrears.

Figure 19.8 shows the diverse information sources and illustrates how the distributions of the Bayesian network are assessed by fusion of multiple information sources. The assessment of conditional probability distributions includes estimation from data, logistic regressions and questionnaires.

Figure 19.9 shows an example of a conditional probability distribution of a financial factor given Health. Health has states one to nine while the financial factor has states one to six. For each state of Health a probability distribution over the states of the financial factor is specified. As the figure shows, the value of the financial factor increases as the value of health increases.

The conditional probability distribution $\mathbb{P}(D \mid H, EA)$ of default given earlier arrears and health is approximated using a logistic regression model:

$$\text{logit}(D_j = 1) = \alpha_{EA} + \beta_{EA} H_j,$$

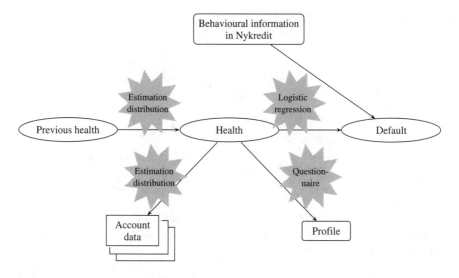

Figure 19.8 The estimation and specification of the conditional probability distribution use diverse information sources.

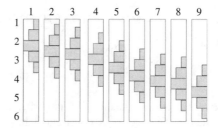

Figure 19.9 The conditional probability distribution of a financial factor given health $\mathbb{P}(FF \mid H)$

where $D_j = 1$ is the event that the customer will default and α_{EA} and β_{EA} are real valued coefficients conditional on earlier arrears and H_j is the state of health. The coefficients of the logistic regression are estimated from data.

19.3.4 BayesCredit model example

Figure 19.10 illustrates the structure of one of the core models of BayesCredit. The model is based on an augmented naïve Bayes structure. The Health variable is the key variable of the model. The model consists of the Health variable, a variable representing the health of the corporation at the previous analysis, a set of financial factors, a variable representing earlier arrears, and a variable representing

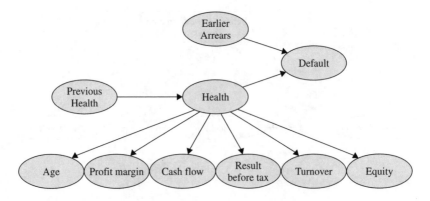

Figure 19.10 A naïve Bayes structure over health and a subset of financial factors augmented with three variables capturing the previous health state, earlier arrears, and default (for illustrative purpose).

the event whether or not the corporate will default within the next time period. The time period is one year.

The model structure shown in Figure 19.10 produce the following factorization over its variables:

$$\mathbb{P}(H, PH, EA, F_1, \ldots, F_n) = \mathbb{P}(PH)\,\mathbb{P}(H\,|\,PH)\,\mathbb{P}(D\,|\,EA, H)\,P(EA)$$

$$\prod_{i=1}^{n}\mathbb{P}(F_i\,|\,\mathrm{pa}(F_i))$$

where H is health, PH is previous health, EA is earlier arrears, D is default, and F_i is a financial factor. The model is constructed based on a naïve Bayes model structure between the health variable and the financial factors. The naïve Bayes model structure is augmented with a variable representing the previous health of the company as this captures the historical performance of the company. The variable representing whether or not the company will default has **Health** and **Earlier Arrears** as parents.

The model supports efficient computation of the probability of default given observations on any subset of variables in the model including the empty set. This flexibility to handle missing observations is one of the key properties of Bayesian networks and was a determining factor in Nykredit's selection of Bayesian networks as the appropriate technology for the advanced internal rating model. Figure 19.11 shows an example of the computation of probability of default given incomplete knowledge about the corporate.

The figure shows the probability of default conditional on equity capital and earlier arrears. The equity capital is DKK 4420–13 500 millions and the earlier arrears is 0 to 1 months. The conditional probability of default is computed to be 1.6%.

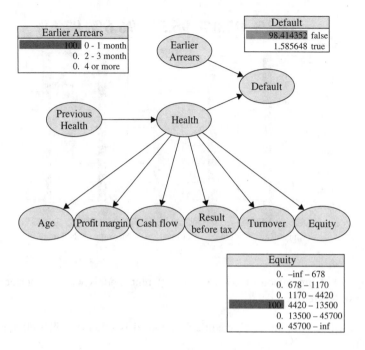

Figure 19.11   Given equity capital of DKK 4420–13 500 millions and earlier arrears is 0 to 1 months, the probability of default is 1.6%.

19.4   Model benchmarking

BayesCredit has been tested by the credit analysis agency Standard & Poor's, where BayesCredit attained a high score compared to other commonly used risk models. The graph in Figure 19.12 presents a comparative measure of the accuracy of the model. In this case, Nykredit's new model is compared with Standard & Poor's benchmark models.

As can be observed, Nykredit's model has a higher accuracy than Standard & Poor's benchmark models. Based on Standard & Poor's own evaluation of BayesCredit, the following conclusions on the performance of the model have been reached:

- Definition of risk profiles: based on the result of extensive tests, we believe that Nykredit's new model is able to differentiate different risk profiles when evaluating the creditworthiness of companies through the probability of default.

- Benchmark evaluation: the overall performance of Nykredit's new model is as good (or better) than that of the benchmark models developed by Standard & Poor's based on the same database.

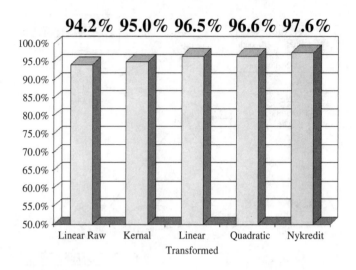

Figure 19.12 Comparison of accuracy of different models with the performance of the new Nykredit model.

Overall, in our opinion, Nykredit's new model is efficient and well suited to the bank.

19.5   Benefits from technology and software

A number of benefits from using Bayesian networks and Hugin software to compute the probability of default have materialized during the development, the integration, and the usage phase. Some of the main benefits are listed below:

- A Bayesian network is a powerful communication tool. One of the benefits of Bayesian networks is that a lot of knowledge is captured in the graphical structure. Without the need to compute probabilities, many important properties of a model can be recognized. This makes Bayesian networks a powerful communication tool during the initial discussion of a problem as well as when explaining results after analysis. Because the Bayesian network is hierarchical with the numbers *hidden* within the nodes, attention is focused on the relationship among variables.

- A Bayesian network combines expert knowledge and data. Because default data is so sparse, it is unlikely that Nykredit can build a strong statistical model based on data alone. During the development of BayesCredit, Nykredit used knowledge from credit specialists to a great extent, and the two sources of information – expert knowledge and data – were combined in a flexible and elegant manner.

- A Bayesian network can efficiently handle missing observations. Hugin software contains algorithms that handle missing observations in an exact way since basic axioms of probability are used to compute the probabilities. The handling of missing observations is crucial for BayesCredit because Nykredit must compute the probability of default even in cases where a part of the financial data is not available.

- The algorithms in Hugin software are highly efficient because they exploit properties of the algebra of multiplication and marginalization of probability tables. This is important for Nykredit, because BayesCredit is applied in real time that necessitates a fast computation of probability of default. BayesCredit is also applied offline on the entire customer database. This too requires efficient calculations.

- Integration and maintenance of models into existing software is straightforward. The integration of the Hugin Decision Engine into the existing mainframe system was simple. Similarly, the maintenance of models and software is straightforward.

In general, a main advantage of Bayesian network technology for handling uncertainty is that the same model, based upon the same fundamental understanding of the domain, is able to solve different tasks, such as identifying, assessing, monitoring, and mitigating risk. The main advantages of the technology include, but are not limited to:

- Probabilistic graphical models are generally easy to understand and interpret as relationships are represented graphically rather than textually.

- Probabilistic graphical models illustrate the cause – effect relations in an efficient, intuitive and compact way. In general the model represents (conditional) independence and dependence relations (for instance, between key risk indicators, key risk drivers, risk factors, losses, activities, business lines, event types, controls, etc.).

- Probabilistic graphical models offer a mathematical coherent and sound handling of reasoning and decision making under uncertainty.

- Probabilistic graphical models may be constructed by fusion of historical data and domain expert knowledge.

- A cost–benefit analysis of controls can be performed using decision theory to identify the optimal controls based on a scenario analysis.

19.6   Conclusion

The framework of Bayesian networks is a valuable tool for handling uncertainty in Nykredit's line of business. Bayesian networks support the integration of multiple data sources and provide a graphical model representation, which is easy to interpret and analyze. Since September 2003 BayesCredit has successfully been used in production at Nykredit. BayesCredit computes the probability of default for each corporate in three different segments (Industry, Commerce and service, and Property) based on internal and external data on the corporate. Due to differences between the types of corporates in the three segments, there is a separate model for each segment.

Not only is BayesCredit efficient with a high performance, the use of Bayesian networks and Hugin software have implied a number of other significant improvements in Nykredit. The integration of models into the existing mainframe system was and still is very efficient and straightforward. Maintenance on and updates of models are simple tasks to perform as they amount only to replacing an existing file containing the outdated model specification with a file containing the new or updated model specification.

In addition to using Bayesian networks for computing the probability of default for different segments of customers, we see a number of opportunities for applying Bayesian networks to different types of problems.

Hugin Expert A/S was a natural choice for Nykredit when the decision to develop internal risk models was made. The company is well established on the market for software for advanced model-based decision-making under uncertainty. Hugin software is known to be very reliable, efficient and user-friendly. Finally, the Hugin software package is equipped with a variety of functionalities that Nykredit has found to be very useful. Data conflict analysis is one such example. Another determining factor in the selection of software provider was that, besides their core technology, Hugin also provides excellent services, such as training, consultancy and effective technical support.

20

Risk management in robotics

Anders L. Madsen
HUGIN Expert A/S, Gasværksvej 5, 9000 Aalborg, Denmark

Jörg Kalwa
Atlas Elektronik GmbH, Germany

and

Uffe B. Kjærulff
Aalborg University, Denmark

20.1 Introduction

One of the main objectives of the ADVOCATE project (acronym for *Advanced On-board Diagnosis and Control of Autonomous Systems*)[1] was to increase the performance of unmanned underwater vehicles in terms of availability, efficiency, and reliability of the systems and in terms of safety for the systems themselves as well as for their environments. The main objective of the ADVOCATE II project[2] was to improve and extend the general-purpose software architecture to cover Autonomous Underwater Vehicles (AUVs) and Autonomous Ground Vehicles (AGVs) and to apply it on three different types of autonomous vehicles. Based on the software architecture the aim is to increase the degree of automation, efficiency, and reliability of the vehicles.

---

[1] ADVOCATE was formed in 1997 under the ESPRIT programme of the European Community.
[2] ADVOCATE II was formed in 2001 under the IST programme of the European Community.

Bayesian Networks: A Practical Guide to Applications   Edited by O. Pourret, P. Naïm, B. Marcot
© 2008 John Wiley & Sons, Ltd

The software architecture is designed in a highly modular manner to support easy integration of diverse artificial intelligence techniques into existing and new control software. The basic idea is that the artificial intelligence techniques will perform diagnosis on the state of the vehicle and in the case of malfunction suggest appropriate recovery actions in order to recover from this situation.

Three end-user partners were involved in the ADVOCATE II project: University of Alcalá designs modules for AGVs for surveillance applications, Ifremer designs AUVs for scientific applications, and Atlas Elektronik designs AUVs and semi-AUVs for industrial applications. Each end-user partner presented diagnosis and control problems related to a specific vehicle.

In this chapter, we focus on the new type of AUV designed by Atlas Elektronik in 2005. The vehicle, called DeepC, is designed to allow unsupervised autonomous mission durations of up to 60 hours. The diagnosis and control problems considered are detection of corrupt sensor signals and identification of corresponding appropriate recovery actions. Corrupt sensor signals can be caused by sensor malfunctions or noise.

The sensor is a sonar (= sound navigation and ranging), which is used to detect obstacles in the vicinity of the underwater vehicle in order to avoid collisions. As a sonar emits and receives sound pulses, any ambient noise might have an influence on the functionality of the system and thus on the safety of the vehicle.

Based on an assessment of properties of the problem, we decided to use probabilistic graphical models to support the detection of failure situations and to suggest appropriate recovery actions in case of (high probability of) a failure situation. The main reasons for this choice are that the problem domain is highly structured and that it is subject to a large amount of uncertainty.

20.2   DeepC

Atlas Elektronik has developed a new type of underwater vehicle operating with autonomous mission durations of up to 60 hours. This vehicle is referred to as DeepC (see Figure 20.1).

Figure 20.1 The DeepC underwater vehicle.

The DeepC vehicle – developed under the support and promotion of the Federal Ministry of Education and Research of Germany 2000–2005 – is a fully autonomous underwater vehicle with specific payload for oceanographic and oceanlogic applications. The vehicle is supposed to operate up to 60 hours without supervision. To master this challenge one of the outstanding features of the DeepC is its reactive autonomy. This property allows situation-adapted mission and vehicle control on the basis of multi-sensor data fusion, image evaluation and higher-level decision techniques. The aim of these active and reactive processes is to achieve a high level of reliability and safety for longer underwater missions in different sea areas and in the presence of different sea bottom topologies.

The long mission durations impose the need for advanced artificial intelligence techniques to detect, avoid, and recover from dysfunctions. Atlas Elektronik was faced with problems, which could not easily be solved by existing systems. The current approach to handle mission faults is to abort the mission, which is, however, very expensive.

As a fully autonomous system, the DeepC vehicle has to rely on its sensors to survive operationally. The DeepC is equipped with an advanced object detection and obstacle avoidance system. The object detection system consists of a mechanically scanning, forward looking sonar and its control electronics. In principle a sonar operates similar to a radar. It emits short pulses of ultrasound and records any echoes from the environment. The intensity of the echoes can be scaled into brightness of a gray-scale image. One axis of the image represents the time or distance from the sonar, the other axis the relative direction into which the sonar is pointing. Any objects are detected by means of image analysis.

This system works well when the sonar image is of sufficient quality. The problem considered is to construct a model for assessing the sonar image quality and for suggesting actions to avoid object collisions in case the quality is degraded.

Figure 20.2 illustrates the problem domain in some detail. The vehicle is equipped with a forward looking sonar located at its nose. The sonar image (when of significant quality) will show possible objects which the vehicle should circumnavigate in order not to damage itself. Noise and reverberation may reduce the quality of the sonar image in which case the object detection system may fail to identify objects. When the sonar image is of insufficient quality this should be detected and appropriate actions to increase the quality of the sonar image should be suggested.

20.2.1 Critical and safe detection distance

The critical detection distance is the specific distance at which an obstacle must be identified. It is the range the AUV needs physically to avoid a collision by changing the course or making a full stop. The critical distance varies with size of the object. The minimal turning radius of DeepC is 30 meters. From this geometric constraint, the critical range for small objects can be computed. It is necessary to add a fixed time tr (e.g., $tr = 6$ seconds) for image analysis and reaction time

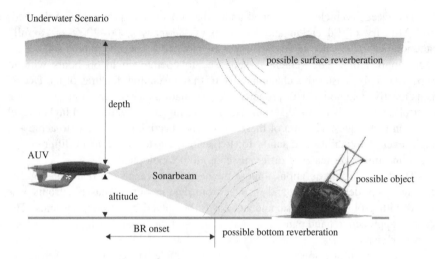

Figure 20.2 An illustration of the problem domain.

(from receiving the image until start of an avoidance manoeuvre) to calculate the speed dependent minimal distance for reaction.

The safe detection distance (SDD) is quite similar to critical detection distance. They may be used synonymously because the safe detection distance is necessary to avoid an infinitely wide obstacle. An operational sonar range less than this distance is considered unsafe as only small obstacles could be avoided. We have assumed the critical detection distance to be equal to the safe detection distance.

The effect of reverberation of sound from the bottom or sea surface can be compared to the effect of fog on visibility: a distant object may be hidden in the reverberation. If the necessary safe detection distance is larger than the distance of reverberation onset, then any objects hidden in the fog are detected too late to be avoided. In general, the larger the SDD, the greater the risk of a collision.

Regarding countermeasures the safe detection distance is fortunately a function of speed (see Figure 20.3, which explains the concepts).

If the distance of reverberation onset is smaller than the safe detection distance, then the speed has to be reduced so that the distance of reverberation onset is equal to or larger than the safe detection distance. Alternatively, the altitude above the sea bottom may be increased to reduce the probability of the presence of obstacles.

For the application the probability of bottom and surface reverberation has to be assessed. The range of bottom reverberation onset (BRO) can be calculated from the vertical opening cone of the sonar which is ± 10 degrees. Figure 20.4 illustrates bottom reverberation onset. The concept of surface reverberation onset is similar.

Note, that the bottom effect may not be present, depending on the physical properties of the sea bottom. The image analysis must assess whether or not reverberation is present. The BRO is an indicator only. Surface reverberation onset (SRO) behaves in the same way as BRO.

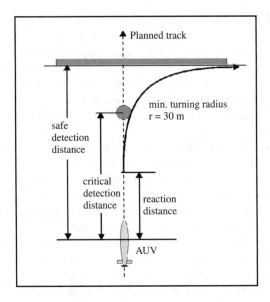

Figure 20.3 Concepts of reaction time, critical detection distance, safe detection distance, and turning radius.

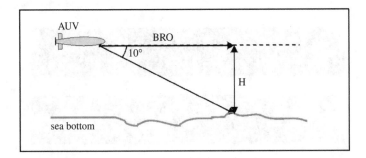

Figure 20.4 An illustration of bottom reverberation onset.

20.2.2 Sonar image generation

The sonar used in this context is a mechanical sector scanning sonar. This means that the emitted and received sonar pulse (ping) covers a very small area – the horizontal beam width is about two degrees and the maximum range is 100 meters. After each ping the sonar head turns around the vertical axis by two degrees and sends its next pulse. The sonar turns and sends ping by ping to cover a wider area. For instance, to cover 40 degrees of the horizon, the head has to send and receive 20 pulses.

After sending the pulse the sonar switches to receive mode and performs an analog-to-digital (A/D) conversion of the received sound intensity at a defined sampling rate. This rate defines the amount of data samples in the radial direction (distance).

The received sound intensity of the single samples is converted into a number with a range of 0 to 255 (one byte). The higher the value the more return signal (echo) has been received. The range and angle parameters of the sonar image span up a matrix: range versus orientation of the sonar head. Typically, there are about 20–40 angular positions of orientation (rows), each of them containing about 650 data samples (columns).

The intensities or echoes in each cell of this matrix, which may be displayed as an image, may have different sources. In the ideal case there is only the echo of objects. But in practice these are overlaid by noise and reverberation from the bottom or sea surface. There is also a volume reverberation, but this can be neglected here as it is of no significance for far-field assessments. There is a wide range of sources of noise. There may be natural sources as, e.g., rain, sound of waves, or the among acoustic scientists well known shrimp-snapping, there may be artificial sources as, e.g., ships, or generated own-noise and there may be noise from electrical influences during the signal processing.

Figure 20.5 shows five examples of typical images. The first exhibits some artefacts due to electrical noise, the second has a closed area object, the third has the same closed area object with bad noise, the fourth displays an object larger than the image, and the fifth has bottom reverberation.

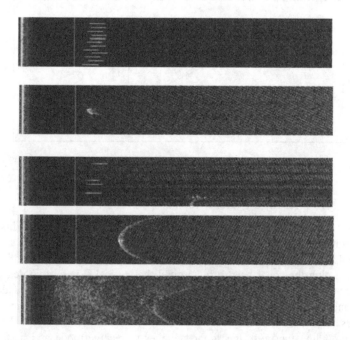

Figure 20.5 Typical examples of five different types of sonar images.

Any high noise intensities are clearly visible as bright spots. The classification into objects, artefacts, or noise is a difficult task even for a skilled sonar operator. But it is possible to recognize even an object hidden in clutter. The human decision is based on presence, distribution and structure of intensities, as well as on contrasts. Therefore it is important to develop indicators for the analysis of sonar images.

20.2.3   Sonar image analysis

The sonar image can be viewed as a typical gray-scale image. Thus, some important quality indicators can be extracted from mathematical features as image mean value, entropy, and substance.

The use of the image mean value is quite clear; it indicates the level of noise, if there are no objects present. If there are objects in the image, the mean value increases dependent on the spatial proportion of the object. The image mean M is defined as

$$M = N^{-1} \sum_j i_j, \qquad (20.1)$$

where i_j denotes the pixel value of pixel j (i.e., i_j is a value from 0 to 255) and N is the number of pixels.

Entropy is a measure of disorder of a system. In our setting image entropy H is related to the amount of information an image contains. A highly ordered system can be described using fewer bits of information than a disordered one. For example, a string containing one million '0' can be described using run-length encoding as [('0', 1E6)] whereas a string of random symbols will be much harder to compress. Hence, high entropy means a uniformly distributed noisy image. The image entropy H can be expressed as

$$H = \sum_{i=1}^{255} p_i \cdot \log(1/p_i), \qquad (20.2)$$

where i denotes the pixel value (i.e., a value from 0 to 255) and p_i is the probability of a pixel value (relative pixel count) within the image. Both symbols can be represented in an image histogram.

The histogram of an image refers to a histogram of the pixel intensity values. This histogram is a graph showing the number of pixels in an image at each different intensity value found in that image. In our case we have an eight-bit gray-scale image. There are 256 different possible intensities, and so the histogram will graphically display 256 numbers showing the distribution of pixels amongst those gray-scale values. See Figure 20.6 for an example of pixel count as a function of pixel value.

As the third image quality indicator, we introduce the notion of image substance S as an expression for weighted high valued echo pixels. The image substance can be expressed as

$$S = \sum_i p_i \cdot i^{\log i}. \qquad (20.3)$$

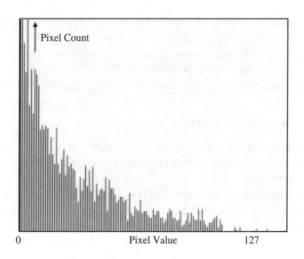

Figure 20.6 Pixel count as a function of pixel value.

Typically, high valued echoes are more sparse in the image than values near the mean. Substance gives clear evidence that there are significant echoes (reverberation or objects) in the image. The higher the number of high valued pixels the larger the substance.

In the assessment of sonar image quality is it important to distinguish between noise and reverberation. Noise may be present in the sonar image due to noise from the vehicle or other sources of noise. Noise from the vehicle may be due to, for instance, payload activity or acoustic noise.

If the sonar beam strikes the ground or the water surface, then there may be a reflection from theses surfaces. This reflection produces reverberation. The intensity of reverberation depends on the roughness of the surface. Whether the condition for a potential reverberation onset is geometrically fulfilled, can be calculated from the sonar parameters and the distance to the surface.

In essence the sonar image assessment problem is a tracking problem in the sense that we want to monitor the time variant quality of the sonar image. In case of degraded image quality an appropriate recovery action shall be determined. To solve the sonar image quality assessment problem and the decision making problem, we developed a probabilistic graphical model.

20.3   The ADVOCATE II architecture

The purpose of a system based on the ADVOCATE II architecture is to assist the operator or a robot's control system (piloting system) in managing fault detection, risk assessment, and recovery plans under uncertainty. One of the design main goals was to allow easy integration of different artificial intelligence techniques into pre-existing systems. The decision to support the simultaneous use of diverse artificial

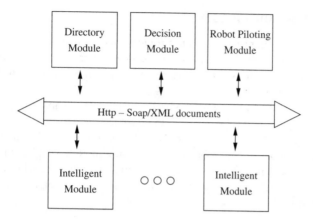

Figure 20.7 The communication architecture.

intelligence techniques was made to allow these techniques to collaborate on the task of reasoning and making decisions under uncertainty. This raises the question of how to most efficiently integrate different artificial intelligence techniques into new and existing systems. We have found that this is most efficiently done through an open and generic architecture with a sophisticated communication interface. The generic communication protocol is based on SOAP/XML technology implementing HTTP for communication between different types of modules (see Figure 20.7).

The architecture is widely generic, open, and modular consisting of four different types of modules.

20.3.1   Robot piloting module

The robot piloting module is the main interface between a robotic system and the ADVOCATE modules. It has access to mission plans, sensor and actuator data. This module also has the permission to interrupt the initially planned behavior of the robot to enforce recovery actions (recovery plans) received from the decision module. A recovery action may be moderate as, e.g., changing some parameter, but it may also be a drastic action as, e.g., an emergency surfacing manoeuvre.

20.3.2   Decision module

As indicated above the decision module communicates with the piloting module, but also with intelligent modules, receiving diagnoses and recovery actions, and the user. Thus, it is the central point of the architecture. The decision module manages the diagnosis and recovery action processes. This includes integration of information provided by different intelligent modules, user validation of diagnosis and recovery actions when required by the system, and translation of recovery actions into recovery plans.

20.3.3   Intelligent module

The role of an intelligent module is to provide possible diagnoses, suggestions for recovery actions, or both. A typical diagnosis on an operational vehicle corresponds to identification of system state while a recovery action corresponds to performing a sequence of actions on the vehicle (e.g., to avoid collision or to recover from a dysfunction). To reach this aim an intelligent module encapsulates a knowledge base to a specific problem domain and an inference engine.

The intelligent module communicates with the robot vehicle piloting module and the decision module. The robot vehicle piloting module supplies the intelligent module with data. These data are used in conjunction with the knowledge base to generate diagnoses and recovery actions. The diagnosis or recovery action is communicated to the decision module.

20.3.4   Directory module

The directory module is the central point of inter-module-communication in the sense that it maintains a list of registered and on-line modules and its addresses. When registering at the directory module, each module obtains this list to initiate communication.

20.4   Model development

The sonar quality assessment problem concerns reasoning and decision making under uncertainty. First, the quality of the sonar image should be assessed based on navigation data and a number of image quality indicators. Second, an appropriate action to reduce noise in the sonar image has to be suggested based on an assessment of the sonar image and the development of the system over time. A single model to assess the sonar image quality and to suggest an appropriate action was to be developed. For this reason we use influence diagrams.

The traditional influence diagram is useful for solving finite, sequential decision problems with a single decision maker who is assumed to be non-forgetful. The limited-memory influence diagram relaxes some of the main constraints of the traditional influence diagram. The limited-memory influence diagram is well suited for representing (and solving) the sonar quality assessment problem.

The Hugin tool [14, 233, 292] was used to construct and deploy the probabilistic graphical model developed.

20.4.1   Limited memory influence diagrams

Probabilistic graphical models are well-suited for reasoning and decision making in domains with inherent uncertainty. The framework of probabilistic graphical models offers a compact, intuitive, and efficient graphical representation of dependencies between entities of a problem domain [249]. The graphical structure makes it easy for nonexperts to understand and interpret this kind of knowledge representation. We use a limited memory influence diagram (LIMID) to describe multistage decision problems in which the traditional non-forgetting assumption of influence diagrams is relaxed [272].

Let V_D be the decision variables and V_C the chance variables of the influence diagram such that $V = V_C \cup V_D$. The influence diagram is a compact representation of a joint expected utility function (EU):

$$EU(V) = \prod_{X \in V_C} P(X \mid pa(X)) \cdot \sum_{Y \in V_U} f(pa(Y)), \tag{20.4}$$

where V_U is the set of utility nodes. The solution to an influence diagram is an (optimal) strategy specifying the (optimal) decision policy for each decision in V_D as a function of each possible observation. In principle it is necessary to solve an influence diagram only once as this produces the (optimal) strategy.

The traditional influence diagram assumes a non-forgetting decision maker and that there exists a linear order on all decisions. The non-forgetting assumption implies that all previous observations are recalled at each decision. The linear order on decisions implies a partial order on chance variables relative to the decisions. The traditional influence diagram is solved by performing a sequence of rollback-and-collapse operations (also known as variable eliminations) where decisions are eliminated by maximization and chance variables are eliminated by summation. The rollback-and-collapse procedure iterates over variables in reverse order of the partial order. In some cases the non-forgetting assumption introduces very large decision policies as each decision is a function of all previous observations and decisions.

The two aforementioned assumptions of the traditional influence diagram are relaxed in the LIMID interpretation. This implies that it is possible to represent and solve decision problems where there is no total order on the decisions and where the decision maker may be assumed to be forgetful. In a LIMID representation it is necessary to specify the information available to the decision maker at each decision.

20.4.2   Knowledge elicitation

The construction of a probabilistic graphical model can be a labor intensive task with respect to both knowledge acquisition and formulation. The resources available for the model development consisted of domain experts with limited knowledge on probabilistic graphical models and knowledge engineers with limited domain knowledge. In this setup, model development usually proceeds as an iterative process where the model is refined in each step of the iteration. The knowledge engineer elicits the necessary information from domain experts and tries to capture the knowledge of the domain expert in a model. At each iteration the knowledge engineer and domain expert assess the quality and performance of the model in order to focus the resources available at the next iteration on the most sensitive parts of the model.

In the ADVOCATE II project the knowledge engineer and domain expert had limited opportunities for face-to-face meetings. For this reason we developed a knowledge acquisition methodology to support the knowledge extraction process [248]. The knowledge elicitation method should not rely on familiarity with probabilistic graphical models and it should be applicable to problems where the knowledge engineer and the domain expert are situated at different locations.

Following the knowledge acquisition approach, we start out by breaking down the overall problem (called *sonar image*) into subproblems/causes, subsubcauses, etc. This gives us a hierarchy of causes, which serves as a basis for identifying the diagnoses (or root causes), which are the leaves of the causes hierarchy, and gives us a structured way of relating available observations and background information to these root causes. In this process it is important to ensure that the subcauses of a given cause are mutually exclusive and exhaustive.

The scheme is based on building a problem hierarchy for an overall problem. The problems (or causes) of the hierarchy relate to the states of the different parts of a vehicle and its environment.

Figure 20.8 shows such a cause hierarchy related to the sonar image assessment problem. The causes of the hierarchy are grouped into causes that qualify as satisfactory explanations of the overall problem and causes that do not. The first group of causes are referred to as *permissible diagnoses*. The subset of these that can actually be identified based on available information are referred to as *possible diagnoses*. Only possible diagnoses are present in Figure 20.8 and they are marked with a '+' symbol.

The cause hierarchy acts as a road-map for describing the relevant diagnostic information and the possible recovery actions. A cause of a subtree of the cause hierarchy that does not contain any possible diagnoses is unlikely to provide

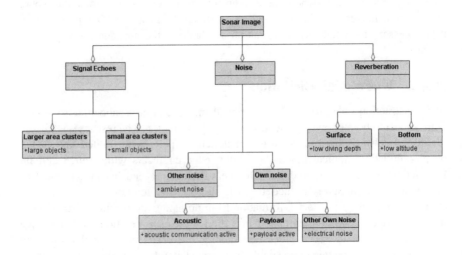

Figure 20.8 Cause hierarchy for the sonar image assessment problem.

relevant diagnostic information or error recovery information. Thus, if there are no observable manifestations of the cause strong enough to identify a possible diagnosis for the cause, we need not worry about it when eliciting the diagnostic and error recovery information.

The domain expert provides the relevant diagnostic information and the recovery actions in matrix form with one row for each cause and one column for each kind of diagnostic information (i.e., background information and symptoms) and one column for possible recovery actions.

The qualitative knowledge elicited following such a scheme provides a sufficient basis for a knowledge engineer to construct the structure of a probabilistic graphical model, on the basis of which the quantitative knowledge can then be elicited.

For a detailed description of the knowledge acquisition scheme, we refer the reader to [248].

The cause hierarchy and its diagnostic information provides a sufficient basis for constructing an initial structure of a model for this problem.

20.4.3 Probabilistic graphical model

It is clear that the amount of disturbance in the sonar image and the presence of objects will be time dependent. Time is discretized into intervals of eight seconds, which is equal to the time the image analysis component needs to analyze a single sonar image. Currently, the model consists of three time-slices, i.e., the suggested recovery actions are based on a 16 seconds look-ahead.

We model the problem as a discrete time, finite horizon partially observed Markov decision process. The model is dynamic in the sense that it models the behavior of the system over time. The state of the system at any given point in time is only partially observed as navigation data and sonar image data are available, but not all entities of the problem domain are observed.

The development of the system is represented as a time-sliced, limited memory influence diagram. As such the model will produce a diagnosis or set of diagnoses and an appropriate recovery action or set of recovery actions given a set of observations on the exact state of the system. The model may produce a diagnosis given any subset of observations whereas a set of recovery actions is determined based on the observations on the navigation data and image quality data. Navigation data consists of vehicle depth, altitude, pitch, and speed, while image quality data consists of pixel intensity mean, substance, and entropy of the digital version of the sonar image.

Figure 20.9 shows the time-sliced model with three time slices where each time slice is represented using an instance of the LIMID class shown in Figure 20.10. In the model, time progresses from left to right. Each instance node is an instantiation of the model representing the system at any given point in time. Thus, the model represents the system at three consecutive time steps separated in time by eight seconds.

Figure 20.9 The top-level LIMID class for sonar image assessment consists of three instances of the model (class) shown in Figure 20.10.

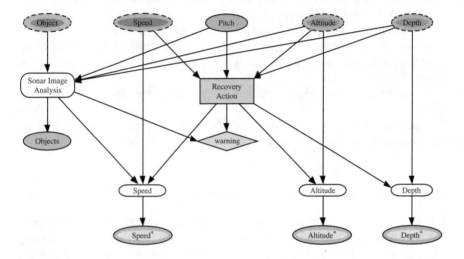

Figure 20.10 The generic time-slice model for sonar image assessment.

The LIMID class shown in Figure 20.10 is a decomposed representation of the system at any given point in discrete time. The nodes with dashed, shaded borders are input nodes, while the nodes with shaded borders are output nodes (input nodes are typically shown at the top or left of a figure and output nodes are typically shown at the bottom or right of a figure). Input and output nodes are used to link time slices together. The output nodes of one time-slice are connected to the input nodes of the subsequent time-slice. An input node of time slice T_{i+1} is equal to (or a placeholder for) the linked output node of time slice T_i. These nodes are used to transfer the belief state of the system from one time slice to the next and are referred to as the interface variables I_i.

In total, when considering all models, there are seven groups of nodes:

1. diagnosis variables;

2. variables representing background information;

3. symptom variables;

4. auxiliary variables;

5. intervention variables;

6. decision variable(s);

7. utility functions.

The diagnosis, background information, symptom, auxiliary, and intervention variables are all discrete random variables (i.e., which can be in one of a finite set of mutually exclusive and exhaustive states). The states of the variables are specified by the domain expert.

The structure of the model specifies that altitude, depth, pitch, and speed are observed prior to the decision recovery action. This implies that the policy for the recovery action is a function of state space configurations of these nodes to the domain of the decision. Notice that no other informational links are present in the graph. Hence, at the next decision only the most recent observations on altitude, depth, pitch, and speed are used.

The decision has a potential impact on speed, altitude, and depth of the vehicle. Deciding on a recovery action changing any of these properties will impact the quality of the next sonar image.

The instance node Sonar Image Analysis, which is an instance of the network class shown in Figure 20.11, models the sonar image assessment process. The three image quality indicators are represented in this class by the nodes Entropy, Mean, and Substance. The quality indicators are influenced by the presence of disturbance or objects in the sonar image. Disturbance may be caused by reverberation or noise.

The sonar image analysis model is based on a naive Bayes structure over image disturbance, image quality indicators and image quality where image disturbance is the hidden variable, the image quality indicators are observed, and image quality is the variable being predicted.

Since the image quality indicators are observed each time a sonar image is analyzed, we need to resolve the LIMID with the observations on the image quality

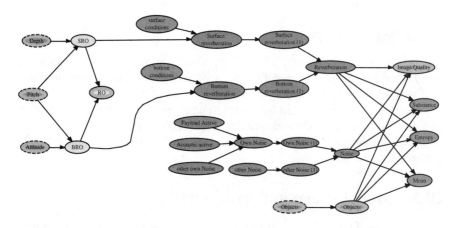

Figure 20.11 The model (class) representing the sonar image assessment process.

indicators entered as evidence in conjunction with the navigation data in order to compute the expected utility of each decision option.

The possible outputs of the model are a set of diagnoses quantified by a probability, a set of recovery actions quantified by an expected utility, a collision risk indicator, and a prediction of the future development of the system under optimal behavior (according to the model). The possible diagnoses are good image quality and bad image quality. The possible recovery actions are no action, warning, change speed, increase altitude, and increase depth. Out of the possible outcomes of the model, only the quantified diagnoses and recovery actions are used.

The conditional probability distributions of the models have been specified based on a combination of assessment by interviewing a domain expert, estimating parameters from observational data, and using mathematical expressions relating configurations of parent variables to states of their common child variable. Some parameters of the sonar image analysis class, shown in Figure 20.11, have been estimated from observational data. That is, the sonar image assessment model has been constructed based on information fusion of domain expert knowledge and observational data. A database of 1048 sonar images have been assembled. Each image has been classified as good or bad by a domain expert.

Utility functions have been assessed from the expert. Utility functions have been assessed based on the cost of damage to the vehicle, the cost of mission abort, and the cost of delays.

The entire model consists of 124 variables with 3556 probability parameters most of which have been assessed from the domain expert.

The total clique size of an optimal junction tree representation of the LIMID is 108 841. This is significantly smaller than the total clique size of an optimal junction tree representation of the model interpreted as a traditional influence diagram. The total clique size of an optimal junction tree representation of the traditional influence diagram is greater than $2^{32} - 1$. Thus, the LIMID representation has made it tractable to represent and solve the sonar image assessment problem as a probabilistic graphical model.

20.4.4 Model verification

The model has developed over a significant number of iterations of structure and parameter assessments. The model has been verified by a domain expert at scheduled technical meetings including reviews.

The model has been verified based on test cases developed by a domain expert, on simulated data, and real world data. The entire ADVOCATE II system including all modules have been tested in a real-life setting at the Atlas Elektronik test pond in Bremen, Germany.

20.5   Model usage and examples

In combination, Figures 20.12–20.14 give an example of how the developed model is used to diagnose a possible malfunction of a sensor, i.e., a bad sonar image, and

Figure 20.12   A bad sonar image with noise.

Policy for Recovery Action (TO_D)											
Pitch (TO_pitch)	...	...						-0.5 - 0.5			
Depth (TO_depth)	...	...	...	...	...			5 - 10			
Altitude (TO_altitude)	...	...	...	...	...	...	...	...	5 - 10		...
Speed (TO_speed)	...	...	...	...	...	...	...	...	0.5 - 1		...
no action	-	-	-	-	-	-	-	-	-25536.24	-	-
warning	-	-	-	-	-	-	-	-	46098.4	-	-
change speed	-	-	-	-	-	-	-	-	39896.96	-	-
increase altitude	-	-	-	-	-	-	-	-	40444.12	-	-
increase depth	-	-	-	-	-	-	-	-	42024.12	-	-

Figure 20.13   Given the observations the optimal decision is to give a warning indicating that noise is present in the sonar image.

to give warnings or suggest recovery actions that will help to improve the image quality or decrease the probability of damage to the vehicle.

Figure 20.12 shows a typical bad image with noise. Figure 20.13 shows the expected utilities given the navigation data and the values of the image quality indicators for each possible recovery action.

Figure 20.14 shows the posterior probability of the variable representing image quality given observations on navigation data and image quality indicators. The altitude and depth of the vehicle both are between 5 and 10 meters while the pixel mean of the image is 12.128, the substance is 59.438, and the entropy is 0.579. The posterior probability of a bad image is high.

The option with the highest expected utility is the (locally) optimal decision given the observations and the model. Thus, given the navigation data and values from the bad quality image, the model suggests giving a warning that there may be noise present in the sonar image. The vehicle is neither close to the surface nor close to the bottom. Hence, the assessment is that the image is most likely bad due to noise.

20.6   Benefits from using probabilistic graphical models

Probabilistic graphical models naturally capture, represent, and take the inherent uncertainty of the problem domain into account. There are a lot of uncertainties

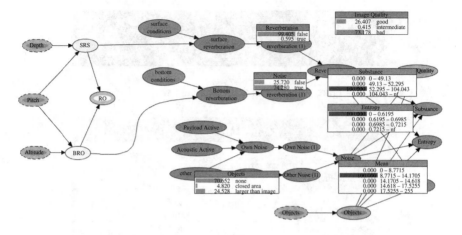

Figure 20.14 Given the values of the image quality indicators the image is classified as *bad*.

present in the domain of autonomous underwater vehicles. It is important to assess and efficiently represent this uncertainty. The graphical structure of a probabilistic graphical model makes the independence assumptions of the model explicit and easy to comprehend for domain experts with limited knowledge of the technology.

The hierarchical construction of the LIMID enforced by the object-oriented paradigm has simplified the knowledge acquisition phase considerably as it is easy to focus on well-defined subparts of the model in isolation.

Using instance nodes, it is a simple task to create and maintain multiple instances of the same LIMID class. Furthermore, it is a simple task to change the class of an instance node to another class. This is particularly useful in the knowledge acquisition phase where each LIMID class has been revised and updated multiple times. This also supports efficient version control.

Probabilistic graphical models is one of three artificial intelligence techniques considered in the ADVOCATE II project. The other two techniques considered are neuro-symbolic systems and fuzzy logic. The decision module of the ADVOCATE II architecture serves the central role of translating and merging the results from different intelligent modules. The artificial intelligence modules need to agree on common scales in order to be able to compare and integrate diagnoses and recovery actions. The probability of an event and the expected utility of a decision option are simple to translate into the common scale used by the decision module.

20.7   Conclusion

The quality assessment of sonar-images has been performed successfully using a probabilistic graphical model. The problem is quite complex; the size of the optimal junction tree of the traditional influence diagram representation is greater than the address space on a 32-bit computer. By the introduction of the limited memory influence diagrams (LIMID) it was possible to reduce the size to a manageable amount.

Furthermore, we have assumed the image indicators not to be available at the next decision in order to reduce the size of the decision policies and to increase the computational efficiency. The cost of this assumption seems to be negligible. But by doing so, it is necessary to resolve the LIMID at each time-step.

Since the technology does not support *infinite horizon* and continuous time decision problems, we have discretized time into three time steps. This enables us to perform an analysis with 16 seconds look-ahead. To perform an analysis with a longer look-ahead amounts to including additional time slices in the model.

Notice that it is not possible to solve the problem using a non time-sliced model.

Acknowledgment

Parts of this chapter are taken from [239] and [293].

We would like to thank the partners of ADVOCATE II: University of Alcalá (Spain), Getronics (France), Technical University of Madrid (Spain), Atlas Elektronik GmbH (Germany), Ifremer (France), Hugin Expert A/S (Denmark), Innova S.p.A (Italy), and e-motive (France).

The European Community supported the ADVOCATE and ADVOCATE II projects under grants ESPRIT 28584 and IST-2001-34508, respectively.

21

Using machine foreknowledge to enhance human cognition

Philippe Baumard

Professor, University Paul Cézanne, France, IMPGT 21, rue Gaston Saporta, 13100 Aix-en-Provence, France

21.1 Introduction

Foreknowledge is the ability to construct meaning and interpretation before fully acquiring a complete set of stimuli. Contrary to its folkloric acceptation, foreknowledge is a usual, though essential, component to all human communications. Human perception is continuous. Most of the filtering that occurs between the reception and the interpretation of stimuli is unknown to individuals; interpretation itself directs which stimuli we perceive or fail to perceive without our full awareness [439, 484]. As where the cursor is placed between applying schemata to stimuli, or vice versa, building schemata from stimuli, has been left to philosophers. Rosset [401] has a definitive answer: human perception is similar to grass: it grows, it transforms, may retract, indistinctively for the human eye.

This issue might be less trivial than it seems when it comes to designing information systems that emulate human communications. As Tversky and Kahneman suggested, the principle of invariance that underlies the rational theory of choice does not genuinely represent the reality of human decision-making [459]. Although preconceived rules are normatively essential, they inherently fail to be descriptively accurate. Hence, individuals create rules on the fly as stimuli emerge, as much as they distort stimuli to fit the newly created rules. Most Bayesian filtering or mining

Bayesian Networks: A Practical Guide to Applications   Edited by O. Pourret, P. Naïm, B. Marcot
© 2008 John Wiley & Sons, Ltd

technologies, on the contrary, tend to artificially separate rules from knowledge, stimuli from framework, hence assuming there exists a normative, rational and stable theoretical corpus that describes human communication behavior.

This chapter describes an experiment that further led to a prototype, and ended in a patented technology.[1] The initial objective of this research, following a work on individual and group tacit knowledge [34] was simply to emulate implicit knowledge within an information system. At first, the finding seemed self-evident as a machine without natural sensors could not generate its own knowledge, and furthermore its own implicit knowing. If one considers that machine-generated learning, based on human rules, with or without human inputs, can be considered 'machine knowledge', then part of this knowledge could be 'unknown' to the machine, either because it has not been yet computed, or it is not currently acted or in use. Hence, machines could hold 'tacit knowledge', i.e., knowledge that they cannot express, in the faults that separate exploration and exploitation, as defined by March [296].

However, and contrary to human beings, machines exploitation doesn't bias their exploration, as exploration has been designed with an exploitative purpose. As Starbuck and Miliken noted, much exploration is conducted within the experimental realm of exploitation, and vice versa [439]. We can therefore assume that the condition of the existence of implicit learning resides in the need of discovery; and consequently, that we need to emulate an *imperfect* discovery process if we want to get close to the simulation of human implicit learning.

21.2   Human foreknowledge in everyday settings

Most human cognition is experiential, either that we look forward, or we look backward. As Gavetti and Levinthal put it [177, p. 113], 'cognition is a forward-looking form of intelligence that is premised on an actor's beliefs about the linkage between the choice of actions and the subsequent impact of those actions on outcome'. Foreknowledge of potential outcomes develop with experience, not related with age, but more so related to similarities in patterns and design of the complexity of the situation being apprehended [158, 246]. People naturally rate the outcomes of their predictions accordingly to their results, either positively or negatively. This experiential wisdom generates itself 'rules of thumb' and experiential memory that can be contrary to rational models or individual mental models [280]. The variability of preferences, the volatility of aspirations over time and according to contexts, make it difficult to generalize *ex ante* normative rules of cognition for an individual. Prior experiences influence how stimuli are selected and perceived, and in turn, new stimuli can revise previous experience. We conclude from these theoretical findings that a Bayesian model aiming at simulation human cognition should have two arrows, i.e., should not prefer the value of models over data. A new and scarce configuration of data may well become the new model, while all sets of previous models can no longer handle new stimuli. Hence, both models and

---

[1]US Patent No 7,251,640. 'Method and system for measuring interest levels of digital messages', published on July 31st, 2007

stimuli should display *mortality*, the latter being either trigger by data or by the model itself.

The valuation of cognition results is itself problematic, as people do not have absolute rules to value knowledge. In fact, most informational behavior of human beings is somehow 'positional', in Hirsch's terms [212]. People seek to gain information that allows them to gain a positional perspective, i.e., to position their own past experience in regard of the new one, while asserting their own position towards the new situation. Information superiority is hence more a theoretical construct than an organizational reality. People do not see information as having absolute superior value, but rather an immediate, practical leverage for their on-going programs. Rather than challenging information being used and articulated to on-going programs, people favor collateral and *ad hoc* organizations, which would leave their main programs untouched, and their deeds unchallenged [510]. For instance, secrecy that can turn banal information into a highly destructive one follows a social psychology that is more embedded into ignorance than malignity [430]. Human inferences are imperfect, and turn genuine interpretations into misleading and contaminating shortcuts [339]. Moreover, most information is unconsciously processed by individuals, so that people would generally not be able to identify how they gathered, assimilated and transformed information that leads to their acts and deeds [282, 340, 386]. People in organizations do not discriminate between *genuine* information and its many doubles. Reality is taken for granted, until proven otherwise. Artifacts are as much instrumental as the real itself, for realities are known as genuine only when they get dismantled [401, 402]. From this, we concur that context is an undifferentiated element of knowing, as it participates both in the filtering and in the construction of knowledge.

Thus, there is an 'information-generating' behavior, as much as there is an 'action-generating' behavior in organizations [438]. People generate information whilst listening upwards; they document ex post an action-generation whose rationale they surely lost; they fine-tune their past commitments with the new criteria for approval; they reuse uncompleted achievements that have bred frustration and translate them into the new set of core beliefs. From these findings, we concur that each individual communication behavior is unique. We hence reject the hypothesis that accuracy of a computer simulation of human cognition can be attained by collaborative filtering. The objective is therefore to capture the incongruities of the communication behavior of an individual, rather than seeking an emulated communication behavior statistically congruent with others.

While decision theorists try to improve and rationalize what they would optimistically call the 'human thinking system,' in real life, 'people do not know all of the sources of stimuli, nor do they necessarily know how to distinguish relevant from irrelevant information' [439, p. 41]. People act according to 'criteria' that they view as important, sometimes quite unable to define why these criteria are important, sometimes totally ignoring, or forgetting why they decided that these criteria were important, for individuals have the tendency 'to deal with a limited set of problems and a limited set of goals' [297, p. 117]. Human beings constantly struggle with incomplete stimuli, uncertain beliefs and limited rationality. Consequently, 'interpretations are generated within the organization in the face of considerable perceptual ambiguity' [298, p. 19]. Robotic research has adequately integrated this dimension while pursuing the goal of minimizing or eradicating classical direct human–machine interface, however stalling on the problem of machine intention [447]. On possible solution has been suggested, and applied, by Brooks [64, p. 291]. It consists of grounding the system design in an evolutionist perspective, i.e., the reasoning system should not have any preconceived determinants or rules, and should therefore be 'born' with the user. Brooks calls it the subsumption architecture, and defines it with three key ideas: 'Recapitulate evolution, or an approximation thereof, as a design methodology (...). Keep each layer added as a short connection between perception and actuation, Minimize interactions between layers' [64, p. 291]. We concur from these findings that our system should not have any preconceived model about human communication, and derives the construction of the individual model from an evolutionist learning of his or her communication behavior. This also implies that the system should have an autonomous *self-designing* capability, hence assuming by itself, with all the replicated weaknesses, biases, discretionary characteristics of the individual cognition it is emulating.

Last but not least, human beings have a skill that does find any competition in machines. Individuals are truly skilled at self-deception, and self-deception plays a major part in the on-going and resilient nature of their perception [186]. When faced with the problem described in the above paragraph, i.e., conflicting between belief and evidence, human beings can lie to themselves about the stimuli as to temporarily protect the defaulted, albeit vital, model. Or, to the contrary, human beings can temporarily alter their core beliefs in order to accept a stimulus that potentially invalidate the model itself. Concurrently, human beings have a built-in system that allows them to escape Laing's paradox, or as he put it himself: 'The range of what we think and do is limited by what *we fail to notice*. And because we fail to notice that we fail to notice, there is little we can do to change, until we notice, how failing to notice shapes our thoughts and deeds' (quoted in [186, p. 24]). Truncated, missing or contradictory stimuli can be replaced on the fly by self-deceptive ones, that do not allow accuracy, but preserves the on-going flow of reasoning. Such a phenomenon is not solely individual, as Ruddick [404] suggested. People with whom we interact do the same, and therefore, self-deceptive stimuli also contribute to maintain the continuity of interaction with others. Thrown into perception [484], we need those real-time and self-deceptive adjustments to continue interaction, without forsaking the goal of achieving sense-making in the longer run. We then later 'unlearn' [208] what we previously accumulated on the sake of interaction continuity (e.g., 'action generators', [438] or internal or psychological consistency [186]. Moreover, self-deceptive stimuli are essential to the processes of exploration and discovery as they constitute the core mechanism behind the discovery of incongruities [236]. We concur from these findings that a system willing to emulate human cognition should display two essential characteristics: first, it should be able to generate, carry, apply and discover rules that may be self-deceptive, i.e., internal logical consistency of the model should not be expected to be related, or influence, an external consistency of the contents. This postulates of course that we should not seek an accuracy in semantics, neither should we try to use contents as a classifier of the Bayesian network. Second, and most importantly, the targeted system should be able to 'unlearn', i.e., temporarily or permanently discard generated rules and/or stimuli, accordingly to the congruity or incongruity of the flow of reasoning in which they are inserted.

21.3   Machine foreknowledge

My first encounter with Bayesian belief networks took place in 1997. The first experiment consisted in emulating very simple strategic planning flows of reasoning: measuring a competitive asymmetry, or the level of a threat of entry on a given market. These first experiments were published with J.A. Benvenuti [35], and some samples are available on the Hugin website.[2] In those attempts, canonic strategic models were drafted from literature, and the operator simply informed the models with data relative to the situation being analyzed. Such traditional Bayesian belief networks do not posses any kind of machine foreknowledge: they can merely apply a truncated body of beliefs, or 'expertise', and improve over time the reliability of the initial models based on incoming data. These first experiments corroborated Druzdzel and Simon's findings that showed that conditional probability tables in a Bayesian network were capable of representing reversible causal mechanisms [138]. The models could accurately suggest initial entry variables values for an expected result, e.g., 'success of entry'.

The idea that a similar design could be applied to human cognition, or communication, was swiftly abandoned. Such a model would require a stability and equilibrium in human communication over time, which was contrary to organizational behavior literature findings. Spirtes et al. [437] suggested imposing externally a value on any node in a graphical model, which implies an exogenous manipulation of the network, e.g., removing the arcs of the network. This issue was detailed by Pearl [357] who suggested that imposing a value on a variable renders the variable independent of its direct causes. Although this strategy corroborates Jones' description of incongruity and self-deception mechanisms [236], it doesn't ponder the existence of the cause. As Druzdzel and Van Leijen put it: 'wearing a raincoat protects one from getting wet but it does not make the rain go away' [137, p. 51].

Inspired by [64], a radically different design was therefore adopted. Starting with a prior model of human communication would impede the sought characteristics of the system. For instance, human motivation to swiftly trash an incoming message does not necessarily relate to a negative opinion of the sender. To the contrary, a very important message that requires discretion can be immediately trashed, the user preferring to use his or her own memory rather than being vulnerable in keeping such data in his or her computer. Bayesian-based filtering systems often failed to recognize such patterns because they are based on preconceived

---

[2]http://www.hugin.com/developer/Samples/Comp_analysis/

'communication theories' that don't take into account the inseparability of contents, context and purpose. The only way around such an obstacle is to ignore all pre-conceived rules or theorems of human communication altogether. Although, if willing to conceive even an auto-generated Bayesian network, we needed an objective classifier as a source of network generation. I started to work with a very simple classifier, that was later revealed as a pertinent intuition. I chose 'intensity of communication' as this classifier, as intensity is always relative to context, time and people. Hence, the value of this classifier is not instructed to the network, but learned over time, by measuring the relative frequency and density of interactions between a pair of individuals, relatively compared to the median intensity of interrelation with all other individuals. This intensity, in turn, is used to build arcs, indifferently between contents (chunk of signals), individuals and contexts.

Chow and Liu [96] suggested a method to learn probabilistic graphical models directly from data. The problem with such a method is to achieve a trade-off between the accuracy of the model and the complexity of the approximation. We substituted this trade-off with a combination of incongruity and mortality. We define mortality of an arc as its net overall contribution to the variance of communication intensity. Like human filtering processes [439], the system should simply be able to disconnect or discard an arc with high mortality, which doesn't mean that this mortality would have the same value with another combination of nodes and arcs. The incongruity measurement was inspired by [236]: Jones defines incongruity as a relative value of a match or mismatch with an individual expectations. Hence, if incongruity is expected, and incongruity is served, then the individual would evaluate the incoming information as 'congruous'. Jones, who was Churchill's advisor for counter-intelligence during World War II, intensively used such a phenomenon to deceit German forces. For our system, it meant that incongruity should be measured in regard of the user's expectations concerning the emitter of information. Here again, the variable had to adopt relative values of incongruity levels with previous messages of the same emitter, compared with the overall median incongruity that the receiver is used to handle. This approach is the exact opposite of seeking Shannon's entropy [424], and then, for instance, comparing entropies of different chunks of information to deduct a level of 'mutual information' [165]. It is not the mutuality of information that is being sought, such as in collaborative filtering systems, but the *mutuality of communication intensity*. The latter can be achieved with completely different sets of information on the receiver and the emitter sides, either because people do not always use the same words to qualify a same object (e.g., 'love' and 'desire'), or because a mutual intensity can be triggered by simultaneous, although different or opposite concepts (e.g., 'love' and 'hate').

The second step was to adopt a recurrent design to allow the self-generation of a Bayesian network based on Chow and Liu's findings [96]. We decided than an evolutionist design should not comprise any form of either handcrafted teaching, nor predefined handcrafted features. The main obstacle was to then design a program that would comprise subroutines for separating knowledge from inferences, therefore allowing the subroutines for auto-generating the Bayesian network. The objective was to design a system that would emulate the user's *foreknowledge* of his or her own interest in an incoming message, or another individual's level of interest in an outgoing message from the user. The level of interest was measured by the predicted level of *communication intensity* that the outgoing or incoming was likely to trigger based on the history of interactions of the two users. The purpose of the Bayesian network is first to self-generate a network to multiple interactions, with multiple signals and multiple individuals based on the *intensity of communication* classifier, and then to be able to predict the most probable receiver of a message or signal in a large population of other interlocutors, and vice versa, the most probable contents or signals for a given population of other interlocutors.

The overall design of the system was founded on a postulate that implicit memory does not exist, following a suggestion of Willingham and Preuss [499]. Based on previous work from Graf and Schacter [188] – that defines implicit memory as a set of memory tasks – Willingham and Preuss [499] assert that 'implicit memory tasks make no reference to the initial encoding episode and are not necessarily associated with awareness of engaging in recall'. Our question was: why do individuals display a higher propensity to communicate with certain individuals rather than others? Previous work on tacit knowledge [34] suggested that people do not need to fully articulate a body of knowledge to act upon it. This suggested that an individual's propensity to communicate with another is most likely embedded in a body of past experiences, knowledge, past stimuli that he or she cannot express. It was then possible that a system that would record and remember over time 'intense' moments of communication, and their related combinations of contents, contexts and individuals, would adequately emulate the functioning of individual tacit knowledge. The assumption was that we tend to forget the combinations of contents, personas and contexts that triggered previous intensity because (a) those experiences are too many, (b) because of our bounded rationality [297] that tend to reduce those past experiences as simpler sets, and (c) a natural 'mortality' of the memory of those events occur over time. The graph of Figure 21.1 (patent excerpt) shows the on-going interaction between contents, context and individuals. Mortality is estimated as the relative probability of (211), knowing the relative conditional probability of (210) in conjunction with the relative probability of (212).

As the graph of Figure 21.1 illustrates on steps (200) and (201), there is no previous knowledge, nor model, of the individual's communication behavior; this includes generic communication theories of rules. The system builds the implicit rules of interpersonal communications of the individual as they are captured and learned. Hence, the second postulate behind the construction of this experimental Bayesian network was that if an individual encounters an *unknown* individual, where past experience and past interaction are absent, he or she would then assess a propensity or willingness to communicate based on previous experiences of other individuals. Therefore, when unknown individuals or stimuli are presented to the Bayesian network, it operates with truncated data, and propagates evidence on all other available variables. For example, the individual may be unknown, but the topic of conversation, i.e., the signals being exchanged, might have been the object of other past interactions with other individuals. In that case, two operations are conducted: the first one is the autonomous teaching of previous arcs and nodes

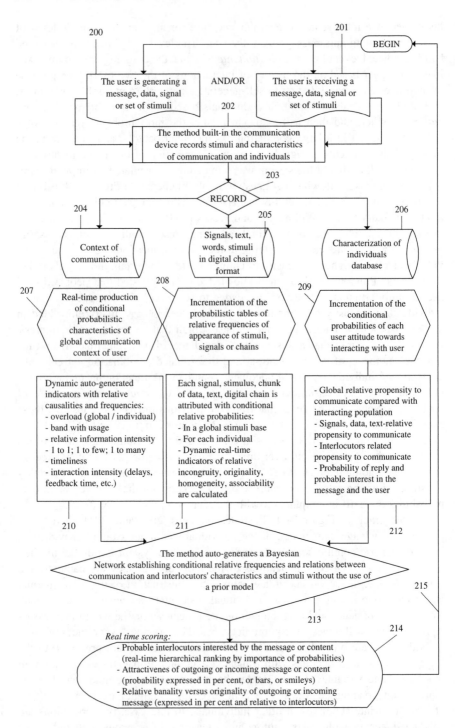

Figure 21.1 Overall design of interest-level of interpersonal communications Bayesian network (from patent).

relating to those new contents, and the second one is the automated generation of arcs and nodes specific to the new individual.

What occurs at the individual level is even more salient at the organizational level. If an individual may possibly remember critical relations between people, contents and contexts, an organization cannot, as those experiences are dispersed between hundreds of individuals. This is a well known obstacle to a theory of 'organization mind' [407]. One of the correlate objectives of the technology becomes the emulation of 'organization mind', given that all employees are equipped with the client-technology, and that legal constraints does not forbid the implementation of a server technology to compile, compare and run all gathered Bayesian networks without the awareness of the users. In such conditions, all individual networks being run 'at a distance' from their users, the technology would allow them to detect high probabilities of willingness to communicate between individuals who ignore each other, and never had a single contact. Conversely, the same interrogation of centrally gathered networks would of course allow to detect individuals who display very high asymmetries in propensities and willingness to communicate. Given the reverse causal probability capacity of the network, it would also allow us to determine groups of individuals more likely to be interested in a given data or combination of signals.

21.4   Current application and future research needs

We crafted a prototype on the postulate that human communication behavior is idiosyncratic and unique to each individual. The prototype is able, for instance, to predict which individual in the network of people of the user, might probably be interested in what is currently typed on the keyboard. We thus achieved a real-time 'scoring' of outgoing communications that can eventually display unforeseen correspondents who are not in the list of addressees, or, in a server version, discover an unknown individual that might display a strong interest in the chunk of knowledge being created.

The current application is to simulate, and eventually replace, an individual's communication behavior, by being able to emulate its reactions, according to the time of the day, his or her personal status of cognitive overload, and his or her communication habits, relatively to other individuals, and relative to exchanged stimuli. One advantage, and simultaneously inconvenient, of the current application is that it is not based on semantics or interpretation of the meaning of the contents, signals, or stimuli exchanged between individuals or machines. The current system has no knowledge of the meaning given to the messages. While it is advantageous for privacy concerns, it also means that the system handles synonyms independently.

Another characteristic of our prototype resides in the deliberate choice of crafting a 'mute' and autonomous technology. Users cannot teach handcrafted features. Such handcrafted features, such as *ex ante* communication rules or values of discrete or continuous variables have also been deliberately discarded from its design. One significant advantage is that the system continuously learn from human behavior, to build its own 'implicit' memory, and record discovered conditional probabilities that relate people, contents and context. One obvious backdrop is that such a system does not benefit from the human capability of radical change. The system does not have any form of psychology, and therefore, is not capable of paradigmatic and brutal change of its own learning (even if this point is still in discussion by behaviorists concerning human beings). It presents a double-edge characteristic of the system. Long forgotten or overlooked intensities between contents and people are signaled by the system, even if such relations have been discarded by the human being. On the contrary, being deprived of semantic learning, the system is currently incapable of self-revising relations based on semantic proximity between chunks of knowledge.

The prototype currently learns every single action and reaction of a single individual in relation with his or her interlocutors. In some sense, it is more efficient than human beings themselves, whose filtering acts on the detection of signals themselves [439]. The current prototype does not discard a stimulus for being opposite – i.e., for triggering a too high value of incongruity – of previously learned relations. However, as intensities of communication vary over time, and with individuals, the filtering is being learned by the Bayesian network through its current classifiers.

Another limitation resides in the fact that the current system is only cognitive and limited to electronic communications. It does not capture, and does not take into account the behavioral attributes of individual preferences of communication with other individuals. It also does not take into account all other forms of interpersonal communication that may occur outside the realm of electronic interchange. However, as further electronics interchanges may reflect experiences that have taken place in the physical world, the technology will of course learn from the improvement or degradation of the physical interaction, if it has consequences on the mutual or asymmetric cognitive perceptions of individuals.

As it is currently designed, the application is capable of absorbing relational databases representing a history of communication in the industry standards. For instance, we ran experiments with existing e-mail software databases that allows the retrospective construction of an initial Bayesian network. After rerunning chronologically all interactions stored between the targeted individual (the user) and all its interlocutors, the prototype becomes immediately operational, and thus able to predict the attractiveness of the next received or emitted message. However, this learning is not complete because individual behaviors regarding storage of previous e-mails vary considerably. Some people are true archivists, and collect an exhaustive treasury of their former interactions; others use electronic interactions as they would do with voice devices, and systematically erase every communication. The absence of initial handcrafted teaching, and further absence of human supervised teaching, is thus double edged. On one hand, without any importation of previous history of interactions, the system learns over time in ideal conditions. In the beginning, its accuracy is poor, due to the scarcity of signals and discovered relations. In the long run, its accuracy strengthens, and allows the discovery of unsuspected interests in a given chunk of knowledge, or inversely, overlooked interlocutors that might show a great interest in a given information. On the other hand, when importing an e-mail, or instant message, database, we consequently import the biases that occurred when the user discarded or retained old interactions. More research is thus needed in improving the importation of historical databases of interactions.

21.5   Conclusion

Tastes of human beings vary according to their humor, the context in which they communicate, and the individuals with whom they interact. The objective of this experiment was to learn in real time all the conditional probabilistic links and nodes between all stimuli, contexts and people, as they interact with the targeted human being. The objective is to tend, over time, to faithfully reproduce in a belief network technology, a complete clone of the communication behavior to the human being observed. The current prototype represents an application of lifelong learning of a predictive user modeling [241]. On the contrary for Kapoor and Horvitz's example of BusyBody, we tried to achieve such lifelong learning without direct supervision, or handcrafted previous teaching and modeling. We learned a few things on the way. First, it is possible to capture the communication behavior of an individual without prior model, prior handcrafted features of teaching. The learning output is idiosyncratic, and cannot be applied to another individual. It is a computer emulation of a given individual's communication behavior, inwards and outwards. We discovered two main advantages.

First, human cognition is limited, whilst the emulated computer implicit memory knows no limit. It does not display any form of creativity, but it is capable of discovery by 'third-party' linking. For instance, it will discover an individual that might contribute to the current generation of an individual knowledge, given the learning system of another individual granted access to its own learning. Future research will therefore focus on agent-learning based on the same technology.

Second, we surely learned that there is much interest in studying intense relations amongst few samples, rather that trying to generalize from numerous relations of weak intensity. As Bayesian technology showed in the past its pertinence for knowledge mining, it should in the future accomplish similar breakthroughs in human communication mining and eventually, cloning. Finally, machine foreknowledge is achievable. Of course, machines are currently incapable of inventing independent reasoning with the same subtlety and speed as human beings. Discovery is achieved by means of self-generating rules on the basis of the observed relative intensities of interaction between contents, people and context. The machine, in producing this foreknowledge, benefits from its intensive capability of exhausting potential relations between stimuli and individuals. In this matter, Bayesian technologies probably achieve results that could not be reached by traditional rule-based AI. While the current prototype shows expected outcomes at the individual level, we believe that the most interesting findings will come from organizational and inter-organizational applications.

22

Conclusion

Patrick Naïm

Elseware, 26–28 rue Danielle Casanova, 75002 Paris, France

A conclusion is usually some sort of prospective analysis. The question we would generally ask is 'where do we go from here?'. This would address actually two separate questions: where do you go from here, and where does the Bayesian network technology go?

We will answer both questions by trying to understand what we exactly do when we use Bayesian networks. This should help you, the reader, to understand whether you should consider this approach for your applications, and also allow us to understand if Bayesian networks are different from other techniques, and how.

22.1   An artificial intelligence perspective

Bayesian networks are part of artificial intelligence. According to AAAI, artificial intelligence is 'the scientific understanding of the mechanisms underlying thought and intelligent behavior and their embodiment in machines' [21].

Artificial intelligence (AI) was 50 years old in 2006,[1] and it can be said at the same time that it exceeded our expectations – if you consider computerized chess programs, pattern recognition applications, and that it failed to fulfill the dreams of pioneers of the field, to build machines behaving in an 'unexpected' way, even to their creators.

This contradiction may have something to do with our expectations for 'magic' in AI. AI is now present everywhere – computer games, credit card authorizations, face recognition, robots, etc. – but none of these applications seems magic. Simply because, as Marvin Minsky put it [319], 'Even though we don't yet understand how

---

[1]The field is considered to have been born at a conference on the campus of Dartmouth College in 1956 [310].

Bayesian Networks: A Practical Guide to Applications   Edited by O. Pourret, P. Naïm, B. Marcot
© 2008 John Wiley & Sons, Ltd

brains perform many mental skills, we can still work toward making machines that do the same or similar things. "Artificial intelligence" is simply the name we give to that research. But as I already pointed out, this means that the focus of that research will keep changing, since as soon as we think we understand one mystery, we have to move on to the next'.

This is the theoretical, scientist perspective on AI. From the point of view of an engineer having used AI for almost 25 years in different applications, here is a simplified perspective. In the 1980s, AI was generally synonymous with expert systems. What we were trying to do was to put experts in computer programs. Experts are supposed to solve problems better and faster than you and me. And, to make a long story short, we, simply, were asking them to explain how they did so. This was incredibly naïve. More or less, we were asking the expert to describe how he or she succeeded in solving a problem, for instance a medical diagnostic problem, or an air traffic control problem, and to write it down. So you had to find 'the expert', to assume he or she would agree to share his or her expertise, and finally to assume that he or she would succeed in doing so.

In most situations, this failed. In the favorable situations where we succeeded to identify an expert, performing the knowledge elicitation under a deterministic constraint was a true challenge. Indeed, expert systems are based on a formal approach of knowledge. Knowledge is seen as a special type of data, handled by a computer program called an inference engine. Expert systems use formal logic of order 0 or 1, but always assume a deterministic causality. Starting from a known situation, or facts P, and rules in the general form *If P, then Q*, an expert system will produce new facts (Q), using the syllogism as an inference engine.

It is really difficult in practice to force an expert to formulate a deterministic rule. Indeed, part of the superiority of experts compared to generic problem solving methods is their ability to handle rules with a limited validity domain. In other words, these rules tolerate exceptions. When prompted to express such rules, an expert is put in a very difficult situation. Either he or she formulates as deterministic and universal a partial and context-dependent rule that would yield false conclusions when used within an inference engine. Or he or she has to identify and describe in full detail the context that would make the rule actually exact and applicable with no exception, which is, in practice, impossible.

In my experience, I remember only one spectacular success of expert systems in an application to commercial airline pilot mission scheduling. The airline expert was regularly doing better that sophisticated optimization programs, and the team building the expert system succeeded in capturing his expertise. In most other nontrivial situations (air-traffic control, defence, civil security), experts were simply unable to express how they would resolve a given problem, given that in most cases, they were asked to describe how they would address a severe, and, generally, unprecedented, situation.

The technology and algorithms used in expert systems, inference engines, have not disappeared. They live now an independent existence with business rules engines. Business rules usually implement organization strategies in detailed tactics, for instance in pricing policies. Business rule-sets may be arbitrarily complex, and subject to frequent revisions. The ability of inference engines to properly separate the inference logic from the rules has been determinant for this type of application. The good news here is that the rules are, by nature, immediately available and explicit. There is no need to torment an expert to collect them. But this is another story, some sort of a happy side-effect of AI research.

Some years later, in the early 1990s, neural networks became the popular alternative in artificial intelligence applications. Neural networks indeed had a decisive advantage compared to expert systems. The knowledge acquisition phase, which caused most expert system-based projects to fail, was nonexistent, or, at least, very simplified. Indeed, instead of asking experts how they actually address a problem, we simply needed to collect cases the expert solved in the past, and the neural network will learn them. By learning the cases, we mean here that the neural network will try to reproduce the input–output association contained in the provided set of resolved cases. For instance, when dealing with a diagnostic problem, the neural network would build on a history of couples (observed symptoms, diagnostic), and try to find a mapping that would fit most expert conclusions. Hopefully this mapping would also show some generalization performance, i.e., will perform reasonably well on new examples.

From my experience again, neural networks applications were usually more successful. This is not due to some sort of superiority of the technique, but simply to the fact that the initial exigence of neural networks was clearer, and was naturally eliminating ill-specified problems. In order to consider applying neural networks to your problem, you had to have a reasonably large set of solved cases at hand, with all the cases represented with the same input–output structure. This worked quite well for expert-based pattern recognition problem, whatever the pattern involved: diagnosis, signature identification, fault detection, etc.

The main problem with neural networks was that they were considered as black boxes. This means that, even if they succeeded to mimic the expert behavior, in most cases it was not possible to undertand how they did so. This was a strong limitation when using them in 'mission-critical' applications. Some attempts have been made to 'open the black box'. This would have been a smart way to overcome both the limitations of neural networks and expert systems: when the expert cannot explicitly state his or her problem-solving method, a neural network will learn to replicate it, and then be translated into an expert system. The neural network would have behaved as an artificial knowledge engineer. Again, in my experience, this did not succeed in practice.

22.2 A rational approach of knowledge

Now we are coming to Bayesian networks. Even if this book demonstrates that Bayesian networks have a broad application scope, it is probably too early to assess the success of this technique in practical engineering problems. It has been exactly 10 years since I first attended UAI in 1997 in Providence, RI (*Uncertainty in Artificial Intelligence*, the 'official' BN conference). I remembered that I was traveling with a friend working with a bank. I used to introduce him as a potential 'user', and he really felt as a strange species among BN researchers. Today, after training hundreds of people to use Bayesian networks, in fields as diverse as risk management, roofing, planning military aircrafts missions, formulating lubricant additives, etc., I believe that BN really have some features that can make them a powerful and practical AI tool.

22.2.1   Using expert knowledge

First, a Bayesian network model uses expert knowledge. The contribution expected from an expert is generally limited to domain knowledge, not problem solving knowledge. In principle, this does not require the same level of introspection required when using expert systems. In other words, Bayesian networks simply require *knowledge on the domain*, while expert systems would usually require *knowledge on the expert* herself.

Expert knowledge can be used to build the causal structure of the model, the graph, or to quantify the probabilities used in the model. In theory, several experts can coexist in a model, at least as far as probability assessment are concerned. There is no need to reach a consensus between experts, or to average their opinion.

22.2.2   Acquiring knowledge from data

A Bayesian network model can also use the knowledge stored in data. Bayesian networks can learn, a feature they share with neural networks. But various forms of learning can be considered. There is no need to structure the problem into a stimulus–response scheme, as in neural networks. Learning can be implemented from the simplest level, to update a probability table of a node from an augmented database, up to a general level, to create the model structure from data. Most of the applications described in this book discuss some type of learning. Parameter learning, i.e., learning conditional probabilities tables from data, is used for instance in the HEPAR network (Chapter 2), or crime risk factor analysis in Chapter 5, or the Nykredit network in Chapter 19. Structure learning – i.e., causal relationship discovery, is used for instance in Chapter 6 for spatial dynamics modeling, and in Chapter 11, for sensor validation.

22.2.3   Assessing uncertainty

A Bayesian network model also recognizes uncertainty, i.e., lack of knowledge. Indeed, assessing probabilities in a model is the most natural way to do so.

From a scientific viewpoint, causation is the foundation of determinism: identifying all the causes of a given phenomenon would allow predicting the occurrence and unfolding of this event. Similarly, probability theory is the mathematical perspective on uncertainty. Even in situations where an event is totally unpredictable, the laws of probability can help to envision and quantify the possible futures. Knowledge is the reduction of uncertainty; when we gain a better and better understanding of a phenomenon, the random part of the outcome decreases compared to the deterministic part. Some authors introduce a distinction between uncertainty and variability, the latter being an intrinsic randomness of a phenomenon that cannot be reduced. In the framework of deterministic physics, there is no such thing as variability, and apparent randomness is only the result of incomplete knowledge. Invoking Heisenberg's uncertainty principle in our discussion may seem disproportionate. But should we do it, we understand the principle as stating that ultimate knowledge is not reachable, rather than that events are random by nature 'In the sharp formulation of the law of causality (if we know the present exactly, we can calculate the future) it is not the conclusion that is wrong but the premise' [209].

Although other artificial intelligence techniques proposed various ways to deal with uncertainty, none used a plain probability assessment. Expert systems used to deal with uncertainty using confidence values for rules, but the way these confidence values were combined during the inference process was not always clear. Usual neural network models (feed-forward, or stimulus–response models) will never give a 100% sure response when shown a specific pattern in a real life application, and the strength of the response can be interpreted as a probabiliy under specific conditions. However neural networks are unable to properly handle uncertain or missing inputs.

22.2.4   Combining knowledge sources

Finally, and this is probably the most important, a Bayesian network allows us to combine different sources of knowledge (multiple experts, data), together with lack of knowledge recognized to a given extent, into a consistent model, and makes it usable for different types of problem solving.

For this reason, Bayesian networks really are a knowledge engineering tool. This is made particularly clear in some applications presented above. For instance, in the clinical decision support application, presented in Chapter 3, the development of TakeHeartII was based on the combination of 'published epidemiological models of CHD, supplemented by medical expertise when [...] unsure of interpretation'.

The complex industrial process operation application, presented in Chapter 18, reports combination of several knowledge sources: expert subjective estimates, scientific theories, fragments of statistical models, based on a large number of cases.

The terrorism risk management application, presented in Chapter 14, shows possibly the most advanced use of this ability of Bayesian networks to assemble knowledge pieces. This is made necessary by the nature of the application, since 'although a limited number of experts may be able to understand and manage a given risk, no human can manage all of the components of thousands of risks simultaneously'. The authors have here developed a specific framework to allow several submodels into a RIN (Risk Influence Network): 'a set of reusable network fragments that could be combined into an asset-threat specific RIN'. Other implementations, not documented in this book, of expert cooperation within a single knowledge models have been developed in military applications (effect based operation[2]) and generalized to decision making involving several expertises in civil contexts.

22.2.5 Putting knowledge to work

A significant part of the applications presented in this book implement more or less a form of diagnosis. Actually the Bayes theorem itself simply relates the probability of a cause given an effect to the probability of an effect given a cause. All the models created are based on causal knowledge. In other words, the initial knowledge of experts is expressed in a forward mode, i.e., from causes to effects. This is therefore not a surprise that the first nontrivial use of this knowledge is to use it backwards, i.e., to identify the likely causes of an observed situation. This is of course the case for the medical diagnosis application presented in Chapter 2, and for several other applications such as student modeling (Chapter 10), or complex industrial process operation (Chapter 18).

When multiple causes interact, Bayesian networks can also be used to perform simulation, i.e., to use knowledge in the forward direction: what would be the typical consequences of a given combination of causes? This type of use is discribed in Chapter 8, on conservation of a threatened bird.

However, Bayesian networks offer more sophisticated types of inference (Table 22.1) to use the knowledge collected in the system. In the medical context, prognostic models also exist, even if they are less represented than diagnostic models. As explained in Chapter 2, 'Medical prognosis attempts to predict the future state of a patient presenting with a set of observed findings and assigned treatment'. A prognostic model would typically include a diagnostic (backward analysis) and a simulation subpart. In medical applications, Bayesian networks can also be used to recommend additional investigations in order to support the diagnosis. In a situation when no clear diagnosis emerge, a Bayesian network model can identify the piece of information (typically, an additional test) that would clearly differentiate the possible causes of the observed symptoms.

Table 22.1   Types of inference.

Simulation	Computation of the probability distribution of an outcome given assumptions on the causes.
Diagnosis	Computation of the probability of possible causes given an observed situation.
Hypothesis evaluation	Comparison of different assumptions given an evidence.
Sensitivity analysis	Evaluation of the impact of an evidence on the conclusion.
Information gain	Expected impact of an additional investigation on the present diagnosis.

---

[2]See for instance the SIAM tool of SAIC: http://www.saic.com/products/software/siam/

In Chapter 7, on inference problems in forensic science, a more subtle type of inference is discussed. The 'likelihood ratio is a consideration of the probability of the evidence (E) given both, competing propositions that are of interest in a particular case typically forwarded by the prosecution and the defense.' This ratio can be computed directly from the model. Bayesian networks offer other different ratios that can support decision, such as sensitivity analysis or information gain.

Inference in Bayesian networks cannot however address directly problems such as planning, i.e., finding an optimal sequence of actions. Specific architectures can be designed for such a task, as shown in Chapter 20, on risk management in robotics, but it must be recognized that other AI techniques such as neural networks are usually better suited to help solving temporal control problems.

22.2.6   Knowing your knowledge

Making knowledge explicit increases knowledge. Using expert knowledge in actual, decision support systems helps reducing approximations and the resulting frustration usually experienced in qualitative discussions with experts. This effect is expected with any knowledge based systems, provided it is not a black box. Learning from data is part of a this knowledge awareness process, which is particularly easy when using Bayesian networks.

For instance, in Chapter 2, a systematic procedure of model validation was applied in order to validate the model. In the application, the model was built from clinical data and revised with experts: 'This analysis has brought several corrections to the model. For example, when observing the variable 'Enlarged spleen', the expert noticed that the presence of the finding does not influence probability distribution of 'Toxic hepatitis' and 'Reactive hepatitis'. Therefore, the arcs from 'Toxic hepatitis' and 'Reactive hepatitis' to 'Enlarged spleen' were drawn.' This particular example show that knowledge used in a Bayesian network can be easily edited, even when learnt from data. Improving expert knowledge from data is also possible. The easiest way to do so is simply to learn a local probability table when the structure has been defined by an expert.

Chapter 6, on spatial dynamics in geography, also exhibits an interesting example of how Bayesian networks can be used to improve knowledge about a given domain, and to discover causal relationships. A database of indicators measured across several communes in the coastal region under study has been learnt by a Bayesian network. The resulting model allows us to interpret the main drivers of urbanization, some of them being unexpected from the author's initial point of view.

22.2.7   Limitations

Unfortunately, and as any other tool, Bayesian networks have limitations. It is not always easy to detect limitations in an application book, since only feasible applications can be described in such a book.

Computational complexity is one of the strongest limitations of Bayesian networks. Bayesian networks algorithms are of nonpolynomial complexity. This means that the computation time grow exponentially as the network complexity grows. Network complexity is not measured by the number of nodes, but by a more technical quantity, which has to do with the connectivity of the network. In practice, however, we can see in the applications presented that most networks had reasonable sizes: 70 nodes for the medical diagnosis network in Chapter 2, about 20 nodes for the different networks used in clinical decision support in Chapter 3, 39 nodes for the spatial dynamics modeling network in Chapter 6, from 35 to 60 nodes in risk influence networks used for terrorism threat assessment in Chapter 14, and 124 nodes in the autonomous vehicle network described in Chapter 20.

This limitation is generally not a difficulty in expert based networks, since most knowledge bases expressed by human expert will generally stay within reasonable size limits. This can be more of a problem when dealing with several networks brought together, such as RIN used in Chapter 14, although in this particular case network sizes remain small. This complexity issue becomes crucial when dealing with temporal reasoning. Applications in robotics for instance require a causal model explicitly dealing with temporal dependencies: information gathered now determines present action, while present action is one of the causes of future external feedbacks. Planning, i.e., optimal selection of a sequence of actions, therefore requires us to unfold the model through time, hence increasing its size, and generally its complexity and connectivity. This type of application – and its limits – is addressed in Chapter 20, on risk management in robotics.

22.3   Future challenges

Although Bayesian networks are certainly not the Holy Grail of artificial intelligence, they definitely are a solid basis for knowledge engineering. They allow us to use various sources of knowledge, even contradicting ones, to make knowledge embedded in data explicit, to use this knowledge for various types of problem solving, and finally to improve it through online learning.

Artificial intelligence remains a challenge for the next decades. Indeed, intelligence cannot be limited to inference and learning, but requires action. Embedding artificial intelligence systems in the real world is probably the next challenge of artificial intelligence, far beyond simply connecting an offline 'artificially intelligent system' to external sensors and actuators.

Bibliography

[1] S. Acid, L. M. de Campos, J. M. Fernández-Luna, and J. F. Huete. An information retrieval model based on simple Bayesian networks. *International Journal of Intelligent Systems*, 18:251–265, 2003.

[2] S. Aeberhard, O. de Vel, and D. Coomans. Comparative analysis of statistical pattern recognition methods in high dimensional settings. *Pattern Recognition*, 27:1065–1077, 1994.

[3] F. Agterberg, G. Bonham-Carter, and D. Wright. Statistical pattern integration for mineral exploration. In *Computer Applications in Resource Estimation Prediction and Assessment for Metals and Petroleum*, pages 1–21. Pergamon Press, Oxford and New York, 1990.

[4] Air Force Instruction 31-210. The Air Force Antiterrorism/Force Protection (AT/FP) Program. 1 July 1977.

[5] J. Aires-de Sousa. Verifying wine origin: A neural network approach. *American Journal of Enology and Viticulture*, 47:410–414, 1996.

[6] C. G. G. Aitken, T. Connolly, A. Gammerman, G. Zhang, D. Bailey, R. Gordon, and R. Oldfield. Statistical modelling in specific case analysis. *Science and Justice*, 36:245–255, 1996.

[7] C. G. G. Aitken and A. Gammerman. Probabilistic reasoning in evidential assessment. *Journal of the Forensic Science Society*, 29:303–316, 1989.

[8] C. G. G. Aitken, A. Gammerman, G. Zhang, T. Connolly, D. Bailey, R. Gordon, and R. Oldfield. Bayesian belief networks with an application in specific case analysis. In A. Gammerman, editor, *Computational Learning and Probabilistic Reasoning*, pages 169–184. John Wiley & Sons, Chichester, 1996.

[9] C. G. G. Aitken and F. Taroni. *Statistics and the Evaluation of Evidence for Forensic Scientists*. John Wiley & Sons, Inc., New York, 2nd edition, 2004.

Bayesian Networks: A Practical Guide to Applications   Edited by O. Pourret, P. Naïm, B. Marcot
© 2008 John Wiley & Sons, Ltd

[10] C. G. G. Aitken, F. Taroni, and P. Garbolino. A graphical model for the evaluation of cross-transfer evidence in DNA profiles. *Theoretical Population Biology*, 63:179–190, 2003.

[11] M. Ajmone Marsan, G. Balbo, G. Conte, S. Donatelli, and G. Franceschini. *Modelling with Generalized Stochastic Petri Nets*. Wiley Series in Parallel Computing, John Wiley & Sons, Ltd, New York, 1995.

[12] S. Alag, A. Agogino, and M. Morjaria. A methodology for intelligent sensor measurement, validation, fusion, and fault detection for equipment monitoring and diagnostics. *Artificial Intelligence for Engineering Design, Analysis and Manufacturing (AIEDAM), Special Issue on AI in Equipment Service*, 15(4):307–319, 2001.

[13] S. Ampuero and J. O. Bosset. The electronic nose applied to dairy products: A review. *Sensors and Actuators B: Chemical*, 94(1):1–12, August 2003.

[14] S. K. Andersen, K. G. Olesen, F. V. Jensen, and F. Jensen. HUGIN – a shell for building bayesian belief universes for expert systems. In *Proceedings of the Eleventh International Joint Conference on Artificial Intelligence*, Detroit, N.S. Sridharan (editor) pages 1080–1085, Morgan Kaufmann 1989.

[15] J. L. Anderson. Embracing uncertainty: The interface of Bayesian statistics and cognitive psychology. *Conservation Ecology*, 2(1):27, 1998.

[16] S. Andreassen, M. Woldbye, B. Falck, and S. K. Andersen. MUNIN – A causal probabilistic network for interpretation of electromyographic findings. In J. McDermott, editor, *Proceedings of the Tenth International Joint Conference on Artificial Intelligence*, pages 366–372, Los Altos, CA, Morgan Kaufmann Publishers, 1987.

[17] Argonne National Laboratory. Argonne software helps emergency responders plan and prepare. `http://www.anl.gov/Media\_Center/News/2003/news030404.htm`.

[18] D. E. Arking, G. Atzmon, A. A. Arking, N. Barzilai, and H. C. Dietz. Association between a functional variant of the KLOTHO gene and high-density lipoprotein cholesterol, blood pressure, stroke, and longevity. *Circulation Research*, 96:412–418, 2005.

[19] G. Arroyo-Figueroa and L. Sucar. A temporal Bayesian network for diagnosis and prediction. In *Proceedings of the Fifteenth Conference on Uncertainty in Artificial Intelligence*, Stockholm, K. B. Laskey, H. Prade (editors) pages 13–20, Morgan Kaufmann, 1999.

[20] G. Assmann, P. Cullen, and H. Schulte. Simple scoring scheme for calculating the risk of acute coronary events based on the 10-year follow-up

of the Prospective Cardiovascular Münster (PROCAM) Study. *Circulation*, 105(3):310–315, 2002.

[21] http://www.aaai.org.

[22] ASTM International. Standard practice for use of scrap tires in civil engineering applications. ASTM Standard Design D6270–98, 2004.

[23] N. O. Attoh-Okine. Valuation-based systems for pavement management decision making. In *Uncertainty Analysis in Engineering and Sciences*, B. M. Ayyab and M. M. Gupta, chapter 11, pages 157–176. Springer, Berlin, 1997.

[24] N. O. Attoh-Okine. Addressing uncertainties in flexible pavement maintenance decisions at project level using Bayesian influence diagrams. In J. L. Gifford, editor, *Infrastructure Planning and Management*, pages 362–366. American Society of Civil Engineers, 1993.

[25] N. O. Attoh-Okine and I. Ahmad. Application of Bayesian influence diagrams to risk analysis of highways construction costs. In, *Proceedings Computing in Civil Engineering*, J. P. Mohsen, editor, ASCE, 1995.

[26] N. O. Attoh-Okine and A. K. Appea. Probabilistic graphical networks in pavement management decision making. In N. O. Attoh-Okine, editor, *Artificial Intelligence and Mathematical Methods in Pavement and Geomechanical Systems*, pages 119–133. A.A. Balkema Publishers, Rotterdam, 1998.

[27] N. O. Attoh-Okine and S. Bowers. A Bayesian belief network model of bridge deterioration. *Bridge Engineering*, 159(2):69–76, 2006.

[28] N. O. Attoh-Okine and W. Roddis. Uncertainties of asphalt layer thickness determination in flexible pavements-influence diagram approach. *Civil Engineering Systems*, 15:107–124, 1998.

[29] AUPA DRE PACA. La métropolisation dans l'espace méditerranéen français. *Les cahiers de la métropolisation Aix-en-Provence*.

[30] B. Baesens, M. Egmont-Petersen, R. Castelo, and J. Vanthienen. Learning Bayesian networks classifiers for credit scoring using Markov chain Monte Carlo search. In *Proceedings of International Congress on Pattern Recognition*, Quebec pages 49–52, Springer-Verlag, Berlin, 2002.

[31] R. Baeza-Yates and B. Ribeiro-Neto. *Modern Information Retrieval*. Addison-Wesley, Harlow, UK, 1999.

[32] V. Bahn. Habitat requirements and habitat suitability index for the Marbled Murrelet (*Brachyramphus marmoratus*) as a management target species in the Ursus Valley, British Columbia, 1998. Diplomarbeit Thesis, Phillips Universität, Marburg, Germany.

[33] C. T. Baldwin, V. G. Nolan, D. F. Wyszynski, Q. L. Ma, P. Sebastiani, S. H. Embury, A. Bisbee, J. Farrell, L. S. Farrer, and M. H. Steinberg. Association of klotho, bone morphogenic protein 6 and annexin a2 polymorphisms with sickle cell osteonecrosis. *Blood*, 106(1):372–375, July 2005.

[34] P. Baumard. *Tacit Knowledge in Organizations*. Sage, London, 1999.

[35] P. Baumard and J. A. Benvenuti. *Compétitivité et systèmes d'information.* InterEditions, Paris, 1998.

[36] Bayesware Limited, London. *Bayesware – User Manual*, 2000.

[37] J. Bechta-Dugan, S. Bavuso, and M. Boyd.   Dynamic fault-tree models for fault-tolerant computer systems.   *IEEE Transactions on Reliability*, 41:363–377, 1992.

[38] J. Bechta-Dugan, K. Sullivan, and D. Coppit. Developing a low-cost high-quality software tool for dynamic fault-tree analysis. *IEEE Transactions on Reliability*, 49(1):49–59, 2000.

[39] A. Becker and P. Naïm. *Les réseaux bayésiens – Modèles graphiques de connaissance*. Eyrolles, Paris, 1999.

[40] I. Beinlich, H. Suermondt, R. Chavez, and G. Cooper. The ALARM monitoring system: A case study with two probabilistic inference techniques for belief networks. In *Proceedings of the Second European Conference on Artificial Intelligence in Medical Care*, London J. Hunta, J. Cookson, J. Wyatt, editors, pages 247–256, Springer-Verlag, Berlin, 1989.

[41] S. R. Beissinger. Population viability analysis: past, present and future. In S. R. Beissinger and D. R. McCullough, editors, *Population Viability Analysis*, pages 5–17. University of Chicago Press, Chicago, Illinois, 2002.

[42] N. H. Beltrán, M. A. Duarte-Mermoud, M. A. Bustos, S. A. Salah, E. A. Loyola, A. I. Peña Neira, and J. W. Jalocha. Feature extraction and classification of Chilean wines. *Journal of Food Engineering*, 1:1–10, July 2006.

[43] N. H. Beltrán, M. A. Duarte-Mermoud, S. A. Salah, M. A. Bustos, A. I. Peña Neira, E. A. Loyola, and J. W. Galocha. Feature selection algorithms using Chilean wine chromatograms as examples. *Journal of Food Engineering*, 67(4):483–490, April 2005.

[44] C. Berzuini, R. Bellazzi, S. Quaglini, and D. J. Spiegelhalter. Bayesian networks for patient monitoring. *Artificial Intelligence in Medicine*, 4:243–260, 1992.

[45] A. Beverina and B. Ware. Method and apparatus for risk management, 2006. US Patent 7130779, filed 14 May 2001, and issued 31 October 2006.

[46] A. Biedermann and F. Taroni. Bayesian networks and probabilistic reasoning about scientific evidence when there is a lack of data. *Forensic Science International*, 157:163–167, 2006.

[47] A. Biedermann and F. Taroni. A probabilistic approach to the joint evaluation of firearm evidence and gunshot residues. *Forensic Science International*, 163:18–33, 2006.

[48] C. Bishop. *Neural Networks for Pattern Recognition*. Oxford University Press, New York, 2002.

[49] A. Bobbio, L. Portinale, M. Minichino, and E. Ciancamerla. Improving the analysis of dependable systems by mapping fault trees into Bayesian networks. *Reliability Engineering and System Safety*, 71:249–260, 2001.

[50] A. Bobbio and D. C. Raiteri. Parametric fault-trees with dynamic gates and repair boxes. In *Proceedings of the Annual Reliability and Maintainability Symposium RAMS2004*, pages 101–106, Los Angeles, 2004.

[51] G. Bonham-Carter. *Geographic Information Systems for Geoscientists: Modeling with GIS*. Pergamon Press, Ontario, 1994.

[52] G. Bonham-Carter and F. Agterberg. Application of a microcomputer-based Geographic Information System to mineral-potential mapping. In *Microcomputer-based Applications in Geology, II, Petroleum*, pages 49–74. Pergamon Press, New York, 1990.

[53] R. Boondao. A Bayesian network model for analysis of the factors affecting crime risk. *WSEAS Transactions on Circuit and Systems*, 3(9): November 2004.

[54] D. Botstein and N. Risch. Discovering genotypes underlying human phenotypes: Past successes for Mendelian disease, future approaches for complex disease. *Nature Genetics supplement*, 33:228–237, 2003.

[55] R. Bouckaert. *Bayesian belief networks: from construction to inference*. PhD thesis, University of Utrecht, 1995.

[56] R. Bouckaert. Bayesian network classifiers in Weka. Technical report, No. 14 University of Waikato, Department of Computer Science, Hamilton, New Zealand, 2004.

[57] H. Boudali and J. Bechta-Dugan. A new Bayesian network approach to solve dynamic fault trees. In *Proceedings of the Annual Reliability and Maintainability Symposium RAMS2005*, Alexandria, USA, pages 451–456, Springer-Verlag, 2005.

[58] M. S. Boyce, E. M. Kirsch, and C. Serveen. Bet hedging applications for conservation. *Journal of Bioscience*, 27:385–392, 2002.

[59] R. Brackett. Statement before the Subcommittee on the Federal Workforce and Agency Organization, Committee on Government Reform, U.S. House of Representatives, 17 May 2005.

[60] P. J. Brantingham and P. L. Brantingham. Notes on the geometry of crime. In P. J. Brantingham and P. L. Brantingham, editors, *Environmental Criminology*, pages 27–54. Sage Publications, Beverly Hills, 1981.

[61] L. Breiman. Random forests. *Machine Learning*, 45:5–32, 2001.

[62] B. W. Brook, M. A. Burgman, H. R. Akcakaya, J. J. O'Grady, and R. Frankman. Critiques of PVA ask the wrong questions: throwing the heuristic baby out with the numerical bath water. *Conservation Biology*, 16:262–263, 2002.

[63] F. J. Brooks. GE gas turbine performance characteristics. Technical Report GER-3567H, GE Power Systems, Schenectady, NY, 2000.

[64] R. A. Brooks. From earwigs to humans. *Robotics and Autonomous Systems*, 20(24):291–304, 1997.

[65] K. Brudzewski, S. Osowski, and T. Markiewicz. Classification of milk by means of an electronic nose and SVM neural networks. *Sensors and Actuators B: Chemical*, 98(2–3):291–298, March 2004.

[66] R. Brunet. Models in geography? A sense to research. *Cybergeo*, 12th European Colloquium on Quantitative and Theorelical Geography, Saint-Valéry-en-Caux, France, 2001.

[67] A. Bryman and D. Cramer. *Quantitative Data Analysis with SPSS for Windows: A Guide for Social Scientists*. Routledge, London, 1997.

[68] D. L. Buckeridge, H. Burkom, M. Campbell, and W. R. Hogan. Algorithms for rapid outbreak detection: a research synthesis. *Journal of Biomedical Informatics*, 38(2):99–113, 2005.

[69] S. G. Bunch. Consecutive matching striation criteria: A general critique. *Journal of Forensic Sciences*, 45:955–962, 2000.

[70] A. Bunt and C. Conati. Probabilistic student modelling to improve exploratory behaviour. *Journal of User Modelling and User-Adapted Interaction*, 13(3):269–309, 2003.

[71] A. E. Burger. Conservation assessment of marbled murrelets in British Columbia: review of the biology, populations, habitat associations, and conservation. Part A of Marbled Murrelet Conservation Assessment, Canada Wildlife Service, Delta, British Columbia, Technical Report Series No. 387, 2002.

[72] A. E. Burger, T. A. Chatwin, S. A. Cullen, N. P. Holmes, I. A. Manley, M. H. Mather, B. K. Schroedor, J. D. Steventon, J. E. Duncan, P. Arcese, and E. Selak. Application of radar surveys in the management of nesting habitat for marbled murrelets in British Columbia. *Marine Ornithology*, 32:1–11, 2004.

[73] A. E. Burger, J. Hobbs, and A. Hetherington. Testing models of habitat suitability for nesting Marbled Murrelets, using low-level aerial surveys on the North Coast, British Columbia. Unpublished report, Ministry of Water, Land and Air Protection, Smithers, B.C., 2005.

[74] C. Burges. *A Tutorial on Support Vector Machines for Pattern Recognition*. Kluwer Academic Publishers, Boston, 2000.

[75] B. E. Burnside, D. L. Rubin, and R. Shachter. A Bayesian network for mammography. Technical report, Stanford Medical Informatics Dept, 2000.

[76] G. Bush. National preparedness. Homeland Security Presidential Directive (HSPD-8), Washington, DC. 17 December, 2003.

[77] Busselton Health Studies Group. The Busselton health study. Website, December 2004.

[78] M. A. Bustos, M. A. Duarte-Mermoud, N. H. Beltrán, S. A. Salah, A. I. Peña Neira, E. A. Loyola, and J. W. Galocha. Clasificación de vinos chilenos usando un enfoque Bayesiano. *Viticultura y Enología Profesional*, 90:63–70, 2004.

[79] M. Cabezudo, M. Herraiz, and E. Gorostiza. On the main analytical characteristics for solving enological problems. *Process Biochemistry*, 18:17–23, 1983.

[80] J. P. Callan, W. B. Croft, and S. M. Harding. The INQUERY retrieval system. In *Proceedings of the 3rd International Conference on Database and Expert Systems Applications*, Valencia, A. Min Tjoa and I. Ramog, editors, volume 1980, pages 78–83. Springer-Verlag, Vienna, 1992.

[81] A. Campbell, V. Hollister, and R. Duda. Recognition of a hidden mineral deposit by an artificial intelligence program. *Science*, 217(3):927–929, 1982.

[82] L. R. Cardon and J. I. Bell. Association study designs for complex diseases. *Nature Reviews Genetics*, 2:91–99, 2001.

[83] Carnegie Mellon Software Engineering Institute. Introduction to the OCTAVE approach, 2003.

[84] E. Carranza and M. Hale. Geologically constrained probabilistic mapping of gold potential, Baguio district, Philippines. *Natural Resources Research*, 9(3):237–253, 2000.

[85] E. Carranza, J. Mangaoang, and M. Hale. Application of mineral deposit models and GIS to generate mineral potential maps as input for optimum land-use planning in the Philippines. *Natural Resources Research*, 8(2):165–173, 1999.

[86] E. Castillo, J. M. Gutierrez, and A. S. Hadi. *Expert Systems and Probabilistic Network Models*. Springer, New York, 1997.

[87] M.-A. Cavarroc and R. Jeansoulin. L'apport des réseaux Bayésiens pour la recherche de causalité spatio-temporelles. In *Assises Cassini 1998*, Marne-la-Vallée, France, 1998.

[88] P. K. Chan, W. Fan, L. A. Prodromidis, and S. J. Stolfo. Distributed data mining in credit card fault detection. *IEEE Intelligent Systems*, 14(6):67–74, 1999.

[89] C. Chang and C. J. Lin. OSU support vector machines (SVMs) toolbox. Technical report, Department of Computer Science and Information Engineering, National Taiwan University, 2002.

[90] E. Charniak. Belief networks without tears. *AI Magazine*, pages 50–62, 1991.

[91] P. Cheeseman and J. Stutz. Bayesian classification (AutoClass): Theory and results. In U. M. Fayyad, G. Piatetsky-Shapiro, P. Smyth, and R. Uthurusamy, editors, *Advances in Knowledge Discovery and Data Mining*, pages 153–180. MIT Press, Cambridge, MA, 1996.

[92] J. Cheng, R. Greiner, J. Kelly, D. Bell, and W. Liu. Learning Bayesian networks from data: an information-theory based approach. *Artificial Intelligence*, 137:43–90, 2002.

[93] Y. Chiaramella. Information retrieval and structured documents. *Lectures Notes in Computer Science*, pages 291–314, 2001.

[94] D. Chickering, D. Geiger, and D. Heckerman. Learning Bayesian networks is NP-hard. Microsoft Research, Microsoft Corporation, 1994. Technical Report MSR-TR-94-l7.

[95] G. Chiola, C. Dutheillet, G. Franceschinis, and S. Haddad. Stochastic well-formed coloured nets for symmetric modelling applications. *IEEE Transactions on Computers*, 42(11):1343–1360, 1993.

[96] C. K. Chow and C. N. Liu. Approximating discrete probability distributions with dependence trees. *IEEE Transactions On Information Theory*, IT-14(11), 1968.

[97] R. T. Clemen and R. L. Winkler. Calibrating and combining precipitation probability forecasts. In R. Viertl, editor, *Probabilities and Bayesian Statistics*, pages 97–110. Plenum, New York, 1987.

[98] C. Conati, A. Gertner, and K. Van Lehn. On-line student modelling for coached problem solving using Bayesian networks. In J. Kay, editor, *Proceedings of the 7th International Conference on User Modelling*, pages 231–242. Banff, Springer-Verlag, 1999.

[99] R. Cook, I. W. Evett, J. Jackson, P. J. Jones, and J. A. Lambert. A hierarchy of propositions: Deciding which level to address in casework. *Science and Justice*, 38:231–240, 1998.

[100] G. Cooper. The computational complexity of probabilistic inference using Bayesian belief networks. *Artificial Intelligence*, 42:393–405, 1990.

[101] G. Cooper and D. Dash. Bayesian biosurveillance of disease outbreak. In *Uncertainty in Artificial Intelligence: Proceedings of the Sixteenth Conference (UAI-2004)*, pages 94–103. Morgan Kaufmann Publishers, San Francisco, CA, 2004. .

[102] G. F. Cooper. *NESTOR: a computer-based medical diagnostic aid that integrates causal and probabilistic knowledge*. PhD thesis, Stanford University, Computer Science Department, 1984.

[103] G. F. Cooper and E. Herskovits. A Bayesian method for the induction of probabilistic networks from data. *Machine Learning*, 9(4):309–348, 1992.

[104] F. Cortijo-Bon. Selección y extracción de características, April 2004. http://www-etsi2.ugr.es/depar/ccia/rf/www/tema5_00-01_www/tema5_00-01_www.html.

[105] R. G. Cowell. FINEX: a probabilistic expert system for forensic identification. *Forensic Science International*, 134:196–206, 2003.

[106] R. G. Cowell, A. P. Dawid, S. L. Lauritzen, and D. J. Spiegelhalter. *Probabilistic Networks and Expert Systems*. Springer, New York, 1999.

[107] F. Crestani, L. M. de Campos, J. M. Fernández-Luna, and J. F. Huete. A multi-layered Bayesian network model for structured document retrieval. *Lecture Notes in Artificial Intelligence*, 2711:74–86, 2003.

[108] F. Crestani, M. Lalmas, C. J. van Rijsbergen, and L. Campbell. Is this document relevant?... probably. A survey of probabilistic models in information retrieval. *ACM Computing Surveys*, 30(4):528–552, 1991.

[109] N. Cristianini and J. Shawe-Taylor. *An Introduction to Support Vector Machines and other Kernel-Based Methods*. Cambridge University Press, Cambridge, 2000.

[110] M. J. Daly, J. Rioux, S. F. Schaffner, T. Hudson, and E. S. Lander. High-resolution haplotype structure in the human genome. *Nature Genetics*, 29:229–232, 2001.

[111] J. N. Darroch, S. L. Lauritzen, and T. P. Speed. Markov fields and log linear models for contingency tables. *Annals of Statistics*, 8:522–539, 1980.

[112] P. Das and T. Onoufriou. Areas of uncertainty in bridge management: framework for research. In *Transport Research Record, National Research Council*, pages 202–209, National Research Council, Washington DC, 2000.

[113] S. Dash and V. Venkatasubramanian. Challenges in the industrial applications of fault diagnostic systems. *Computers and Chemical Engineering*, 24(2–7):785–791, 2000.

[114] A. P. Dawid, J. Mortera and P. Vicard, Object-oriented Bayesian networks for complex forensic DNA profiling problems, Forensic Science International, 169:169–205, 2007.

[115] A. P. Dawid and I. W. Evett. Using a graphical method to assist the evaluation of complicated patterns of evidence. *Journal of Forensic Sciences*, 42:226–231, 1997.

[116] A. P. Dawid and S. L. Lauritzen. Hyper Markov laws in the statistical analysis of decomposable graphical models. *Annals of Statistics*, 21:1272–1317, 1993, correction, 23(5): 1864, 1995.

[117] A. P. Dawid, J. Mortera, V. L. Pascali, and D. van Boxel. Probabilistic expert systems for forensic inference from genetic markers. *Scandinavian Journal of Statistics*, 29:577–595, 2002.

[118] L. M. de Campos, J. M. Fernández-Luna, and J. F. Huete. The BNR model: foundations and performance of a Bayesian network-based retrieval model. *International Journal of Approximate Reasoning*, 34:265–285, 2003.

[119] L. M. de Campos, J. M. Fernández-Luna, and J. F. Huete. Bayesian networks and information retrieval. An introduction to the special issue. *Information Processing and Management*, 40(5):727–733, 2004.

[120] L. M. de Campos, J. M. Fernández-Luna, and J. F. Huete. Using context information in structured document retrieval: An approach using influence diagrams. *Information Processing and Management*, 40(5):829–847, 2004.

[121] L. M. de Campos, J. M. Fernández-Luna, and J. F. Huete. Improving the context-based influence diagram model for structured document retrieval: removing topological restrictions and adding new evaluation methods. *Lecture Notes in Computer Science*, 3408:215–229, 2005.

[122] M. Deb. VMS deposits: geological characteristics, genetic models and a review of their metallogenesis in the Aravalli range, NW India. In M. Deb, editor, *Crustal Evolution and Metallogeny in the Northwestern Indian Shield*, pages 217–239. Narosa Publishing House, New Delhi, 2000.

[123] Decision Focus Inc. *INDIA, Influence Diagram Software.* Boston, MA, 1991.

[124] F. Decoupigny. Modélisation d'un déplacement sur une double échelle. In P. Mathis, editor, *Graphes et réseaux – modélisation multiniveau*, pages 77–91. Hermès Sciences, Paris, 2003.

[125] F. Dellaert. The Expectation Maximization Algorithm. Technical Report GIT-GVU 02-02, College of Computing, Georgia Institute of Technology, 2002.

[126] R. Descartes. Discourse on the method of rightly conducting the reason in the search for truth in the sciences, 1637. Freely downloadable at www. gutenberg.org.

[127] L. Diappi, P. Bolchi, and L. Franzini. Improved understanding of urban sprawl using neural networks. In J. Van Leeuwen and J. Timmermans, editors, *Recent Advances in Design and Decision Support Systems*. Kluwer Academic Publishers, Dordrecht, 2004.

[128] F. Díez. Teaching probabilistic medical reasoning with the Elvira software. In R. Haux and C. Kulikowski, editors, *IMIA Yearbook of Medical Informatics*, pages 175–180. Schattauer, Stuttgart, 2004.

[129] F. J. Díez and M. J. Druzdzel. Canonical probabilistic models for knowledge engineering. Technical Report CISIAD-06-01, UNED, Madrid, Spain, 2006.

[130] F. J. Díez, J. Mira, E. Iturralde, and S. Zubillaga. DIAVAL, a Bayesian expert system for echocardiography. *Artificial Intelligence in Medicine*, 10:59–73, 1997.

[131] Direction Régionale de l'Équipement Marseille. Carte d'occupation du sol de la région Provence-Alpes-Côte d'Azur. CD-ROM, 1999.

[132] P. Domingos and M. Pazzani. Beyond independence: Conditions for optimality of the simple Bayesian classifier. In *Proceedings of the Thirteenth International Conference on Machine Learning*, pages 105–112. Morgan Kaufmann, San Francisco, CA, 1996.

[133] P. Domingos and M. Pazzani. On the optimality of the simple Bayesian classifier under zero-one loss. *Machine Learning*, 29:103–130, 1997.

[134] B. Downs. Safely securing U.S. ports: The port security assessment program. Proceedings (US Coast Guards), Spring 2006.

[135] M. Druzdzel and L. van der Gaag. Building probabilistic networks: Where do the numbers come from – a guide to the literature, guest editors introduction. *IEEE Transactions in Knowledge and Data Engineering*, 12:481–486, 2000.

[136] M. J. Druzdzel and F. J. Díez. Combining knowledge from different sources in causal probabilistic models. *Journal of Machine Learning Research*, 4:295–316, 2003.

[137] M. J. Druzdzel and H. V. Leijen. Causal reversibility in Bayesian networks. *Journal of Experimental and Theoretical Artificial Intelligence*, 13(1):45–62, 2001.

[138] M. J. Druzdzel and H. A. Simon. Causality in Bayesian belief networks. In *UAI93 – Proceedings of the Ninth Conference on Uncertainty in Artificial Intelligence*, Providence, D. Heckerman and E. H. Marndani, editors, pages 3–11. Morgan Kaufmann, 1993.

[139] R. Duda and P. Hart. *Pattern Classification and Scene Analysis*. John Wiley & Sons, Inc., New York, 1973.

[140] R. Duda, P. Hart, N. Nilsson, and G. Sutherland. Semantic network representations in rule-based interference systems. In *Pattern-Directed Inference Systems*, pages 203–221. Academic Press, 1978.

[141] G. Dupuy. *La dépendance automobile: symptômes, diagnostics, traitement*. Anthropos, Paris, 1999.

[142] Y. Dutuit and A. Rauzy. A linear-time algorithm to find modules of fault tree. *IEEE Transactions on Reliability*, 45:422–425, 1996.

[143] Y. Dutuit and A. Rauzy. Efficient algorithms to assess components and gates importances in fault tree analysis. *Reliability Engineering and System Safety*, 72:213–222, 2000.

[144] W. Edwards. Influence diagrams, Bayesian imperialism, and the *Collins* case: An appeal to reason. *Cardozo Law Review*, 13:1025–1074, 1991.

[145] P. Etievant and P. Schilich. Varietal and geographic classification of French red wines in terms of mayor acids. *Agriculture Food and Chemistry*, 47:421–498, 1989.

[146] ETS. *Electronic Sensor Technology 7100 Fast GC Analyzer: Operation Manual*, 1999.

[147] J. Evert, E. Lawler, H. Bogan, and T. T. Perls. Morbidity profiles of centenarians: Survivors, delayers, and escapers. *Journals of Gerontology Series A: Biological Sciences and Medical Sciences*, 58:232–237, 2003.

[148] I. W. Evett, P. D. Gill, G. Jackson, J. Whitaker, and C. Champod. Interpreting small quantities of DNA: the hierarchy of propositions and the use of Bayesian networks. *Journal of Forensic Sciences*, 47:520–530, 2002.

[149] I. W. Evett, J. A. Lambert, and J. S. Buckleton. A Bayesian approach to interpreting footwear marks in forensic casework. *Science and Justice*, 38:241–247, 1998.

[150] N. Fenton and M. Neil. The 'jury observation fallacy' and the use of Bayesian networks to present probabilistic legal arguments. *Mathematics Today – Bulletin of the IMA*, 36:180–187, 2000.

[151] F. Ferrazzi, P. Sebastiani, I. S. Kohane, M. Ramoni, and R. Bellazzi. Dynamic Bayesian networks in modelling cellular systems: a critical appraisal on simulated data. In *19th IEEE International Symposium on Computer-Based Medical Systems (CBMS 2006), 22–23 June 2006, Salt Lake City, Utah, USA*, pages 544–549. IEEE Computer Society, 2006.

[152] C. Flint, M. Harrower, and R. Edsall. But how does place matter? Using Bayesian networks to explore a structural definition of place. In *New Methodologies for Social Sciences*, Boulder, CO, March 2000. University of Colorado.

[153] P. Foley. Predicting extinction times from environmental stochasticity and carrying capacity. *Conservation Biology*, 8:124–137, 1994.

[154] http://www.fortiusone.com.

[155] Foundation for Neural Networks and University Medical Centre Utrecht. 'PROMEDAS' a probabilistic decision support system for medical diagnosis. Technical report, SNN Foundation for Neural Networks and University Medical Centre Utrecht, Nijmegen, Netherlands, 2002.

[156] C. Fraley and A. E. Raftery. Model-based clustering, discriminant analysis, and density estimation. *Journal of the American Statistical Association*, 97:611–631, 2002.

[157] R. C. G. Frankin, D. J. Spiegelhalter, F. MacCartney, and K. Bull. Combining clinical judgments and clinical data in expert systems. *Internation Journal of Clinical Monitoring and Computing*, 6:157–166, 1989.

[158] D. Freudenthal. The role of age, foreknowledge and complexity in learning to operate a complex device. *Behaviour and Information Technology*, 20(1):23–135, 2001.

[159] W. Friedewald, R. Levy, and D. Fredrickson. Estimation of the concentration of low-density lipoprotein cholesterol in plasma, without use of the preparative ultracentrifuge. *Clinical Chemistry*, 18:499–502, June 1972.

[160] J. H. Friedman. On bias, variance, 0/1-loss and the curse-of-dimensionality. *Data Mining and Knowledge Discovery*, 1(1):55–77, 1997.

[161] N. Friedman, D. Geiger, and M. Goldszmidt. Bayesian network classifiers. *Machine Learning*, 29(2–3):131–163, 1997.

[162] N. Friedman and M. Goldszmidt. Learning Bayesian network from data. Technical report, Berkeley, University of California, 1998. http://www.cs.berkeley.edu/~nir/Tutorial [Accessed 5 January 2004].

[163] R. D. Friedman. A close look at probative value. *Boston University Law Review*, 66:733–759, 1986.

[164] R. D. Friedman. A diagrammatic approach to evidence. *Boston University Law Review*, 66:571–622, 1986.

[165] J. Fritz. Distribution-free exponential error bound for nearest neighbor pattern classification. *IEEE Transactions on Information Theory*, IT-21:552–557, 1975.

[166] K. Fukunaga and R. Hayes. Estimation of classifier performance. *IEEE Transactions on Pattern Analysis and Machine Intelligence*, 11:1087–1101, 1989.

[167] G. Fusco. Looking for sustainable urban mobility through Bayesian networks. *Scienze Regionali / Italian Journal of Regional Sciences*, 3:87–106, 2003. Also in Cybergeo No. 292. http://www.cybergeo.presse.fr.

[168] G. Fusco. La mobilité quotidienne dans les grandes villes du monde: application de la théorie des réseaux bayésiens. *Cybergeo*, 260, 2004. http://www.cybergeo.presse.fr.

[169] G. Fusco. Un modèle systémique d'indicateurs pour la durabilité de la mobilité urbaine. PhD thesis, University of Sophia-Antipolis, Nice (France) and Politecnico di Milano (Italy), 2004. Published by Chambre de Commerce Italienne de Nice, 2 volumes + CD-ROM.

[170] S. Gabriel, S. Schaffner, H. Nguyen, J. Moore, J. Roy, B. Blumenstiel, J. Higgins, M. DeFelice, A. Lochner, M. Faggart, S. N. Liu-Cordero, C. Rotimi, A. Adeyemo, R. Cooper, R. Ward, E. Lander, M. Daly, and D. Altshuler. The structure of haplotype blocks in the human genome. *Science*, 296, 2002.

[171] M. Gaeta, M. Marsella, S. Miranda, and S. Salerno. Using neural networks for wine identification. In *Proceedings from IEEE International Joint Symposia on Intelligence and Systems*, Washington DC, pages 418–421, IEEE Computer Society, 1998.

[172] S. Galan, F. Aguado, F. J. Díez, and J. Mira. NasoNet: joining Bayesian networks and time to model nasopharyngeal cancer spread. In S. Quaglini, P. Barahona, and S. Andreassen, editors, *Artificial Intelligence in Medicine, Lecture Notes in Computer Science Subseries, LNAI 2101*, pages 207–216. Springer-Verlag, Heidelberg, 2001.

[173] P. Garbolino. Explaining relevance. *Cardozo Law Review*, 22:1503–1521, 2001.

[174] P. Garbolino and F. Taroni. Evaluation of scientific evidence using Bayesian networks. *Forensic Science International*, 125:149–155, 2002.

[175] M. García, M. Aleixandre, J. Gutiérrez, and M. Horrillo. Electronic nose for wine discrimination. *Sensors and Actuators B: Chemical*, 113(2):911–916, February 2006.

[176] J. Gardner and P. Bartlett. A brief history of electronic noses. *Sensors and Actuators B: Chemical*, 18(1–3):210–211, March 1994.

[177] G. G. Gavetti and D. D. Levinthal. Looking forward and looking backward: Cognitive and experiential search. *Administrative Science Quarterly*, 45(1):113–137, 2000.

[178] D. Geiger and D. Heckerman. Knowledge representation and inference in similarity networks and Bayesian multinets. *Artificial Intelligence*, 82(1–2):45–74, 1996.

[179] D. Geiger and D. Heckerman. A characterization of Dirichlet distributions through local and global independence. *Annals of Statistics*, 25:1344–1368, 1997.

[180] J. Gemela. Financial analysis using Bayesian networks. *Applied Stochastic Models in Business and Industry*, 17:57–67, 2001.

[181] Geological Survey of India. Total intensity aeromagnetic map and map showing the magnetic zones of the Aravalli region, southern Rajasthan and northwestern Gujarat, India, 1981. Hyderabad, India, 4 sheets (1:250 000).

[182] Géoméditerranée. Analyse de l'évolution de l'occupation du sol, 2003. Étude réalisée pour le Conseil Régional de Provence-Alpes-Côte d'Azur, Sophia Antipolis.

[183] L. Getoor, J. T. Rhee, D. Koller, and P. Small. Understanding tuberculosis epidemiology using structured statistical models. *Artificial Intelligence in Medicine*, 30:233–256, 2004.

[184] J. Ghosh and A. Nag. An overview of radial basis functions networks. In *Radial Basis Function Neural Network Theory and Applications*, R. J. Howlerv and L. C. Jain, editors, Physica-Verlag, Heidelberg, 2000.

[185] C. Glymour, R. Scheines, P. Spirtes, and K. Kelly. *Discovering Causal Structure: Artificial Intelligence, Philosophy of Science, and Statistical Modeling*. Academic Press, San Diego, CA, 1987.

[186] D. Goleman. *Vital Lies, Simple Truths: The Psychology of Self-Deception.* Simon & Schuster, New York, 1985.

[187] W. Goodfellow. Attributes of modern and ancient sediment-hosted, seafloor hydrothermal deposits. In *Proceedings of an International Workshop on Sediment-hosted Lead-Zinc Deposits in the Northwestern Indian Shield, New Delhi and Udaipur, India*, pages 1–35, UNESCO, New Dehli, 2001.

[188] P. Graf and D. L. Schacter. Implicit and explicit memory for new associations in normal and amnesic subjects. *Journal of Experimental Psychology: Learning, Memory, and Cognition*, 11:501–518, 1985.

[189] A. Graves and M. Lalmas. Video retrieval using an MPEG-7 based inference network. In *Proceedings of the 25th ACM–SIGIR conference*, Tampere, Finland, M. Bealieu et al. editors, pages 339–346, 2002.

[190] A. Grêt-Regamey and D. Straub. Spatially explicit avalanche risk assessment linking Bayesian networks to a GIS. *Natural Hazards and Earth System Sciences*, 6:911–926, 2005.

[191] V. Grimm and C. Wissell. The intrinsic mean time to extinction: a unifying approach to analyzing persistence and viability of populations. *Oikos*, 105:501–511, 2004.

[192] R. Gulati and J. Bechta-Dugan. A modular approach for analyzing static and dynamic fault trees. In *Proceedings of the Annual Reliability and Maintainability Symposium (RAMS 1997)*, Philadelphia, pages 1–7, 1997.

[193] K. L. Gunderson, F. J. Steemers, G. Lee, and L. G. Mendoza. A genome-wide scalable SNP genotyping assay using microarray technology. *Nature Genetics*, 37:549–554, 2005.

[194] S. Gupta, Y. Arora, R. Mathur, Iqballuddin, B. Prasad, T. Sahai, and S. Sharma. Lithostratigraphic map of Aravalli region, 1995. 1:250 000 map, Geological Survey of India, Calcutta, India, 4 sheets.

[195] S. Gupta, Y. Arora, R. Mathur, Iqballuddin, B. Prasad, T. Sahai, and S. Sharma. Structural map of the Precambrian of Aravalli region, 1995. 1:250 000 map, Geological Survey of India, Calcutta, India, 4 sheets.

[196] S. Gupta, Y. Arora, R. Mathur, Iqballuddin, B. Prasad, T. Sahai, and S. Sharma. The Precambrian geology of the Aravalli Region. *Memoir Geological Survey of India*, 123, 1997.

[197] S. Haldar. Grade-tonnage model for lead-zinc deposits of Rajasthan, India. In *International Workshop on Sediment-hosted Lead-Zinc Deposits in the Northwestern Indian Shield*, pages 153–160. New Delhi and Udaipur, India, 2001.

[198] D. J. Hand, H. Mannila, and P. Smyth. *Principles of Data Mining*. MIT Press, Cambridge, MA, 2001.

[199] T. Hastie, R. Tibshirani, and J. Friedman. *The Elements of Statistical Learning, Data Mining, Inference, and Prediction*. Springer-Verlag, New York, 2001.

[200] S. Haykin. *Neural Networks: A Comprehensive Foundation*. Macmillan College Publishing Company, New York, 1994.

[201] D. Heckerman. Probabilistic similarity networks. *Networks*, 20:607–636, 1990.

[202] D. Heckerman. Bayesian networks for knowledge discovery. In *Advances in Knowledge Discovery and Data Mining*, pages 153–180. MIT Press, Cambridge, MA, 1996.

[203] D. Heckerman. Bayesian networks for data mining. *Data Mining and Knowledge Discovery*, 1:79–119, 1997.

[204] D. Heckerman. A tutorial on learning with Bayesian networks. In M. Jordan, editor, *Learning in Graphical Models*, pages 301–354. MIT Press, Cambridge, MA, 1998.

[205] D. Heckerman, J. S. Breese, and K. Rommelse. Decision-theoretic troubleshooting. *Communications of the ACM*, 38(3):49–56, 1995. Special issue on real-world applications on Bayesian networks.

[206] D. Heckerman, D. Geiger, and D. M. Chickering. Learning Bayesian networks: The combinations of knowledge and statistical data. *Machine Learning*, 20:197–243, 1995.

[207] D. E. Heckerman, E. J. Horvitz, and B. N. Nathwani. Toward normative expert systems: Part I. The Pathfinder project. *Methods of Information in Medicine*, 31:90–105, 1992.

[208] B. Hedberg. How organizations learn and unlearn. In P. Nystrom and W. Starbuck, editors, *Handbook of Organizational Design*, pages 1–27. Oxford University Press, 1981.

[209] W. Heisenberg. Über den anschaulichen Inhalt der quantentheoretischen Kinematik und Mechanik. *Zeitschrift für Physik*, 43:172–198, 1927.

[210] M. Henrion. Some practical issues in constructing belief networks. In L. Kanal, T. Levitt, and J. Lemmer, editors, *Uncertainty in Artificial Intelligence 3*, pages 161–173. Elsevier Science Publishers, Amsterdam, 1989.

[211] A. Heron. The geology of central Rajputana. *Memoir Geological Survey of India*, 79(1), 1953.

[212] F. Hirsch. *Social Limits to Growth*. Routledge & Kegan Paul, London, 1977.

[213] J. Hoh and J. Ott. Mathematical multi-locus approaches to localizing complex human trait genes. *Nature Reviews Genetics*, 4:701–709, 2003.

[214] K. E. Holbert, A. S. Heger, and N. K. Alang-Rashid. Redundant sensor validation by using fuzzy logic. *Nuclear Science and Engineering*, 118(1):54–64, 1994.

[215] E. E. Holmes. Beyond theory to application and evaluation: diffusion approximations for population viability analysis. *Ecological Applications*, 14:1272–1293, 2004.

[216] A. Holst. The use of Bayesian neural network model for classification tasks. PhD thesis, Department of Numerical Analysis and Computer Science, Royal Institute of Technology, Stockholm, Sweden, 1997.

[217] L. Hope and K. Korb. A Bayesian metric for evaluating machine learning algorithms. In G. Webb and X. Yu, editors, *Proceedings of the 17th Australian Joint Conference on Advances in Artificial Intelligence (AI'04)*, volume 3229 of *LNCS/LNAI Series*, pages 991–997. Springer-Verlag, Berlin, Germany, 2004.

[218] L. Hope, A. Nicholson, and K. Korb. TakeheartII: A tool to support clinical cardiovascular risk assessment. Technical Report TR 2006/209, Clayton School of Information Technology, Monash University, 2006.

[219] L. R. Hope. Information Measures for Causes Explanation and Prediction. PhD thesis, Clayton School of Information Technology, Monash University, 2007.

[220] R. A. Howard. Decision analysis: Practice and promise. *Management Science*, 34:679–695, 1988.

[221] C. Hsu, C. Chang, and C. Lin. A practical guide to support vector classification. Technical report, Department of Computer Science and Information Engineering, National Taiwan University, 2003.

[222] D. N. Humphreys and L. E. Katz. Five year field study of the water quality effects of tire shreds placed above the water table. In *79th Annual Meeting of the Transportation Research Board, No. 00-0892*, TRB, National Research Council, Washington DC., 2000.

[223] P. H. Ibargüengoytia, S. Vadera, and L. Sucar. A probabilistic model for information and sensor validation. *Computer Journal*, 49(1):113–126, January 2006.

[224] http://www.imapdata.com.

[225] INSEE. Communes profils – recensement de la population 1999, 2003. INSEE Paris, CD-ROM.

[226] INSEE PACA. La polarisation de l'espace régional par l'emploi, 1998. Sud INSEE dossier 3 INSEE Provence-Alpes-Côte d'Azur-Marseille.

[227] Institut Géographique National. BD carto, 2001. édition 2, IGN Paris, CD-ROM.

[228] International Hapmap Consortium. A haplotype map of the human genome. *Nature*, 429:1300–1320, 2005.

[229] G. Jackson, S. Jones, G. Booth, C. Champod, and I. W. Evett. The nature of forensic science opinion – a possible framework to guide thinking and practice in investigations and in court proceedings. *Science and Justice*, 46:33–44, 2006.

[230] L. Janer-García. Transformada wavelet aplicada a la extracción de información en señales de voz. PhD thesis, Universidad Politécnica de Cataluña, 1998.

[231] C. R. Jeffery. *Crime Prevention through Environmental Design*. Sage Publications, Beverly Hills, 1971.

[232] F. Jensen. *Bayesian Networks and Decision Graphs*. Springer-Verlag, New York, 2001.

[233] F. Jensen, U. B. Kjærulff, M. Lang, and A. L. Madsen. HUGIN – the tool for Bayesian networks and influence diagrams. In *First European Workshop on Probabilistic Graphical Models*, Cuenca, Spain, J. A. Gámez and A. Salmerón, editors, pages 212–221. 2002.

[234] N. P. Jewell. *Statistics for Epidemiology*. CRC/Chapman and Hall, Boca Raton, 2003.

[235] Joint Pub 3-07.2. Joint tactics, techniques, and procedures for antiterrorism. DTIC, March 1998.

[236] R. V. Jones. The theory of practical joking – an elaboration. *Journal of the Institute of Mathematics and its Applications*, January/February: 11(2)10–17, 1975.

404   BIBLIOGRAPHY

[237] J. B. Kadane and D. A. Schum. *A Probabilistic Analysis of the Sacco and Vanzetti Evidence*. John Wiley & Sons Inc., New York, 1996.

[238] J. Kalagnanam and M. Henrion. A comparison of decision analysis and expert rules for sequential analysis. In *Uncertainty in Artificial Intelligence 4*, pages 271–281. North-Holland, New York, 1990.

[239] J. Kalwa and A. L. Madsen. Sonar image quality assessment for an autonomous underwater vehicle. In *Proceedings of the 10th International Symposium on Robotics and Applications*, Seville, Spain, M. Jarnshidi, editor, TSI Press Albuquerque, 2004.

[240] W. B. Kannel, J. D. Neaton, D. Wentworth, H. E. Thomas, J. Stamler, and S. B. Hulley. Overall and coronary heart disease mortality rates in relation to major risk factors in 325,348 men screened for the MRFIT. *American Heart Journal*, 112:825–836, 1986.

[241] A. Kapoor and E. Horvitz. Principles of lifelong learning for predictive user modeling. In *Proceedings of the Eleventh Conference on User Modeling*, Corfu, Greece, C. Conati, K. McCoy, G. Palioras, editors, Springer-Verlag, Berlin, June 2007.

[242] R. E. Kass and A. Raftery. Bayes factors. *Journal of the American Statistical Association*, 90:773–795, 1995.

[243] S. Katz. Emulating the Prospector expert system with a raster GIS. *Computers and Geosciences*, 17:1033–1050, 1991.

[244] R. Keeney and D. von Winterfeldt. Eliciting probabilities from experts in complex technical problems. *IEEE Transactions on Engineering Management*, 3:191–201, 1991.

[245] L. Kemp, G. Bonham-Carter, G. Raines, and C. Looney. *Arc-SDM: ArcView extension for spatial data modeling using weights of evidence, logistic regression, fuzzy logic and neural network analysis*, www.ige.unicamp.br/sdm/ 2001.

[246] D. E. Kieras and S. Bovair. The role of a mental model in learning to operate a device. *Cognitive Science*, 8:255–273, 1984.

[247] J. H. Kim and J. Pearl. CONVINCE: A conversational inference consolidation engine. *IEEE Transactions on Systems, Man, and Cybernetics*, 17(2):120–132, 1983.

[248] U. B. Kjærulff and A. L. Madsen. A methodology for acquiring qualitative knowledge for probabilistic graphical models. In *Proceedings of the*

Fifteenth International Conference on Information Processing and Management of Uncertainty in Knowledge-Based Systems (IPMU), Perugia, Italy, pages 143–150. 2004.

[249] U. B. Kjærulff and A. L. Madsen. Probabilistic networks for practitioners – a guide to construction and analysis of Bayesian networks and influence diagrams. Unpublished manuscript, 2006.

[250] M. Knuiman. Personal communication, 2005.

[251] M. Knuiman and H. Vu. Multivariate risk estimation for coronary heart disease: the Busselton health study. *Australian and New Zealand Journal of Public Health*, 22:747–753, 1998.

[252] D. Koller. Probabilistic relational models. In *9th International Workshop, Inductive Logic Programming*, Bled, Slovenia, S. Dzeroski and P. Flach, editors, Springer-Verlag, Berlin, pages 3–13. 1999.

[253] D. Koller and A. Pfeffer. Object-oriented Bayesian networks. In *Uncertainty in Artificial Intelligence: Proceedings of the Thirteenth Conference*, pages 302–313. Morgan Kaufmann. San Francisco, CA, 1997.

[254] I. Kononenko. Semi-naive Bayesian classifier. In Y. Kodratoff, editor, *Proceedings of the European Working Session on Learning : Machine Learning (EWSL-91)*, volume 482 of *LNAI*, pages 206–219. Springer-Verlag, Porto, Portugal, March 1991.

[255] K. B. Korb, L. R. Hope, A. E. Nicholson, and K. Axnick. Varieties of causal intervention. In *PRICAI'04 – Proceedings of the 8th Pacific Rim International Conference on Artificial Intelligence*, pages 322–331, Auckland, New Zealand, 2004.

[256] K. B. Korb and A. E. Nicholson. *Bayesian Artificial Intelligence*. Computer Science and Data Analysis. Chapman & Hall / CRC, Boca Raton, 2004.

[257] M. Kubat and S. Matwin. Addressing the curse of imbalanced training sets: One-sided selection. In *Proceedings of the Fourteenth International Conference on Machine Learning, ICML'97*, Nashville, TN, pages 179–186. D. H. Fisher, Morgan Kaufmann, 1997.

[258] C. Lacave and F. J. Díez. A review of explanation methods for Bayesian networks. *Knowledge Engineering Review*, 17(2):107–127, 2002.

[259] C. Lacave and F. J. Díez. A review of explanation methods for heuristic expert systems. *Knowledge Engineering Review*, 19:133–146, 2004.

[260] C. Lacave, M. Luque, and F. J. Díez. Explanation of Bayesian networks and influence diagrams in Elvira. Submitted to *IEEE Transactions on Systems, Man and Cybernetics–Part B: Cybernetics*, 19:730–738, 2006.

[261] C. Lacave, A. Oniśko, and F. J. Díez. Use of Elvira's explanation facility for debugging probabilistic expert systems. *Knowledge-Based Systems*, 19:730–738, 2006.

[262] M. Lalmas and I. Ruthven. Representing and retrieving structured documents with Dempster-Shafer's theory of evidence: Modelling and evaluation. *Journal of Documentation*, 54(5):529–565, 1998.

[263] E. S. Lander and N. J. Schork. Genetic dissection of complex traits. *Science*, 265:2037–2048, 1994.

[264] P. Langley, W. Iba, and K. Thompson. An analysis of Bayesian classifiers. In *Proceedings of the National Conference on Artificial Intelligence (AAAI '92)*, San Jose, CA, W. R. Swartout, editor, pages 223–228. 1992.

[265] P. Langley and S. Sage. Induction of selective Bayesian classifiers. In *Proceedings of the Tenth Conference on Uncertainty in Artificial Intelligence (UAI '94)*, Seattle, WA, R. L. de Mántaras and D. Poole, pages 399–406. Morgan Kaufmann, San Francisco, CA, 1994.

[266] H. Langseth and L. Portinale. Bayesian networks in reliability. *Reliability Engineering and System Safety*, 92(1):92–108, 2007.

[267] Los Angeles Police Department website. http://www.lapdonline.org/emergency_services_division/content_basic_view/33044.

[268] K. B. Laskey. MEBN: A first-order Bayesian logic. *Artificial Intelligence*, 172(1):1–68, January 2008.

[269] K. B. Laskey and P. Costa. Of starships and klingons: Bayesian logic for the 21st century. In *Uncertainty in Artificial Intelligence: Proceedings of the 21st Conference*. Edinburgh, AUAI Press, 2005.

[270] K. B. Laskey and S. M. Mahoney. Network fragments: Representing knowledge for constructing probabilistic models. In *Uncertainty in Artificial Intelligence: Proceedings of the Thirteenth Conference*, Morgan Kaufmann, San Francisco, CA, 1997.

[271] S. L. Lauritzen. *Graphical Models*. Oxford University Press, Oxford, 1996.

[272] S. L. Lauritzen and D. Nilsson. Representing and solving decision problems with limited information. *Management Science*, 47:1238–1251, 2001.

[273] S. L. Lauritzen and D. J. Spiegelhalter. Local computations with probabilities on graphical structures and their application to expert systems (with discussion). *Journal of the Royal Statistic Society*, 50:157–224, 1988.

[274] K. LeBeau and S. Wadia-Fascetti. A fault tree model of bridge deterioration. In *Eighth ASCE Speciality Conference on Probabilistic Mechanics and Structural Reliability*, Notre Dame, 2000.

[275] C. Lee and K. Lee. Application of Bayesian network to the probabilistic risk assessment of nuclear waste disposal. *Reliability Engineering and System Safety*, 91(5):515–532, 2006.

[276] D. C. Lee and B. E. Rieman. Population viability assessment of salmonids by using probabilistic networks. *North American Journal of Fish Management*, 17:1144–1157, 1997.

[277] T.-Y. Leong. Multiple perspective dynamic decision making. *Artificial Intelligence*, 105(1-2):209–261, 1998.

[278] D. Leung and J. Romagnoli. Dynamic probabilistic model-based expert system for fault diagnosis. *Computers and Chemical Engineering*, 24:2473–2492, 2000.

[279] D. Leung and J. Romagnoli. An integration mechanism for multivariate knowledge-based fault diagnosis. *Journal of Process Control*, 12:15–26, 2002.

[280] B. Levitt and J. G. March. Organizational learning. *Annual Review of Sociology*, 14:319–340, 1998.

[281] T. S. Levitt and K. B. Blackmond Laskey. Computational inference for evidential reasoning in support of judicial proof. *Cardozo Law Review*, 22:1691–1731, 2001.

[282] P. Lewicki. *Non Conscious Social Information Processing*. Academic Press, New York, 1986.

[283] D. Liew and S. Rogers. Personal communication, September 2004.

[284] Y. J. Lin, H. R. Guo, Y. H. Chang, M. T. Kao, H. H. Wang, and R. I. Hong. Application of the electronic nose for uremia diagnosis. *Sensors and Actuators B: Chemical*, 76(1-3):177–180, June 2001.

[285] P. J. F. Lucas, H. Boot, and B. G. Taal. Computer-based decision-support in the management of primary gastric non-Hodgkin lymphoma. *Methods of Information in Medicine*, 37:206–219, 1998.

[286] P. J. F. Lucas, N. de Bruijn, K. Schurink, and A. Hoepelman. A probabilistic and decision-theoretic approach to the management of infectious disease at the ICU. *Artificial Intelligence in Medicine*, 19(3):251–279, 2000.

[287] P. J. F. Lucas, R. Segaar, and A. Janssens. HEPAR: An expert system for diagnosis of disorders of the liver and biliary tract. *Liver*, 9:266–275, 1989.

[288] P. J. F. Lucas, L. van der Gaag, and A. Abu-Hanna. Bayesian networks in biomedicine and health-care. *Artificial Intelligence in Medicine*, 30:201–214, 2004.

[289] D. Ludwig. Is it meaningful to estimate a probability of extinction? *Ecology*, 80:298–310, 1999.

[290] R. Luk, H. Leong, T. Dillon, A. Chan, W. Croft, and J. Allan. A survey in indexing and searching XML documents. *Journal of the American Society for Information Science and Technology*, 53(6):415–437, 2002.

[291] K. L. Lunetta, L. B. Hayward, J. Segal, and P. V. Eerdewegh. Screening large-scale association study data: Exploiting interactions using random forests. *BMC Genetics*, 5:32, 2004.

[292] A. Madsen, M. Lang, U. B. Kjærulff, and F. Jensen. The Hugin tool for learning Bayesian networks. In *Proceedings of the Seventh European Conference on Symbolic and Quantitative Approaches to Reasoning and Uncertainty*, Bonn, W. Prinz, et al., editors, pages 594–605, Springer-Verlag, Berlin, 2003.

[293] A. L. Madsen and U. B. Kjærulff. *Applications of HUGIN to Diagnosis and Control of Autonomous Vehicles*. Studies in Fuzziness and Soft Computing. Springer, Berlin, 2007.

[294] S. M. Mahoney and K. B. Laskey. Network engineering for agile belief network models. *IEEE Transactions in Knowledge and Data Engineering*, 12, 2000.

[295] R. Manian, D. Coppit, K. Sullivan, and J. Dugan. Bridging the gap between systems and dynamic fault tree models. In *Proceedings IEEE Annual Reliability and Maintainability Symposium*, Washington DC, pages 105–111. IEEE, 1999.

[296] J. March. Exploration and exploitation in organizational learning. *Organization Science*, 2(1):71–87, 1991.

[297] J. G. March and R. M. Cyert. *A Behavioral Theory of the Firm*. Prentice Hall, Englewood Cliffs, NJ, 1963.

[298] J. G. March and J. P. Olsen. *Ambiguity and Choice in Organizations*. Universitets Forlaget, Bergen, 1976.

[299] J. Marchini, P. Donnelly, and L. R. Cardon. Genome-wide strategies for detecting multiple loci that influence complex diseases. *Nature Genetics*, 37:413–417, 2005.

[300] B. G. Marcot. Characterizing species at risk I: modeling rare species under the Northwest Forest Plan. *Ecology and Society*, 11(2):10, 2006.

[301] B. G. Marcot, P. A. Hohenlohe, S. Morey, R. Holmes, R. Molina, M. C. Turley, M. H. Huff, and J. A. Laurence. Characterizing species at risk II: Using Bayesian belief networks as decision support tools to determine species conservation categories under the Northwest Forest Plan. *Ecology and Society*, 11(2), 2006.

[302] B. G. Marcot, R. S. Hothausen, M. G. Raphael, M. M. Rowland, and M. J. Wisdom. Using Bayesian belief networks to evaluate fish and wildlife population viability under land management alternatives from an environmental impact statement. *Forest Ecology and Management*, 153:29–42, 2001.

[303] B. G. Marcot, J. D. Steventon, G. D. Sutherland, and R. K. McCann. Guidelines for developing and updating Bayesian belief networks applied to ecological modeling and conservation. *Canadian Journal of Forest Research*, 36:3063–3074, 2006.

[304] E. R. Martin, M. D. Ritchie, L. Hahn, S. Kang, and J. H. Moore. A novel method to identify gene-gene effects in nuclear families: The MDR-PDT. *Genetic Epidemiology*, 30:111–123, 2006.

[305] J. Martin and K. Van Lehn. Student assessment using Bayesian nets. *International Journal of Human-Computer Studies*, 42:575–591, 1995.

[306] Maryland Emergency Management Agency (MEMA). Critical asset & portfolio risk analysis (CAPRA), University of Maryland, MD, 2005.

[307] T. Masters. *Practical Neural Network Recipes in C++*. Academic Press, New York, 1993.

[308] D. L. Mattern, L. C. Jaw, T.-H. Guo, R. Graham, and W. McCoy. Using neural networks for sensor validation. In *Proceedings of the 34th Joint Propulsion Conference*, Cleveland, Ohio, July 1998. AIAA, ASME, SAE, and ASEE.

[309] M. Mayo and A. Mitrovic. Optimising ITS behaviour with Bayesian networks and decision theory. *International Journal of Artificial Intelligence in Education*, 12(2):124–153, 2001.

[310] J. McCarthy, M. L. Minsky, N. Rochester, and C. E. Shannon. A proposal for the Dartmouth Summer Research Project on Artificial Intelligence.

Report (manuscript), Massachusetts Institute of Technology, A.I. Lab., Cambridge, MA, Aug. 1955.

[311] R. S. McNay, B. G. Marcot, V. Brumovsky, and R. Ellis. A Bayesian approach to evaluating habitat for woodland caribou in north-central British Columbia. *Canadian Journal of Forest Research*, 36:3117–3133, 2006.

[312] J. J. McNeil, A. Peeters, D. Liew, S. Lim, and T. Vos. A model for predicting the future incidence of coronary heart disease within percentiles of coronary heart disease risk. *Journal of Cardiovascular Risk*, 8:31–37, 2001.

[313] D. Meadows. *Indicators and Information Systems for Sustainable Development*. The Sustainability Institute, Hartland Four Corners, VT, 1998.

[314] D. Meidinger and J. Pojar. Ecosystems of British Columbia, 1991. B.C. Ministry of Forests, Special Report Series No. 6. Victoria, BC.

[315] L. Meshkat and J. Dugan. Dependability analysis of systems with on-demand and active failure modes using Dynamic Fault Trees. *IEEE Transactions on Reliability*, 51(2):240–251, 2002.

[316] H. Mhaskar and M. Michelli. Approximation by superposition of sigmoidal and radial basis functions. *Advances in Applied Mathematics*, 13:350–373, 1992.

[317] D. Michie, J. Spiegelhalter, and C. Taylor. *Machine Learning, Neural and Statistical Classification*. Prentice Hall, Upper Saddle River, New Jersey, 1994.

[318] B. Middleton, M. Shwe, D. Heckerman, M. Henrion, E. Horvitz, H. Lehmann, and G. Cooper. Probabilistic diagnosis using a reformulation of the INTERNIST–1/QMR knowledge base: II. Evaluation of diagnostic performance. *Methods of Information in Medicine*, 30(4):256–267, 1991.

[319] M. L. Minsky. The age of intelligent machines: thoughts about Artificial Intelligence. http://www.kurzweilai.net/meme/frame.html?main=/articles/art0100.html.

[320] S. Montani, L. Portinale, and A. Bobbio. Dynamic Bayesian networks for modeling advanced fault tree features in dependability analysis. In K. Kolowrocki, editor, *Proceedings of the European Safety and Reliability Conference (ESREL) 2005, Tri City, June 2005*, pages 1414–1422. Balkema Publishers, Leiden, 2005.

[321] S. Montani, L. Portinale, A. Bobbio, and D. Codetta-Raiteri. RADYBAN: a tool for reliability analysis of dynamic fault trees through conversion into dynamic Bayesian networks. *Reliability Engineering and System Safety*, 2008. [Electronic publication ahead of print].

[322] S. Montani, L. Portinale, A. Bobbio, M. Varesio, and D. Codetta-Raiteri. A tool for automatically translating Dynamic Fault Trees into Dynamic Bayesian Networks. In *Proceedings of the Annual Reliability and Maintainability Symposium RAMS2006*, Newport Beach, USA, pages 434–441, 2006.

[323] A. Moore and M. Lee. Efficient algorithms for minimizing cross validation error. In *Proceedings of the 11th International Conference on Machine Learning*. Morgan Kaufmann, San Francisco, 1994.

[324] J. H. Moore, J. C. Gilbert, C. T. Tsai, F. T. Chiang, T. Holden, N. Barney, and B. C. White. A flexible computational framework for detecting, characterizing, and interpreting statistical patterns of epistasis in genetic studies of human disease susceptibility. *Journal of Theoretical Biology*, 21:252–261, 2006.

[325] M. G. Morgan and M. Henrion. *Uncertainty: A Guide to Dealing with Uncertainty in Quantitative Risk and Policy Analysis*. Cambridge University Press, New York, 1990.

[326] J. Mortera, A. P. Dawid, and S. L. Lauritzen. Probabilistic expert systems for DNA mixture profiling. *Theoretical Population Biology*, 63:191–205, 2003.

[327] K. Murphy. Dynamic Bayesian networks: representation, inference and learning. PhD thesis, University of California, Berkeley, 2002.

[328] R. Murray, R. Janke, and J. Uber. The threat ensemble vulnerability assessment program for drinking water distribution system security. In *Proceedings of the ASCE/EWRICongress*, Salt Lake City, 2004.

[329] W. Murray. A practical approach to Bayesian student modeling. In B. Goettl, H. Halff, C. Redfield, and V. Shute, editors, *Proceedings of the 4th International Conference of Intelligent Tutoring Systems*, San Antonio, TX, pages 424–433. Springer-Verlag, London, 1998.

[330] R. L. Nagel. Pleiotropic and epistatic effects in sickle cell anemia. *Current Opinion in Hematology*, 8:105–110, 2001.

[331] P. Naïm, P.-H. Wuillemin, P. Leray, O. Pourret, and A. Becker. Réseaux bayésiens. *Eyrolles*, 2007.

[332] National Domestic Preparedness Coalition. HLS_CAM: Homeland security comprehensive assessment model. http://www.ndpci.us/hls_cam.html.

[333] National Institute of Standards and Technology. NIST e-handbook. http://www.itl.nist.gov/div989/handbook, 2007.

[334] D. Nauck, F. Klawonn, and R. Kruse. *Foundations of Neuro-Fuzzy Systems*. John Wiley & Sons, Ltd., Chichester, 1997.

[335] Naval Facilities Engineering Command. Mission dependency index. http: //www.nfesc.navy.mil/shore/mdi/index.html, 2004.

[336] R. Neapolitan. *Learning Bayesian Networks*. Prentice Hall, Upper Saddle River, NJ, 2004.

[337] P. Newman and J. Kennworthy. *Sustainability and Cities: Overcoming Automobile Dependence*. Island Press, Washington DC, 1998.

[338] A. C. Newton, E. Marshall, K. Schreckenberg, D. Golicher, D. W. te Velde, F. Edouard, and E. Arancibia. Bayesian belief network to predict the impacts of commercializing non-timber forest products on livelihoods. *Ecology and Society*, 11(2):24, 2006.

[339] R. E. Nisbett and L. Ross. *Human Inferences: Strategies and Shortcomings of Social Judgment*. Prentice Hall, Englewood Cliffs, NJ, 1980.

[340] R. E. Nisbett and T. D. Wilson. Telling more than we know: Verbal reports on mental processes. *Psychological Review*, 84:231–259, 1977.

[341] J. Noguez and L. Sucar. Intelligent virtual laboratory and project oriented learning for teaching mobile robotics. *International Journal of Engineering Education*, 22(4):743–757, 2006.

[342] http://www.norsys.com.

[343] J. B. Nyberg, B. G. Marcot, and R. Sulyma. Using Bayesian belief networks in adaptive management. *Canadian Journal of Forest Research*, 36:3104–3116, 2006.

[344] http://www.nykredit.dk.

[345] G. Oatley and B. Ewart. Crime analysis software: 'pins in maps', clustering and Bayes net prediction. *Expert Systems with Applications*, 25:569–588, 2003.

[346] R. O'Donnell, A. E. Nicholson, B. Han, K. B. Korb, M. J. Alam, and L. Hope. Causal discovery with prior information. In A. Sattar and B. H. Kang, editors, *AI 2006: Advances in Artificial Intelligence, (Proceedings of the 19th Australian Joint Conference on Advances in Artificial Intelligence [AI'06], Hobart, Australia, 4–8 December 2006)*, LNAI Series, pages 1162–1167. Springer-Verlag, Berlin, 2006.

[347] Office for Domestic Preparedness. Texas domestic preparedness assessment, 2002.

[348] A. Oniśko, M. Druzdzel, and H. Wasyluk. A probabilistic model for diagnosis of liver disorders. In *Proceedings of the Seventh Symposium on Intelligent Information Systems (IIS-98)*, Bystra, Poland, M. Klopstock, M. Michalewicz, S. T. Wierzchoń, editors, pages 379–387. Physica-Verlag Heidelberg, 1998.

[349] A. Oniśko and M. J. Druzdzel. Effect of imprecision in probabilities on Bayesian network models: An empirical study. In *Working Notes of the European Conference on Artificial Intelligence in Medicine (AIME-03): Qualitative and Model-based Reasoning in Biomedicine*, Protaras, Cyprus, October 2003.

[350] A. Oniśko, M. J. Druzdzel, and H. Wasyluk. Extension of the Hepar II model to multiple-disorder diagnosis. In S. W. M. Kłopotek and M. Michalewicz, editors, *Intelligent Information Systems,* Advances in Soft Computing *Series*, pages 303–313. Physica-Verlag, Heidelberg, 2000.

[351] A. Oniśko, M. J. Druzdzel, and H. Wasyluk. Learning Bayesian network parameters from small data sets: Application of Noisy-OR gates. *International Journal of Approximate Reasoning*, 27(2):165–182, 2001.

[352] A. Oniśko, M. J. Druzdzel, and H. Wasyluk. An experimental comparison of methods for handling incomplete data in learning parameters of Bayesian networks. In M. Kłopotek, M. Michalewicz, and S. T. Wierzchoń, editors, *Intelligent Information Systems,* Advances in Soft Computing *Series*, Physica-Verlag, Heidelberg, pp. 351–360, 2002.

[353] A. Oniśko, P. Lucas, and M. J. Druzdzel. Comparison of rule-based and Bayesian network approaches in medical diagnostic systems. In S.A.S. Quaglini, P. Barahona, S. Andreasson, editors, *Artificial Intelligence in Medicine, Berlin,* Lecture Notes in Computer Science Subseries, pages 281–292. Springer, Berlin, 2001.

[354] D. Pan. Digital audio compression. *Digital Technical Journal*, 5(2):28–33 1993.

[355] A. Pansini and K. Smalling. *Guide to Electric Power Generation*. Marcel Dekker, New York, 2nd. edition, 2002.

[356] J. Pearl. *Probabilistic Reasoning in Intelligent Systems: Networks of Plausible Inference*. Morgan Kaufmann, San Mateo, CA, 1988.

[357] J. Pearl. Causal diagrams for empirical research. *Biometrika*, 82:669–710, 1995.

[358] J. Pearl and S. Russell. Bayesian networks. Technical Report R-277, University of California, Los Angeles, 2000.

[359] R. Peimer. Target analysis, November 2006. Government Technology's Emergency Management.

[360] M. Peot and R. Shachter. Learning from what you don't observe. In *Proceedings of the Fourteenth Annual Conference on Uncertainty in Artificial Intelligence (UAI–98)*, pages 439–446. Morgan Kaufmann, San Francisco, CA, 1998.

[361] T. T. Perls. The different paths to age one hundred. *Annals of the New York Academy of Science*, 1055:13–25, 2005.

[362] T. T. Perls, L. Kunkel, and A. Puca. The genetics of aging. *Current Opinion in Genetics and Development*, 12:362–369, 2002.

[363] T. T. Perls, M. Shea-Drinkwater, J. Bowen-Flynn, S. B. Ridge, S. Kang, E. Joyce, M. Daly, S. J. Brewster, L. Kunkel, and A. A. Puca. Exceptional familial clustering for extreme longevity in humans. *Journal of the American Geriatrics Society*, 48:1483–1485, 2000.

[364] T. T. Perls and D. Terry. Understanding the determinants of exceptional longevity. *Annals of Internal Medicine*, 139(5):445–449, 2003.

[365] F. Pettijohn. *Sedimentary Rocks*, 3rd edition, Harper-Row, New York, 1975.

[366] A. Pike. Modelling water quality violations with Bayesian networks. PhD thesis, Pennsylvania State University, Pittsburgh, USA, 2001.

[367] P. Pileri. *Interpretare l'ambiente – Gli indicatori di sostenibilità per il governo del territorio*. Alinea Editrice, Firenze, Italy, 2002.

[368] B. Piwowarski and P. Gallinari. A Bayesian network model for page retrieval in a hierarchically structured collection. In *Proceedings of the XML Workshop of the 25th ACM–SIGIR Conference*, Tampere, Finland, M. Beaulieu, et al., editors, 2002.

[369] A. Porwal. Mineral potential mapping with mathematical geological models. PhD thesis, University of Utrecht, Netherlands, 2006.

[370] A. Porwal, E. Carranza, and M. Hale. Extended weights-of-evidence modeling for predictive mapping of base metal deposit potential, Aravalli Province, India. *Exploration and Mining Geology Journal*, 10(4):273–287, 2001.

[371] A. Porwal, E. Carranza, and M. Hale. Bayesian network classifiers. *Computers and Geosciences*, 32(1):1–16, 2006.

[372] A. Porwal, E. J. M. Carranza, and M. Hale. Tectonostratigraphy and base-metal mineralization controls, Aravalli province (western India): new interpretations from geophysical data. *Ore Geology Reviews*, 29(3–4):287–396, 2006.

[373] M. Pradhan, M. Henrion, G. Provan, B. del Favero, and K. Huang. The sensitivity of belief networks to imprecise probabilities: An experimental investigation. *Artificial Intelligence*, 85(1–2):363–397, Aug. 1996.

[374] J. K. Pritchard and N. J. Cox. The allelic architecture of human disease genes: Common disease-common variant. Or not? *Human Molecular Genetics*, 11:2417–2422, 2002.

[375] http://www.promedas.nl.

[376] J. R. Quinlan. *C4.5: Programs for Machine Learning*. Morgan Kaufmann, San Francisco, CA, 1993.

[377] R Development Core Team. *R: A Language and Environment for Statistical Computing*. Vienna, Austria, 2006.

[378] J. Ragazzo-Sánchez, P. Chalier, D. Chevalier, and C. Ghommidh. Electronic nose discrimination of aroma compounds in alcoholised solutions. *Sensors and Actuators B: Chemical*, 2005. Available online.

[379] D. C. Raiteri. The conversion of dynamic fault trees to stochastic Petri nets, as a case of graph transformation. *Electronic Notes on Theoretical Computer Science*, 127(2):45–60, 2005.

[380] M. Ramoni and P. Sebastiani. Bayesian methods. In M. Berthold and D. Hand, editors, *Intelligent Data Analysis: An Introduction*, pages 129–166. Springer, New York, 1999.

[381] M. Ramoni and P. Sebastiani. Learning conditional probabilities from incomplete data: An experimental comparison. In *Proceedings of the Seventh International Workshop on Artificial Intelligence and Statistics*, pages 260–265. Morgan Kaufmann, San Francisco, CA, 1999.

[382] M. Ramoni and P. Sebastiani. *Bayesian Knowledge Discoverer System*, 2000. Knowledge Media Institute, The Open University, UK.

[383] M. G. Raphael, M. J. Wisdom, M. M. Rowland, R. S. Holthausen, B. C. Wales, B. G. Marcot, and T. D. Rich. Status and trends of habitats of terrestrial vertebrates in relation to land management in the interior Columbia River Basin. *Forest Ecology and Management*, 153:63–87, 2001.

[384] A. Rauzy. New algorithms for fault trees analysis. *Reliability Engineering and System Safety*, 05(59):203–211, 1993.

[385] T. R. C. Read and N. A. C. Cressie. *Goodness-of-Fit Statistics for Discrete Multivariate Data*. Springer-Verlag, New York, 1988.

[386] A. S. Reber. Implicit learning and tacit knowledge: an essay on the cognitive unconscious. In Oxford psychology series, volume 19. Oxford University Press, Oxford, 1993.

[387] K. H. Reckhow. Water quality prediction and probability network models. *Canadian Journal of Fisheries and Aquatic Sciences*, 56:1150–1158, 1999.

[388] R. Reddy, G. Bonham-Carter, and A. Galley. Developing a geographic expert system for regional mapping of VMS deposit potential. *Nonrenewable Resources*, 1(2):112–124, 1992.

[389] M. Redmayne. *Expert Evidence and Criminal Justice*. Oxford University Press, Oxford, 2001.

[390] D. H. Reed, J. J. O'Grady, B. W. Brook, J. D. Ballou, and R. Frankham. Estimates of minimum viable population sizes for vertebrates and factors influencing those estimates. *Biological Conservation*, 113:23–34, 2003.

[391] S. Reese. FY2006 Homeland Security Grant Guidance Distribution Formulas: Issues for the 109th Congress. CRS Report RS22349, 9 December, 2005.

[392] Réseau Ferré de France. Contexte et enjeux de la ligne nouvelle en Provence-Alpes-Côte d'Azur, 2005. Marseille, 2 volumes.

[393] W. M. Rhodes and C. Conly. *The Criminal Commute: A Theoretical Perspective*. Wavelend Press Inc., Environmental Criminology, San Francisco USA, 1991.

[394] B. A. Ribeiro-Neto and R. R. Muntz. A belief network model for IR. In *Proceedings of the 19th International ACM–SIGIR Conference*, Zurich H.-P. Frei et al., editors, pages 253–260. ACM, 1996.

[395] O. Rioul and M. Vetterli. Wavelets and signal processing. *IEEE Signal Processing Magazine*, pages 14–38, October 1991.

[396] B. Ripley. *Pattern Recognition and Neural Networks*. Cambridge University Press, Cambridge, 1996.

[397] M. D. Ritchie, L. W. Hahn, N. Roodi, L. R. Bailey, W. D. Dupont, F. F. Parl, and J. H. Moore. Multifactor-dimensionality reduction reveals high-order interactions among estrogen-metabolism genes in sporadic breast cancer. *American Journal of Human Genetics*, 69:138–147, 2001.

[398] T. Robison. Application of risk management to the state & local communities. Los Alamos National Laboratory, 20 March 2006.

[399] V. Robles, P. Larrañaga, J. M. Peña, E. Menasalvas, M. S. Pérez, V. Herves, and A. Wasilewska. Bayesian network multi-classifiers for protein secondary structure prediction. *Artificial Intelligence in Medicine*, 31:117–136, 2004.

[400] Rockit. Roc analysis, 2004. Available from: `http://xray.bsd.uchicago.edu/cgi-bin/roc\_software.cgi` [Accessed 9 September 2004].

[401] C. Rosset. *Le réel: Traité de l'idiotie*. Editions de Minuit, Paris, 1977.

[402] C. Rosset. *Le réel et son double: Essai sur l'illusion*. Gallimard, Paris, 1985.

[403] A. Roy. Stratigraphic and tectonic frame work of the Aravalli mountain range. In A. Roy, editor, *Precambrian of the Aravalli Mountain, Rajasthan, India*. Geological Society of India, Memoir 7, pages 3–31, 1988.

[404] W. Ruddick. Social self-deceptions. In B. McLaughlin and A. Oksenberg Rorty, editors, *Perspectives on Self-Deception*, pages 380–389. University of California Press, Berkeley, 1988.

[405] T. Saaty. *The Analytic Hierarchy Process for Decisions in a Complex World*. McGraw-Hill, New York, 1980.

[406] G. Salton and M. J. McGill. *Introduction to Modern Information Retrieval*. McGraw-Hill, 1983.

[407] L. E. Sandelands and R. E. Stablein. The concept of organization mind. In S. Bacharach and N. DiTomaso, editors, *Research in the Sociology of Organizations*, volume 5, pages 135–161. JAI Press, Greenwich, CT, 1987.

[408] L. Sanders. L'analyse statistique des données en géographie. *RECLUS*, 1989.

[409] L. Sanders and F. Durand-Dastès. L'effet régional: les composantes explicatives dans l'analyse spatiale. *RECLUS*, Montpellier, 1985.

[410] Sandia National Laboratory. Risk assessment methodology for dams (RAMD). `http://ipal.sandia.gov/`.

[411] Sandia National Laboratory. Risk assessment methodology for dams (RAMD) fact sheet. `http://www.sandia.gov/ram/RAMD\%20Fact\%20Sheet.pdf`.

[412] J. Santos, J. Lozano, H. Vásquez, J. Agapito, M. Martín, and J. González. Clasificación e identificación de vinos mediante un sistema de estado sólido. In *Proceedings of the XXI Jornadas de Automática*, Sevilla, Spain, 2000.

[413] M. Santos. Construction of an artificial nose using neural networks. PhD thesis, Centre of Informatics, Federal University of Pernambuco, Brazil, 2000.

[414] S. Sarkar and R. S. Sriram. Bayesian models for early warning of bank failures. *Management Science*, 47(11):1457–1475, 2001.

[415] F. Scarlatti. Statistiche testuali, mappe concettuali, reti Bayesiane e valutazioni del paesaggio. In *XXIII Conferenza Italiana di Scienze Regionali*, Reggio Calabria, Italy, AISRE (Associazione Italiana di Scienze Regionali), CD-ROM, October 2002.

[416] W. G. Schneeweiss. *The Fault Tree Method*. LiLoLe Verlag, Hagen, Germany 1999.

[417] J. T. Schnute, A. Cass. and L. J. Richards. A Bayesian decision analysis to set escapement goals for Fraser River sockeye salmon (oncorhynchus nerka). *Canadian Journal of Fisheries and Aquatic Sciences*, 8(3):219–283, 2000.

[418] D. A. Schum. *Evidential Foundations of Probabilistic Reasoning*. John Wiley & Sons, New York, 1994.

[419] J. Schurmann. *Pattern Classification: A Unified View of Statistical and Neural Approaches*. John Wiley & Sons, New York, 1996.

[420] P. Sebastiani, L. Wang, V. G. Nolan, E. Melista, Q. Ma, C. T. Baldwin and M. H. Steinberg. Fetal hemoglobin in sickle cell anemia: Bayesian modeling of genetic associations. American Journal of Hematology. 2007 Oct 4. [Electronic publication ahead of print].

[421] P. Sebastiani, R. Lazarus, S. T. Weiss, L. M. Kunkel, I. S. Kohane, and M. F. Ramoni. Minimal haplotype tagging. *Proceedings of the National Academy of Science of the United States of America*, 100:9900–9905, 2003.

[422] P. Sebastiani, M. Ramoni, V. Nolan, C. Baldwin, and M. Steinberg. Genetic dissection and prognostic modeling of overt stroke in sickle cell anemia. *Nature Genetics*, 37(4):435–440, 2005.

[423] R. D. Shachter. Evaluating influence diagrams. *Operations Research*, 36(4):589–604, 1986.

[424] C. Shannon. A mathematical theory of communication. *The Bell System Technical Journal*, 27:379–390, 1948.

[425] C. Shannon and W. Weaver. *The Mathematical Theory of Communication*. University of Illinois Press, Urbana, Illinois, 1949.

[426] C. Shenoy and P. P. Shenoy. Bayesian network models of portfolio risk and return, 1999. https://kuscholarworks.ku.edu/dspace/bitstream/1808/161/1/CF99.pdf.

[427] P. P. Shenoy. A valuation-based language for expert systems. *International Journal of Approximate Reasoning*, 9:383–411, 1989.

[428] M. Shwe,   B. Middleton,   D. Heckerman,   M. Henrion,   E. Horvitz, H. Lehmann, and G. Cooper.   Probabilistic diagnosis using a reformulation of the INTERNIST–1/QMR knowledge base: I. The probabilistic model and inference algorithms.   *Methods of Information in Medicine*, 30(4):241–255, 1991.

[429] B. Sierra and P. Larrañaga. Predicting survival in malignant skin melanoma using Bayesian networks automatically induced by genetic algorithms. An empirical comparison between different approaches. *Artificial Intelligence in Medicine*, 14:215–230, 1998.

[430] G. Simmel. The sociology of secrecy and of secret societies. In *Soziologie*, chapter 5, pages 337–402. Duncker & Humblot, Berlin, 1908.

[431] M. Sloth, J. S. Jensen, and M. Faber. Bridge management using Bayesian condition indicators in Bayesian probabilistic network. In *Proceedings 1st International Conference on Bridge Maintenance, Safety and Management*, Barcelona, J. R. Casar, D. M. Frangopol, A. S. Nowak, editors, pages 75–76, 2002.

[432] V. Soto. Neural aromatic classification of Chilean wines. Master's thesis, Department of Electrical Engineering, University of Chile, July 2004.

[433] D. Spiegelhalter, N. Best, B. Carlin, and A. Van der Linde. Bayesian measures of model complexity and fit. *Journal of the Royal Statistical Society Series B*, 64(Part 4):583–639, 2002.

[434] D. J. Spiegelhalter. A statistical view of uncertainty in expert systems. In W. A. Gale, editor, *Artificial Intelligence and Statistics*, pages 17–55. Addison Wesley, Reading, MA, 1986.

[435] D. J. Spiegelhalter, A. P. Dawid, S. L. Lauritzen, and R. G. Cowell. Bayesian analysis in expert systems (with discussion).   *Statistical Science*, 8:219–247, 1993.

[436] D. J. Spiegelhalter and S. L. Lauritzen. Sequential updating of conditional probabilities on directed graphical structures. *Networks*, 20:157–224, 1990.

[437] P. Spirtes, C. Glymour, and R. Scheines. *Causation, Prediction, and Search*. Springer-Verlag, New York, 1993.

[438] W. H. Starbuck. Organizations as action generators. *American Sociological Review*, 48:91–102, Feb. 1983.

[439] W. H. Starbuck and F. J. Miliken. Executives' perceptual filters: What they notice and how they make sense. In D. Hambrick, editor, *The Executive Effect: Concepts and Methods for Studying Top Managers*, pages 35–65. JAI Press, Greenwich, CT, 1988.

[440] A. Stassopoulou, M. Petrou, and J. Kittler. Application of a Bayesian Network in a GIS based Decision Making System. Working paper, Department of Electronics and Electrical Engineering University of Surrey-Guildford, 1996.

[441] H. Steck. Constrained-based structural learning in Bayesian networks using finite data sets. PhD thesis, Institut für Informatik der Technischen Universität München, Germany, 2001.

[442] M. H. Steinberg. Predicting clinical severity in sickle cell anaemia. *British Journal of Haematology*, 129:465–481, 2005.

[443] J. D. Steventon. Environmental risk assessment, base-line scenario: Marbled murrelet. North Coast Land and Resource Management Plan. http://ilmbwww.gov.bc.ca/lup/lrmp/coast/ncoast/docs/reports/era/NC_Mamu_Base%-Line_ERA.pdf, 2003. British Columbia Ministery of Sustainable Ressource Management, Skeena Region, Smithers, BC, Canada.

[444] J. D. Steventon, G. D. Sutherland, and P. Arcese. Long-term risks to marbled murrelet (*Brachyramphus marmoratus*) populations: assessing alternative forest management policies in coastal British Columbia. http://www.for.gov.bc.ca/hfd/pubs/Docs/Tr/Tr012.pdf, 2003. Research Branch, Ministry of Forests, Victoria, British Columbia, Technical Report 012.

[445] J. D. Steventon, G. D. Sutherland, and P. Arcese. A population-viability based risk assessment of marbled murrelet nesting habitat policy in British Columbia. *Canadian Journal of Forest Research*, 36:3075–3086, 2006.

[446] T. Sugden, M. Deb, and B. Windley. The tectonic setting of mineralization in the Proterozoic Aravalli–Delhi orogenic belt, NW India. In S. Naqvi, editor, *Precambrian Continental Crust and its Economic Resources*, pages 367–390. Elsevier, Amsterdam, 1990.

[447] K. A. Tahboub. Intelligent human–machine interaction based on dynamic Bayesian networks probabilistic intention recognition. *Journal of Intelligent and Robotic Systems*, 45(1):31–52, 2006.

[448] F. Taroni, C. G. G. Aitken, G. Garbolino, and A. Biedermann. *Bayesian networks and probabilistic inference in forensic science*. John Wiley & Sons, Chichester, 2006.

[449] F. Taroni, A. Biedermann, P. Garbolino, and C. G. G. Aitken. A general approach to Bayesian networks for the interpretation of evidence. *Forensic Science International*, 139:5–16, 2004.

[450] T. H. Tear, P. Kareiva, P. L. Angermmeier, P. Comer, B. Czech, R. Kaultz, L. Landon, D. Mehlman, K. Murphy, M. Ruckelshaus, J. M. Scott, and G. Wilhere. How much is enough? The recurrent problem of setting measurable objectives in conservation. *BioScience*, 55(10):835–849, 2005.

[451] P. Thagart. Why wasn't O.J. convicted? Emotional coherence and legal inference. *Cognition and Emotion*, 17:361–383, 2003.

[452] The Genome International Sequencing Consortium. Initial sequencing and analysis of the human genome. *Nature*, 409:860–921, 2001.

[453] A. Thomas, D. J. Spiegelhalter, and W. R. Gilks. BUGS: A program to perform Bayesian inference using Gibbs sampling. In J. Bernardo, J. Berger, A. P. Dawid, and A. F. M. Smith, editors, *Bayesian Statistics 4*, pages 837–842. Oxford University Press, Oxford, 1992.

[454] J. Torres-Toledano and L. Sucar. Bayesian networks for reliability analysis of complex systems. In *Lecture Notes in Artificial Intelligence*, volume 1484, pages 195–206. Springer-Verlag, Berlin, 1998.

[455] A. Trotman. Searching structured documents. *Information Processing and Management*, 40:619–63, 2004.

[456] H. R. Turtle. Inference networks for document retrieval. PhD thesis, University of Massachusetts, 1990.

[457] H. R. Turtle and W. B. Croft. A comparison of text retrieval models. *Computer Journal*, 35(3):279–290, 1992.

[458] A. Tversky and D. Kahneman. Judgment under uncertainty: Heuristics and biases. *Science*, 185:1124–1130, 1974.

[459] A. Tversky and D. Kahneman. Rational choice and the framing of decisions. *The Journal of Business*, 59(4):251–278, 1986.

[460] C. Twardy, A. Nicholson, and K. Korb. Knowledge engineering cardiovascular Bayesian networks from the literature. Technical Report TR 2005/170, Clayton School of IT, Monash University, 2005.

[461] C. R. Twardy, A. E. Nicholson, K. B. Korb, and J. McNeil. Epidemiological data mining of cardiovascular Bayesian networks. *Electronic Journal of Health Informatics*, 1(1):1–13, 2006. http://www.ejhi.net.

[462] U.S. Department of Defense. USS Cole Commission Report. 9 January 2001.

[463] U.S. Department of Homeland Security. CIP/DSS National Infrastructure Interdependency Model Documentation. 31 May 2004.

[464] U.S. Department of Homeland Security. Office of Grants and Training. *Fiscal Year 2006 Homeland Security Grant Program: Program Guidance and Application Kit*, 2005.

[465] U.S. Department of Homeland Security. Office of Grants and Training. *FY2006 HSGP Fact Sheet: Risk Analysis*, May 2005.

[466] U.S. Department of Homeland Security. Special Needs Jurisdiction Tool Kit, 2005.

[467] U.S. Department of Homeland Security, prepared by ASME. Guidance on Risk Analysis and Management for Critical Asset Protection (RAMCAP), 2004.

[468] U.S. Department of the Army. FM 34-36: Special operations forces intelligence and electronic warfare operations, September 1991.

[469] L. C. van der Gaag, S. Renooij, C. L. M. Witteman, B. M. P. Aleman, and B. G. Taal. Probabilities for a probabilistic network: A case-study in oesophageal cancer. *Artificial Intelligence in Medicine*, 25(2):123–148, 2002.

[470] V. Vapnik. *Statistical Learning Theory*. John Wiley & Sons, New York, 1998.

[471] V. Vapnik. *The Nature of Statistical Learning Theory*. Springer-Verlag, New York, 2001.

[472] O. Varis. Bayesian decision analysis for environmental and resource management. *Environmental Modelling and Software*, 12:177–185, 1997.

[473] O. Varis and S. Kuikka. Learning Bayesian decision analysis by doing: lessons from environmental and natural resources management. *Ecological Modelling*, 119:177–195, 1999. available online.

[474] P. Veltz. *Mondialisation, villes et territoires*. PUF, Paris, 2005.

[475] Viterbi School of Engineering, University of Southern California. Assessment guidelines for counter-terrorism: CREATE terrorism modeling system (CTMS). 2 June 2005.

[476] P. Waddell. Towards a behavioural integration of land use and transportation modelling. In *9th International Association for Travel Behaviour Research Conference*, Gold Coast, Queensland, Australia, January 2001, D. A. Hensher, editor, Elsevier/Pergamon, Oxford, 2001.

[477] S. Wadia-Fascetti. Bidirectional probabilistic bridge load rating model. In *Transportation Research Board Annual Meeting*, Washington DC, 2006.

[478] B. Wagner. From computer based teaching to virtual laboratories in automatic control. In *29th ASEE/IEEE Frontiers in Education Conference*, San Juan, Puerto Rico, 1999.

[479] C. Wallace, K. Korb, R. O'Donnell, L. R. Hope, and C. R. Twardy. CaMML. `http://www.datamining.monash.edu.au/software/camml`, 2005.

[480] C. S. Wallace and K. B. Korb. Learning linear causal models by MML sampling. In A. Gammerman, editor, *Causal Models and Intelligent Data Management*. Springer-Verlag, Heidelberg, 1999.

[481] G. Warnick, R. Knopp, V. Fitzpatrick, and L. Branson. Estimating low-density lipoprotein cholesterol by the Friedewald equation is adequate for classifying patients on the basis of nationally recommended cutpoints. *Clinical Chemistry*, 36(1):15–19, 1990.

[482] A. Webb. *Statistical Pattern Recognition*. John Wiley & Sons, Chichester, 2002.

[483] P. Weber and L. Jouffe. Reliability modelling with dynamic Bayesian networks. In *SafeProcess 2003, 5th IFAC Symposium on Fault Detection, Supervision and Safety of Technical Processes*, Washington DC, 2003.

[484] K. Weick. *Sensemaking in Organizations*. Sage Publications, London, 1995.

[485] G. Weidl. Root cause analysis and decision support on process operation. PhD thesis, Department of Public Technology, Mälardalen University, Sweden, 2002.

[486] G. Weidl, A. Madsen, and S. Israelsson. Applications of object-oriented Bayesian networks for condition monitoring, root cause analysis and decision support on operation of complex continuous processes: Methodology and applications. *Computer and Chemical Engineering*, 29:1996–2009, 2005.

[487] G. Weidl, A. L. Madsen, and E. Dahlquist. Condition monitoring, root cause analysis and decision support on urgency of actions. *In Book Series FAIA (Frontiers in Artificial Intelligence and Applications), Soft Computing Systems – Design, Management and Applications*, 87:221–230, 2002.

[488] G. Weidl, A. L. Madsen, and S. Israelsson. Object-oriented Bayesian networks for condition monitoring, root cause analysis and decision support on operation of complex continuous processes: methodology and applications. Technical report 2005-1, University of Stuttgart, 2005.

[489] M. Wellman, J. Breese, and R. Goldman. From knowledge bases to decision models. *Knowledge Engineering Review*, 7(1):35–53, 1992.

[490] M. P. Wellman. Fundamental concepts of qualitative probabilistic networks. *Artificial Intelligence*, 44:257–303, 1990.

[491] L. K. Westin. Receiver Operating Characteristic (ROC) analysis. Technical report, Umea University, Sweden, 2002.

[492] J. Whittaker. *Graphical Models in Applied Multivariate Statistics*. John Wiley & Sons, Inc., New York, 1990.

[493] M. Wickerhauser. Acoustic signal compression with wavelet packets. Technical report, Yale University, New Haven, CT, August 1989.

[494] M. Wiel. La transition urbaine. *Mardaga Lièges*, 1999.

[495] J. H. Wigmore. The problem of proof. *Illinois Law Review*, 8:77–103, 1913.

[496] P. Wijayatunga, S. Mase, and M. Nakamura. Appraisal of companies with Bayesian networks. *International Journal of Business Intelligence and Data Mining*, 1(3):329–346, 2006.

[497] *Wikipedia*. 'body mass index' entry, 2007.

[498] S. S. Wilks. Multivariate statistical outliers. *Sankhya*, 25(4):407–426, 1963.

[499] D. B. Willingham and L. Preuss. The death of implicit memory. *Psyche*, 2(15), 1995.

[500] I. Witten and E. Frank. *Data Mining: Practical Machine Learning Tools and Techniques*. Morgan Kaufmann, San Francisco, 2nd edition, 2005.

[501] I. H. Witten and E. Frank. *Data Mining: Practical Machine Learning Tools and Techniques with Java Implementations*. Morgan Kaufmann, San Francisco, 2000.

[502] I. H. Witten, A. Moffat, and T. C. Bell. *Managing Gigabytes*. Morgan Kaufmann, San Francisco, 1999.

[503] E. Wright and K. B. Laskey. Credibility models for multi-source fusion. In *Proceedings of the 9th International Conference on Information Fusion*, Florence, 2006.

[504] S. Wright. The theory of path coefficients: a reply to Niles' criticism. *Genetics*, 8:239–255, 1923.

[505] S. Wright. The method of path coefficients. *Annals of Mathematical Statistics*, 5:161–215, 1934.

[506] Y. Xiang and K.-L. Poh. Time-critical dynamic decision modeling in medicine. *Computers in Biology and Medicine*, 32:85–97, 2002.

[507] A. Yamazaki and T. B. Ludermir. Classification of vintages of wine by an artificial nose with neural networks. In *Tercer Encuentro Nacional de Inteligencia Artificial*, Fortaleza, Brazil, 2001.

[508] C. Yoo and G. F. Cooper. An evaluation of a system that recommends microarray experiments to perform to discover gene-regulation pathways. *Artifical Ingelligence in Medicine*, 31:169–182, 2004.

[509] A. Zagorecki and M. Druzdzel. Knowledge engineering for Bayesian networks: How common are Noisy-MAX distributions in practice? In G. Brewka, S. Coradeschi, A. Perini, and P. Traverso, editors, *The Seventeenth European Conference on Artificial Intelligence (ECAI'06)*, pages 482–489. IOS Press, Amsterdam, 2006.

[510] D. E. Zand. *Information, Organization and Power: Effective Management in the Knowledge Society*. McGraw-Hill, New York, 1981.

[511] M. Zedda and R. Singh. Neural-network-based sensor validation for gas turbine test bed analysis. In *Proceedings of the I MECH E Part I Journal of Systems & Control Engineering*, volume 215, pages 47–56. Professional Engineering Publishing, London, February 2001.

[512] K. T. Zondervan and L. R. Cardon. The complex interplay among factors that influence allelic association. *Nature Reviews Genetics*, 5:89–100, 2004.

Index

absorption (of nodes), 36, 136
accuracy, 25, 83
Akaike information criterion, 61,
    290
Analytica, 309
artificial intelligence, 377–379

BayesiaLab, 111
Bayesian classifier
    naive, 47, 264, 270, 334
    selective naïve, 151
    tree-augmented naive, 288
Bayesian information criterion, 61
Bayesian information reward, 46
Bayesian network
    definition, 12
    dynamic, 32, 226
    naive, see Bayesian classifier
    object-oriented, 319–320
binary decision diagram, 232
black box, 12, 315, 379

canonical model, 25, 210
conditional independence, 9
confusion matrix, 138
cross-validation
    k-fold, 64, 70, 106, 156, 164
    leave-one-out, 267, 280

decision tree, 302
deviance information criterion, 61
diagnosis, 16
Discoverer, 69, 98

discretization, 20, 37, 52, 63, 98,
    232, 276
dynamic Bayesian network, see
    Bayesian network
dynamic fault tree, see fault tree

elicitation
    of probabilities, 24–26, 55,
        75–76, 118–119, 252, 306
    of structure, 20–24, 116–118,
        355–357
Elvira, 18
entropy, 195, 351, 370
expectation maximization, see
    learning algorithms
expert, 2, 378

fault tree, 226, 245, 307
    dynamic, 226
Fourier transform, 285
fuzzy logic, 189, 315, 362

Genie, 18
geographic information system, 97,
    109, 156, 245
Gibbs sampling, 26
GIS, see geographic information
    system

Heisenberg's uncertainty principle,
    381
Hugin, 78, 111, 320, 335, 354, 369
hyper-Dirichlet distribution, 60

Bayesian Networks: A Practical Guide to Applications Edited by O. Pourret, P. Naïm, B. Marcot
© 2008 John Wiley & Sons, Ltd

STATISTICS IN PRACTICE

Human and Biological Sciences

Berger – Selection Bias and Covariate Imbalances in Randomized Clinical Trials
Brown and Prescott – Applied Mixed Models in Medicine, Second Edition
Chevret (Ed) – Statistical Methods for Dose-Finding Experiments
Ellenberg, Fleming and DeMets – Data Monitoring Committees in Clinical Trials: A Practical Perspective
Hauschke, Steinijans & Pigeot – Bioequivalence Studies in Drug Development: Methods and Applications
Lawson, Browne and Vidal Rodeiro – Disease Mapping with WinBUGS and MLwiN
Lui – Statistical Estimation of Epidemiological Risk
Marubini and Valsecchi – Analysing Survival Data from Clinical Trials and Observation Studies
Molenberghs and Kenward – Missing Data in Clinical Studies
O'Hagan, Buck, Daneshkhah, Eiser, Garthwaite, Jenkinson, Oakley & Rakow – Uncertain Judgements: Eliciting Expert's Probabilities
Parmigiani – Modeling in Medical Decision Making: A Bayesian Approach
Pintilie – Competing Risks: A Practical Perspective
Senn – Cross-over Trials in Clinical Research, Second Edition
Senn – Statistical Issues in Drug Development, Second Edition
Spiegelhalter, Abrams and Myles – Bayesian Approaches to Clinical Trials and Health-Care Evaluation
Whitehead – Design and Analysis of Sequential Clinical Trials, Revised Second Edition
Whitehead – Meta-Analysis of Controlled Clinical Trials
Willan and Briggs – Statistical Analysis of Cost Effectiveness Data
Winkel and Zhang – Statistical Development of Quality in Medicine

Earth and Environmental Sciences

Buck, Cavanagh and Litton – Bayesian Approach to Interpreting Archaeological Data
Glasbey and Horgan – Image Analysis in the Biological Sciences
Helsel – Nondetects and Data Analysis: Statistics for Censored Environmental Data
Illian, Penttinen, Stoyan, H and Stoyan D–Statistical Analysis and Modelling of Spatial Point Patterns
McBride – Using Statistical Methods for Water Quality Management
Webster and Oliver – Geostatistics for Environmental Scientists, Second Edition
Wymer (Ed) – Statistical Framework for Recreational Water Quality Criteria and Monitoring

Aitken – Statistics and the Evaluation of Evidence for Forensic Scientists, Second Edition

Balding – Weight-of-evidence for Forensic DNA Profiles

Brandimarte – Numerical Methods in Finance and Economics: A MATLAB-Based Introduction, Second Edition

Brandimarte and Zotteri – Introduction to Distribution Logistics

Chan – Simulation Techniques in Financial Risk Management

Coleman, Greenfield, Stewardson and Montgomery (Eds) – Statistical Practice in Business and Industry

Frisen (Ed) – Financial Surveillance

Fung and Hu – Statistical DNA Forensics

Lehtonen and Pahkinen – Practical Methods for Design and Analysis of Complex Surveys, Second Edition

Ohser and Mücklich – Statistical Analysis of Microstructures in Materials Science

Pourret, Naim & Marcot (Eds) – Bayesian Networks: A Practical Guide to Applications

Taroni, Aitken, Garbolino and Biedermann – Bayesian Networks and Probabilistic Inference in Forensic Science